2010 YEAR BOOK OF DIAGNOSTIC RADIOLOGY®

The 2010 Year Book Series

Year Book of Anesthesiology and Pain Management™: Drs Chestnut, Abram, Black, Gravlee, Lien, Mathru, and Roizen

Year Book of Cardiology®: Drs Gersh, Cheitlin, Elliott, Gold, Graham, and Sundt

Year Book of Critical Care Medicine®: Drs Dellinger, Parrillo, Balk, Dorman, Dries, and Zanotti-Cavazzoni

Year Book of Dermatology and Dermatologic Surgery™: Dr Del Rosso

Year Book of Diagnostic Radiology®: Drs Osborn, Abbara, Birdwell, Elster, Manaster, Oestreich, Offiah, Rosado de Christenson, and Walker

Year Book of Emergency Medicine®: Drs Hamilton, Bruno, Handly, Mullin, Quintana, and Ramoska

Year Book of Endocrinology®: Drs Schott, Apovian, Clarke, Eugster, Ludlam, Meikle, Ovalle, Schinner, Schteingart, and Toth

Year Book of Gastroenterology™: Drs Talley, Dempsey, Harnois, Lange, Pearson, Picco, Rombeau, and Scolapio

Year Book of Hand and Upper Limb Surgery®: Drs Yao and Steinmann

Year Book of Medicine®: Drs Barker, Berney, Garrick, Gersh, Khardori, LeRoith, Talley, and Thigpen

Year Book of Neonatal and Perinatal Medicine®: Drs Fanaroff, Benitz, Neu, and Papile

Year Book of Neurology and Neurosurgery®: Drs Klimo and Rabinstein

Year Book of Obstetrics, Gynecology, and Women's Health®: Drs Dungan and Shulman

Year Book of Oncology®: Drs Thigpen, Arceci, Bauer, Byhardt, Gordon, and Lawton

Year Book of Ophthalmology®: Drs Rapuano, Cohen, Eagle, Flanders, Hammersmith, Myers, Nelson, Penne, Sergott, Shields, Tipperman, and Vander

Year Book of Orthopedics®: Drs Morrey, Beauchamp, Huddleston, Swiontkowski, and Trigg

Year Book of Otolaryngology-Head and Neck Surgery®: Drs Sindwani, Balough, Franco, Gapany, and Mitchell

Year Book of Pathology and Laboratory Medicine®: Drs Raab, Parwani, Bejarano, and Bissell

Year Book of Pediatrics®: Dr Stockman

Year Book of Plastic and Aesthetic Surgery™: Drs Miller, Gosain, Gurtner, Gutowski, Ruberg, Salisbury, and Smith

Year Book of Psychiatry and Applied Mental Health®: Drs Talbott, Ballenger, Buckley, Frances, Krupnick, and Mack

Year Book of Pulmonary Disease®: Drs Barker, Jones, Maurer, Raza, Tanoue, and Willsie

Year Book of Sports Medicine®: Drs Shephard, Cantu, Feldman, Jankowski, Khan, Nieman, Pierrynowski, and Rowland

Year Book of Surgery®: Drs Copeland, Behrns, Daly, Eberlein, Fahey, Huber, Jones, Mozingo, and Pruett

Year Book of Urology®: Drs Andriole and Coplen

Year Book of Vascular Surgery®: Drs Moneta, Gillespie, Starnes, and Watkins

2010 The Year Book of DIAGNOSTIC RADIOLOGY®

Editor-in-Chief

Anne G. Osborn, MD

University Distinguished Professor and Professor of Radiology, University of Utah School of Medicine, Salt Lake City, Utah

ELSEVIER
MOSBY

ELSEVIER
MOSBY

Vice President, Continuity: John A. Schrefer
Editor: Ruth Malwitz
Supervisor, Electronic Year Books: Donna M. Skelton
Electronic Article Manager: Jennifer C. Pitts
Illustrations and Permissions Coordinator: Dawn Vohsen

2010 EDITION

Composition by TnQ Books and Journals Pvt Ltd, India

Editorial Office:
Elsevier
Suite 1800
1600 John F. Kennedy Blvd.
Philadelphia, PA 19103-2899

International Standard Serial Number: 0098-1672
International Standard Book Number: 978-0-323-06828-4

Associate Editors

Suhny Abbara, MD
Associate Professor, Harvard Medical School; Director, Cardiovascular Imaging, Massachusetts General Hospital, Boston, Massachusetts

Robyn L. Birdwell, MD
Associate Professor, Harvard Medical School; Section Head, Division of Breast Imaging, Brigham and Women's Hospital, Boston, Massachusetts

Allen D. Elster, MD, FACR
Professor and Chair, Division of Radiologic Sciences, Wake Forest University Health Sciences, Winston-Salem, North Carolina

B. J. Manaster, MD, PhD
Professor of Radiology, University of Utah School of Medicine, Salt Lake City, Utah

Alan E. Oestreich, MD
Professor Emeritus of Radiology, University of Cincinnati College of Medicine; Pediatric Radiologist Emeritus, Cincinnati Children's Hospital Medical Center, Cincinnati, Ohio

Amaka C. Offiah, BSc, MBBS, MRCP, FRCR, PhD
HEFCE Clinical Senior Lecturer, Academic Unit of Child Health, Sheffield Children's NHS Foundation Trust, Western Bank, Sheffield, United Kingdom

Melissa L. Rosado de Christenson, MD, FACR
Section Chief, Thoracic Imaging, St. Luke's Hospital of Kansas City; Clinical Professor of Radiology, University of Missouri–Kansas City, Kansas City, Missouri

T. Gregory Walker, MD
Associate Radiologist, Harvard Medical School; Associate Director Fellowship, Section of Vascular Imaging and Intervention, Massachusetts General Hospital, Boston, Massachusetts

Table of Contents

Journals Represented

Journals represented in this YEAR BOOK are listed below.

Academic Emergency Medicine
Acta Paediatrica
AJR American Journal of Roentgenology
AJNR American Journal of Neuroradiology
Allergy
American Journal of Sports Medicine
Annals of Emergency Medicine
Annals of Internal Medicine
Annals of Surgery
Annals of Surgical Oncology
Archives of Disease in Childhood Fetal and Eeonatal Edition
Arthritis & Rheumatism
Arthroscopy
British Journal of Radiology
Cardiovascular and Interventional Radiology
Chest
Circulation
Clinical Radiology
European Heart Journal
European Journal of Nuclear Medicine and Molecular Imaging
European Journal of Paediatric Neurology
European Journal of Radiology
European Respiratory Journal
Headache
Journal of Bone and Joint Surgery American
Journal of Bone and Joint Surgery British
Journal of Cardiovascular Computed Tomography
Journal of Clinical Oncology
Journal of Computer Assisted Tomography
Journal of Hand Surgery
Journal of Neurosurgery
Journal of Neurosurgery Pediatrics
Journal of Neurosurgery Spine
Journal of Nuclear Medicine
Journal of Oral and Maxillofacial Surgery
Journal of Pediatric Surgery
Journal of Rheumatology
Journal of the American College of Cardiology
Journal of the American College of Radiology
Journal of the American College of Surgeons
Journal of the American Medical Association
Journal of the American Society of Echocardiography
Journal of Thoracic Oncology
Journal of Ultrasound in Medicine
Journal of Urology
Journal of Vascular and Interventional Radiology
Journal of Vascular Surgery

Laryngoscope
Neurology
Neuroradiology
Neurosurgery
New England Journal of Medicine
Pediatric Emergency Care
Pediatric Infectious Disease Journal
Pediatric Radiology
Pediatrics
Radiology
Skeletal Radiology
Spine
Stroke
Surgery
Surgical Neurology
Thorax
Thyroid
Urology

Standard Abbreviations

The following terms are abbreviated in this edition: acquired immunodeficiency syndrome (AIDS), cardiopulmonary resuscitation (CPR), central nervous system (CNS), cerebrospinal fluid (CSF), computed tomography (CT), deoxyribonucleic acid (DNA), electrocardiography (ECG), health maintenance organization (HMO), human immunodeficiency virus (HIV), intensive care unit (ICU), intramuscular (IM), intravenous (IV), magnetic resonance (MR) imaging (MRI), ultrasound (US), and ribonucleic acid (RNA).

Note

The Year Book of Diagnostic Radiology® is a literature survey service providing abstracts of articles published in the professional literature. Every effort is made to assure the accuracy of the information presented in these pages. Neither the editors nor the publisher of the Year Book of Diagnostic Radiology® can be responsible for errors in the original materials. The editors' comments are their own opinions. Mention of specific products within this publication does not constitute endorsement.

To facilitate the use of the Year Book of Diagnostic Radiology® as a reference tool, all illustrations and tables included in this publication are now identified as they appear in the original article. This change is meant to help the reader recognize that any illustration or table appearing in the Year Book of Diagnostic Radiology® may be only one of many in the original article. For this reason, figure and table numbers will often appear to be out of sequence within the Year Book of Diagnostic Radiology®.

Introduction

I'd like to begin this year's introduction to our annual "best of class" selections for the 2010 YEAR BOOK OF DIAGNOSTIC RADIOLOGY by thanking Dr Robyn Birdwell for her outstanding service over 11 years as our editor for the breast imaging section. Dr Tanya Stephens from M. D. Anderson Hospital in Houston will be taking over for the 2011 edition, and she will continue Dr Birdwell's tradition of selecting highly useful and informative articles on breast imaging. Dr Birdwell's selections this year are very timely, given the current controversy raging on whether screening mammography should—or shouldn't—be routinely performed. Should we be doing less? Doing more? The American College of Radiology has an "Image Gently" campaign going to reduce radiation in imaging. That's great. They also weighed in strongly in the Great Mammography Controversy: Just do it! A personal note: I've been having screening mammograms yearly and plan to continue. It isn't my favorite day of the year when I have it done, but I think it's the Right Thing To Do. Thanks, Robyn—and welcome, Tanya!

With this year's selections, Drs Offiah and Oestreich have addressed two burgeoning areas of interest in pediatric radiology: antenatal and forensic imaging. They also selected articles in the more traditional areas: trauma and child abuse and the reluctance versus overzealousness of some radiologists to testify in these cases. They, too, address the issue of unnecessary or excessive radiation (again, the "Image Gently" discussion that needs to be in the forefront of what we do—and when we do it).

Alan Elster, as always, weighs in with pertinent and timely articles on economics, research, education, and quality of care. Cost-effectiveness and radiation dose are major themes of his selections. The jarring (and highly public) incident of a misprogrammed CT scanner subjecting patients to 8-10 times already-high doses for brain perfusion studies was a wake-up call. Again, the "Image Gently" campaign comes as a highly timely (and already under way) focus for all of us.

What's new in thoracic imaging? Dr Rosado de Christenson focused on the new staging system for lung cancer, officially adopted in August 2009, so it is really "hot off the press." What to do in the evaluation and management of pulmonary nodules as well as senescent lung changes in asymptomatic patients are timely articles as well.

Our other editors have also selected a tantalizing potpourri of timely and practical articles for your delectation and delight. So enjoy our latest—and most timely—selection of articles. And as you go forward in your radiology practice, remember: Image Gently!

Anne G. Osborn, MD
Editor-in-Chief

1 Thoracic Radiology

Introduction

I am pleased to present the final selections of thoracic imaging articles from the 2009 literature. Selected articles largely represent the United States radiology literature, but also included are articles from the European radiologic literature and articles from the medical literature.

Several selected articles are of particular importance. Detterbeck and colleagues outline the new staging system for lung cancer that was officially adopted in August of 2009. Godoy and Naidich present interim guidelines for the evaluation and management of subsolid pulmonary nodules. Copley and colleagues report senescent changes in the lungs of asymptomatic subjects. Finally, the articles in the pulmonary embolism section address various topics including dose reduction, optimization of intravascular enhancement, evaluation of pulmonary thromboembolic disease in pregnancy, and practice patterns in ordering CT pulmonary angiograms in the emergency department.

I hope the YEAR BOOK readers find these articles useful as they address issues that arise in the daily interpretation of thoracic imaging studies.

Melissa L. Rosado de Christenson, MD, FACR

Lung Cancer

Initial Findings and Progression of Lung Adenocarcinoma on Serial Computed Tomography Scans

Saito H, Yamada K, Hamanaka N, et al (Kanagawa Cancer Ctr, Yokohama, Japan)

J Comput Assist Tomogr 33:42-48, 2009

To study the initial findings of lung adenocarcinoma revealed by computed tomography (CT) scanning and observe tumor progression and elucidate appropriate follow-up schedule of tumor diagnosis via CT findings of suspected lung adenocarcinoma.

Method.—We studied 59 patients who had undergone CT scanning twice or more at intervals of 3 months or longer before surgery. We evaluated the initial CT findings as well as all subsequent changes. The rate of tumor growth was estimated by tumor volume doubling time, using the original method of Schwartz. The histological classifications were evaluated according to the criteria of Noguchi et al.

Result.—The initial appearances of lung adenocarcinoma were divided into 4 types: (1) ground-glass opacity–like lesions, (2) bubble-like appearance, (3) small nodules, and (4) scar-like lesions. Ground-glass opacity–like lesions tended to increase in size over the years, with solid parts appearing in some lesions during follow-up examinations. Bubble-like appearance displayed characteristic CT findings and tended to increase over the years from the time of initial diagnosis, and we therefore tended to consider them as old inflammatory lesions. Small nodules tended to increase in size over the months more rapidly than in other types. Scar-like lesions tended to exist mainly in the lungs already damaged by lung fibrosis and/or emphysema and therefore were difficult to detect on initial CT scans.

Conclusion.—We categorized 4 types of initial findings of lung adenocarcinomas detected by CT. We determined that each type of lesion had its own unique characteristic growth patterns and required varying follow-up periods.

▶ To study the initial imaging findings of adenocarcinomas and to devise an appropriate follow-up strategy, the authors retrospectively reviewed the records and CT studies of 59 Japanese patients with excised peripheral adenocarcinomas found between the years 1993 and 2003. The patients were 30 men and 29 women (mean age, 65 years; age range 43-78 years) and had their lung lesions initially detected by X-ray photography and CT screening (31) or on chest imaging for the assessment of other conditions (28).

The lesions were initially characterized based on their CT appearance as ground glass opacity lesions (28 cases), small nodules (11), scar-like lesions (11), and bubble-like appearance lesions (9); the latter exhibited intrinsic air bronchiolograms. The largest lesions at presentation were bubble-like appearance lesions (mean diameter 27 mm ± 6 mm; range 19-37 mm), followed by scar-like lesions (mean diameter 15 ± 10 mm; range 4-33 mm), ground glass opacity lesions (mean diameter 14 ± 7 mm; range 6-35 mm), and small nodules (mean diameter 8 ± 3 mm; range 4-13 mm). The authors then used serial CT imaging to evaluate the tumor doubling time of the lesions.

The authors report the growth and morphologic characteristics of the lesions:

• Mean tumor volume doubling times (shown below) showed that the small nodules were the fastest growing lesions and the bubble-like lesions were the slowest.

o 238 ± 264 days for small nodules

o 348 ± 297 days for scar-like lesions

o 770 ± 914 days for ground glass opacity lesions

o 1363 ± 1141 days for bubble-like appearance lesions

• Ground glass opacity lesions grew slowly over the course of years and became denser with development of intrinsic solid components over time (Fig 2 in the orginal article). The latter were observed to grow rapidly.

- Bubble-like lesions were in some cases initially thought to represent old tuberculosis or organizing pneumonia.
- Scar-like lesions were associated with lung fibrosis, emphysema, old tuberculosis and/or bullae.

The authors present the varying morphologic features of adenocarcinomas of the lung on CT imaging. Based on their observations, they recommend follow-up of ground glass lesions at 6-month to 12-month intervals with excision or shorter interval evaluation when solid components appear. They recommend short interval follow-up chest CT (1-2 months) for ground glass opacity lesions and bubble like lesions to exclude infectious or inflammatory processes. A relatively short interval of follow-up chest CT (3-4 months) is recommended for lesions manifesting as small lung nodules and scar-like lesions. They also recommend an assessment of the patient's history to exclude an inflammatory process as the etiology of bubble-like appearance lesions.

It should be noted that the current recommendations for the CT assessment of small lung nodules used in the United States are those published by the Fleischner Society in 2005.[1] These recommendations are based on the initial size of the nodule and on the patient's risk factors, which include exposure to carcinogens and history of lung cancer in a first degree relative. Follow-up and management guidelines for ground glass opacities and other subsolid nodules have until now remained less clear. However, 2 articles by Godoy et al and Infante et al address this important topic.

M. L. Rosado de Christenson, MD

Reference

1. MacMahon H, Austin JHM, Gamsu G, et al. Guidelines for management of small pulmonary nodules detected on CT scans: a statement from the Fleischner Society. *Radiology*. 2005;237:395-400.

Use of a computer-aided detection system to detect missed lung cancer at chest radiography

White CS, Flukinger T, Jeudy J, et al (Univ of Maryland School of Medicine, Baltimore)
Radiology 252:273-281, 2009

Purpose.—To study the ability of a computer-aided detection (CAD) system to detect lung cancer overlooked at initial interpretation by the radiologist.

Materials and Methods.—Institutional review board approval was given for this study. Patient consent was not required; a HIPAA waiver was granted because of the retrospective nature of the data collection. In patients with lung cancer diagnosed from 1995 to 2006 at two institutions, each chest radiograph obtained prior to tumor discovery was evaluated by two radiologists for an overlooked lesion. The size and location of the nodules were documented and graded for subtlety (grades

1-4, 1 = very subtle). Each radiograph with a missed lesion was analyzed by a commercial CAD system, as was the follow-up image at diagnosis. An age- and sex-matched control group was used to assess CAD false-positive rates.

Results.—Missed lung cancer was found in 89 patients (age range, 51-86 years; mean age, 65 years; 80 men, nine women) on 114 radiographs. Lesion size ranged from 0.4 to 5.5 cm (mean, 1.8 cm). Lesions were most commonly peripheral (n = 63, 71%) and in upper lobes (n = 67, 75%). Lesion subtlety score was 1, 2, 3, or 4 on 43, 49, 17, and five radiographs, respectively. CAD identified 53 (47%) and 46 (52%) undetected lesions on a per-image and per-patient basis, respectively. The average size of lesions detected with CAD was 1.73 cm compared with 1.85 cm for lesions that were undetected (P = .47). A significant difference (P = .017) was found in the average subtlety score between detected lesions (score, 2.06) and undetected lesions (score, 1.68). An average of 3.9 false-positive results occurred per radiograph; an average of 2.4 false-positive results occurred per radiograph for the control group.

Conclusion.—CAD has the potential to detect approximately half of the lesions overlooked by human readers at chest radiography.

▶ The authors highlight known pitfalls of interpretation of chest radiographs in the detection of subtle pulmonary abnormalities, including the superimposition of numerous structures of various sizes and of varying radiographic opacity. The poor conspicuity and small size of many lung lesions contribute to the problem of missed lung cancer on radiography, the second-leading cause of malpractice actions against radiologists, according to the authors. Other factors contributing to missed lung cancer on radiography, listed by the authors include observer error, lesion characteristics (conspicuity and size), technical factors related to image acquisition, failure to compare with previous radiographs, and lack of awareness of clinical information.

The authors set out to evaluate the chest radiographic computer aided (CAD) system (OnGuard 3.0; Riverain Medical) in the analysis of chest radiographs of patients with known missed lung cancers. While the study was supported by a grant from Riverain Medical, the authors indicate that the coauthors who were not corporate employees of Riverain Medical had control of the data and the material submitted for publication.

The authors performed a retrospective review of the cancer registries of a university hospital and a veteran's hospital from 1998-2006 and 1995-2006 respectively, to search for patients with lung cancer. They identified each chest radiograph in which a lung cancer was first detected and reviewed previous radiographs to assess whether the lesions were present retrospectively to identify the missed lung cancers that formed the basis of the study. Most affected patients were men, likely reflecting the patient population at the veteran's hospital. Lesions were characterized as central if they overlapped the hilum. Fig 1 in the original article shows the locations of the 89 lung cancers missed at chest radiography. The radiographs were analyzed using CAD, and a control group of 89 normal radiographs of patients of similar demographic characteristics were also analyzed.

In each of the 89 patients a single lesion had been overlooked on a total of 114 radiographs, most of which were PA chest radiographs (81%). These lesions were overlooked by radiologists without subspecialty training in 88% of cases, thoracic radiologists in 9%, and thoracic imaging fellows in 3%. The lesions were more likely to be in the right lung (60%), and most were in the upper lobes (75%) and in peripheral locations (71%).

There was a significant difference in the location of the CAD detected lesions with most being in the middle lobe or lingula (82%). There was an average of 3.9 false-positive results per image. However, the authors state that if erroneous detection of monitoring devices, rib crossings, and costochondral junctions were excluded, the false-positive rate would have been 2.0. Interestingly, on the control cases, the false-positive rate was only 2.4. The authors did not find a significant difference in the detection of lung cancer nodules based on size. The CAD system identified approximately 50% of the missed lung cancers, many of which were subtle. It should be noted that the image reviewers were experienced thoracic radiologists that were able to detect CAD findings that were likely to be false-positive and differentiate these from the missed lung cancers. However, the authors question whether practicing radiologists reviewing these CAD markings would have been able to perform as well.

M. L. Rosado de Christenson, MD

The New Lung Cancer Staging System

Detterbeck FC, Boffa DJ, Tanoue LT (Yale Univ School of Medicine, New Haven, CT)

Chest 136:260-271, 2009

The International Association for the Study of Lung Cancer (IASLC) has conducted an extensive initiative to inform the revision of the lung cancer staging system. This involved development of an international database along with extensive analysis of a large population of patients and their prognoses. This article reviews the recommendations of the IASLC International Staging Committee for the definitions for the TNM descriptors and the stage grouping in the new non-small cell lung cancer staging system.

▶ This article presents the latest tumor, node, metastasis (TNM) staging system recommendations for lung cancer proposed by the International Association for the Study of Lung Cancer (IASLC) and accepted by the Union Internationale Contre le Cancer (UICC) and the American Joint Committee on Cancer (AJCC) to be published as the seventh edition of the lung cancer staging system in 2009. This staging system is based on 81 015 lung cancer cases collected in 20 countries and 4 continents. Most cases (58%) originate from Europe followed by cases from North America, Asia, and Australia. The

TABLE 3.—Definitions for T, N, M Descriptors

Descriptors	Definitions	Subgroups*
T	**Primary tumor**	
T0	No primary tumor	
T1	Tumor ≤ 3 cm,† surrounded by lung or visceral pleura, not more proximal than the lobar bronchus	
T1a	Tumor ≤ 2 cm†	T1a
T1b	Tumor > 2 but ≤ 3 cm†	T1b
T2	Tumor > 3 but ≤ 7 cm† or tumor with any of the following‡:	
	Invades visceral pleura, involves main bronchus ≥ 2 cm distal to the carina, atelectasis/obstructive pneumonia extending to hilum but not involving the entire lung	
T2a	Tumor > 3 but ≤ 5 cm†	T2a
T2b	Tumor > 5 but ≤ 7 cm†	T2b
T3	Tumor > 7 cm;	$T3_{>7}$
	or directly invading chest wall, diaphragm, phrenic nerve, mediastinal pleura, or parietal pericardium;	$T3_{Inv}$
	or tumor in the main bronchus < 2 cm distal to the carina§;	$T3_{Centr}$
	or atelectasis/obstructive pneumonitis of entire lung;	$T3_{Centr}$
	or separate tumor nodules in the same lobe	$T3_{Satell}$
T4	Tumor of any size with invasion of heart, great vessels, trachea, recurrent laryngeal nerve, esophagus, vertebral body, or carina;	$T4_{Inv}$
	or separate tumor nodules in a different ipsilateral lobe	$T4_{Ipsi\ Nod}$
N	**Regional lymph nodes**	
N0	No regional node metastasis	
N1	Metastasis in ipsilateral peribronchial and/or perihilar lymph nodes and intrapulmonary nodes, including involvement by direct extension	
N2	Metastasis in ipsilateral mediastinal and/or subcarinal lymph nodes	
N3	Metastasis in contralateral mediastinal, contralateral hilar, ipsilateral or contralateral scalene, or supraclavicular lymph nodes	
M	**Distant metastasis**	
M0	No distant metastasis	
M1a	Separate tumor nodules in a contralateral lobe;	$M1a_{Contr\ Nod}$
	or tumor with pleural nodules or malignant pleural dissemination‖	$M1a_{Pl\ Dissem}$
M1b	Distant metastasis	M1b
Special situations		
TX, NX, MX	T, N, or M status not able to be assessed	
Tis	Focus of *in situ* cancer	Tis
T1§	Superficial spreading tumor of any size but confined to the wall of the trachea or mainstem bronchus	$T1_{SS}$

*These subgroup labels are not defined in the IASLC publications[2–5] but are added here to facilitate a clear discussion.
†In the greatest dimension.
‡T2 tumors with these features are classified as T2a if ≤ 5 cm.
§The uncommon superficial spreading tumor in central airways is classified as T1.
‖Pleural effusions are excluded that are cytologically negative, nonbloody, transudative, and clinically judged not to be due to cancer.

recommendations are based on a large international database and backed by careful validation and statistical analysis.

Differences between current and former staging systems are for the most part differences in the categorization of the extent of the primary tumor (T descriptor). Most of the differences relate to tumor size (Table 3). The T3 descriptor now includes satellite nodules in the primary tumor lobe as these patients have a relatively good survival. The T4 descriptor now includes separate tumor nodules in a different but ipsilateral lung lobe. The authors state that the analysis of impact of the latter is based primarily on cases from Asia and Australia representing a potentially skewed population.

The classification of the extent of metastatic disease (M descriptor) now includes the M1a descriptor to designate separate tumor nodules in a contralateral lobe and malignant pleural nodules. The latter are known to afford a worse survival than that of T4 lesions. However, resected patients with pleural dissemination had an unexpected relatively good prognosis, a fact that may have an impact on future treatment. According to the authors, the prognostic impact of having 1 or more than 1 contralateral tumor nodules could not be assessed from the data.

While the N descriptor (extent of involvement of regional lymph nodes) has not changed, there are new anatomic definitions for the different nodal stations.[1] To address the data, the committee defined and used several nodal zones: upper zone (stations 1-4), aortopulmonary zone (stations 5-6), subcarinal zone (station 7), lower zone (stations 8-9), a hilar zone (stations 10-11), and a peripheral zone (stations 12-14). The authors found that the number of involved nodal zones appeared to have a prognostic impact.

The authors clarify the difference between clinical and pathologic staging. Clinical staging is the stage determined after using all information available before any treatment. Imaging findings (including findings at positron emission tomography [PET]-CT) and findings at mediastinoscopy in patients with lung cancer constitute clinical staging of the disease. Pathologic staging is determined after a resection of the neoplasm has been carried out. The authors do acknowledge that while pathologic staging would seem preferable, clinical staging is more practical as it guides management decisions made shortly after diagnosis and before treatment.

The authors state that although it would be hoped that the staging system would provide an algorithm to guide the treatment of patients with lung cancer, a single staging system may not be sufficient to provide such guidance. The authors state that treatment algorithms will have to be defined by clinical trials. However, the new system does address the goal of providing a common nomenclature to be used internationally and the goal of defining prognosis.

Future research directions may lead to a better understanding of the biologic behavior of lung cancer to decide which tumors have a potential to spread to regional nodes, produce local invasion, develop multicentric cancer foci, and which have a propensity for systemic dissemination.

M. L. Rosado de Christenson, MD

Reference

1. Rusch VW, Asamura H, Watanabe H, Giroux DJ, Rami-Porta R, Goldstraw P. The IASLC lung cancer staging project: a proposal for a new international lymph node map in the forthcoming seventh edition of the TNM classification for lung cancer. *J Thorac Oncol.* 2009;4:568-577.

The IASLC Lung Cancer Staging Project: A Proposal for a New International Lymph Node Map in the Forthcoming Seventh Edition of the TNM Classification for Lung Cancer

Rusch VW, on Behalf of the Members of the IASLC Staging Committee (Memorial Sloan-Kettering Cancer Ctr, NY; et al)

J Thorac Oncol 4:568-577, 2009

The accurate assessment of lymph node involvement is an important part of the management of lung cancer. Lymph node "maps" have been used to describe the location of nodal metastases. However, discrepancies in nomenclature among maps used by Asian and Western countries hinder analyses of lung cancer treatment outcome. To achieve uniformity and to promote future analyses of a planned prospective international database, the International Association for the Study of Lung Cancer proposes a new lymph node map which reconciles differences among currently used maps, and provides precise anatomic definitions for all lymph node stations. A method of grouping lymph node stations together into "zones" is also proposed for the purposes of future survival analyses.

▶ The lung cancer staging project was established in 1998 and led to the development of an international lung cancer database. However, there were discrepancies between the lymph node station nomenclatures used in Europe/America and in Japan. Thus, members of the International Association for the Study of Lung Cancer (IASLC) staging committee were charged with the development of a revised lymph node map to reconcile existing differences between the currently used lymph node maps and provide uniformity in the prospective international lung cancer data collection beginning in 2009. The committee included a thoracic radiologist who reviewed the definitions of the different nodal stations to ensure that they would be applicable to clinical staging of lung cancer with cross-sectional imaging. The recommendations were reviewed by the entire IASLC staging committee, which included thoracic surgeons, medical and radiation oncologists, pulmonary medicine physicians, radiologists, pathologists, epidemiologists, and data managers. The proposed lymph node map was also presented at the Japanese thoracic meetings for discussion and approval.

The new system:

- assigns supraclavicular and sternal notch lymph nodes to station 1, separated into right and left by the midline of the trachea
- provides specific upper and lower anatomic boundaries for stations 1 through 10 lymph nodes with specific definitions for right and left lymph nodes where appropriate
- provides new well-defined anatomic landmarks for differentiating station 4 and 10 lymph nodes on the right, station 5 and 10 lymph nodes on the left, and station 10 and 11 lymph nodes bilaterally
- provides anatomic boundaries that are more reliably identified on imaging, on endoscopy, and at surgery

- recognizes that lymphatic drainage in the superior mediastinum primarily occurs to the right paratracheal area past the tracheal midline and incorporates the left lateral wall of the trachea as the boundary between right and left station 2 and 4 lymph nodes
- avoids overlap of nodal stations

The authors discuss previous studies that showed metastases from right upper lobe tumors occur predominantly in right paratracheal lymph nodes; those from left upper lobe tumors affect predominantly the peri- and subaortic lymph nodes, and those from middle and lower lobe tumors affect the subcarinal followed by the right paratracheal lymph nodes. They also highlight evidence of involvement of the mediastinal nodes without involvement of intervening hilar and interlobar lymph nodes, so-called skip metastases in up to 25% of lung specimens injected experimentally and reported in 7% to 26% of resected lung cancers. These are felt to be more frequent in upper lobe lung cancers and in adenocarcinoma cell types.

Continuing discussion will address controversies such as differences in survival between patients with intra versus extranodal disease, single versus multiple lymph node station disease, specific sites of nodal metastases in relationship to the location of the primary lung cancer, the significance of skip metastases, and the role of systematic lymph node dissection. The data suggest that left upper lobe tumors with AP zone skip metastases (stations 5 and 6) were associated with a more favorable prognosis than those involving other N2 nodal stations.

M. L. Rosado de Christenson, MD

Malignant Neoplasia

Is investigation of patients with haemoptysis and normal chest radiograph justified?

Thirumaran M, Sundar R, Sutcliffe IM, et al (Mid Yorkshire Hosps NHS Trust, West Yorkshire, UK)
Thorax 64:854-856, 2009

Background.—Haemoptysis is a common clinical symptom. A small proportion of patients present with haemoptysis and normal chest radiograph. The investigation strategy for this group of patients is unclear. The aim of this study is to see whether further investigations for this group of patients are justified.

Methods.—A retrospective analysis was conducted of consecutive patients presenting with haemoptysis and normal chest radiograph over a period of 56 months irrespective of their smoking status. These patients were investigated by CT of the thorax and fibreoptic bronchoscopy.

Results.—275 episodes of haemoptysis with normal chest radiograph were investigated further in 270 patients (60% males). The median age was 60 years. Twenty-six patients were diagnosed to have respiratory tract malignancies (larynx, 1; trachea, 1; lung, 22; carcinoid, 1; and leiomyoma, 1). Eight (31%) of the 26 patients with respiratory tract malignancy had radical treatment. Fibreoptic bronchoscopy was diagnostic of cancer in 14 (54%) of the 26 patients with malignancy. CT of the thorax was suggestive of cancer in 24 (96%) of the 25 patients with malignancy.

Conclusion.—It is concluded that further investigation of haemoptysis in smokers is justified regardless of the amount or frequency of haemoptysis based on a 9.6% rate of malignancy in this consecutive series. It is recommended that these patients are investigated with CT of the thorax followed by fibreoptic bronchoscopy.

▶ Hemoptysis has varied etiologies, including benign and malignant conditions. Although most cases of hemoptysis are due to benign diseases, this symptom often raises the possibility of pulmonary malignancy. Most patients with hemoptysis who present for medical attention undergo chest radiography.

The authors (who appear to be pulmonary medicine specialists) performed a retrospective review of 270 consecutive patients presenting with 275 episodes of hemoptysis and a normal chest radiograph (as reported by a radiologist or as documented by a chest specialist) between May 2001 and December 2005. The patients were identified using the authors' urgent lung cancer referral and fiberoptic bronchoscopy databases. The authors admit to the possibility of selection bias in this United Kingdom population, as it is not known whether all patients who presented with hemoptysis to this institution were referred to them for further evaluation, as is the local practice. Patients with hemoptysis irrespective of quantity or frequency and regardless of patients' smoking status were evaluated with fiberoptic bronchoscopy (269) and/or chest CT (257). The patients were predominantly men (85%) with a median age of 69 years and an age range of 33 years to 84 years. Approximately 90% were current or former smokers.

The authors found 26 patients (9.6%) with respiratory tract malignancy and 2 additional patients with incidental non-respiratory tract malignancies, 1 with papillary carcinoma of the thyroid and one with lymphoma. Although it has been stated that the diagnostic yield for malignancy increases in patients with > 30 ml of hemoptysis, only 5 patients with pulmonary malignancy had this degree of hemoptysis and just over 40% presented with mild non-recurrent hemoptysis.

In this study, CT had a much higher yield for the diagnosis of respiratory tract malignancy than did bronchoscopy (see abstract). Of 57 patients with an abnormal chest CT inconsistent with malignancy, 30 showed pulmonary pathology that could explain the hemoptysis, including bronchiectasis (16), pneumonia (11), leiomyoma (1), pulmonary embolism and bronchiectasis (1), and tuberculosis (1). Based on the diagnostic yields of chest CT and bronchoscopy, and the findings in their study, the authors recommend that patients presenting with hemoptysis and a normal chest radiograph undergo chest CT followed by bronchoscopy.

M. L. Rosado de Christenson, MD

Lung Nodules - Volumetric Assessment and Computer-Aided Detection

Smooth or attached solid indeterminate nodules detected at baseline CT screening in the NELSON study: cancer risk during 1 year of follow up

Xu DM, van der Zaag-Loonen HJ, Oudkerk M, et al (Univ Med Ctr Groningen, The Netherlands)

Radiology 250:264-272, 2009

Purpose.—To retrospectively determine whether baseline nodule characteristics at 3-month and 1-year volume doubling time (VDT) are predictive for lung cancer in solid indeterminate noncalcified nodules (NCNs) detected at baseline computed tomographic (CT) screening.

Materials and Methods.—The study, conducted between April 2004 and May 2006, was institutional review board approved. Patient consent was waived for this retrospective evaluation. NCNs between 5 and 10 mm in diameter (n = 891) were evaluated at 3 months and 1 year to assess growth (VDT < 400 days). Baseline assessments were related to growth at 3 months and 1 year by using chi(2) and Mann-Whitney U tests. Baseline assessments and growth were related to the presence of malignancy by using univariate and multivariate logistic regression analyses.

Results.—At 3 months and at 1 year, 8% and 1% of NCNs had grown, of which 15% and 50% were malignant, respectively. One-year growth was related to morphology (P < .01), margin (P < .0001), location (P < .001), and size (P < .01). All cancers were nonspherical and purely intraparenchymal, without attachment to vessels, the pleura, or fissures. In nonsmooth unattached nodules, a volume of 130 mm(3) or larger was the only predictor for malignancy (odds ratio, 6.3; 95% confidence interval [CI]: 1.7, 23.0). After the addition of information on the 3-month VDT, large volume (odds ratio, 4.9; 95% CI: 1.2, 20.1) and 3-month VDT (odds ratio, 15.6; 95% CI: 4.5, 53.5) helped predict malignancy. At 1 year, only the 1-year growth remained (odds ratio, 213.3; 95% CI: 18.7, 2430.9) as predictor for malignancy.

Conclusion.—In smooth or attached solid indeterminate NCNs, no malignancies were found at 1-year follow-up. In nonsmooth purely intraparenchymal NCNs, size is the main baseline predictor for malignancy. When follow-up data are available, growth is a strong predictor for malignancy, especially at 1-year follow-up.

▶ Pulmonary nodules are a very common finding in both thoracic and abdominal CT studies performed for other reasons. Imaging follow up of these nodules is typically performed to exclude the possibility of early primary lung cancer.

The authors report the results of a retrospective study performed to determine, whether baseline nodule characteristics and 3-month and 1-year volume doubling times (VDT) are predictive for lung cancer in solid indeterminate lung nodules, detected at baseline CT screening. The authors used data collected in the Dutch-Belgian randomized lung cancer screening trial (NELSON study).

Nodule volumes in this study were determined using three dimensional volumetric computer software (Lungcare). The following definitions were provided:

Spherical nodule–Maximal diameter less than twice the minimal diameter (otherwise the nodule was classified as nonspherical).

Attached nodule–length of contact with another pulmonary structure of more than 50% of the diameter of the nodule (otherwise the nodule was classified as intraparenchymal). Attached nodules were attached to:

- an interlobar fissure
- the pleural surface (subpleural or per the authors "pleural-based")
- juxtavascular (except when attached to the tip of a pulmonary vessel).

Nodules were further classified as smooth, lobulated, spiculated, or irregular.

The authors summarize the published literature with regard to the risk of cancer in pulmonary nodules based on size and morphologic features.

Size

The risk of malignancy increases with increasing nodule size, according to previously published data by Midthun et al.[1]

- < 3 mm in diameter—0.2% probability of malignancy
- Between 4 mm and 7 mm—0.9% probability of malignancy
- Between 8 mm and 20 mm—18% probability of malignancy
- > 20 mm—50% probability of malignancy

Morphology

- Non-solid nodules—34% probability of malignancy
- Part-solid nodules—40% to 50% probability of malignancy
- Solid nodules—15% probability of malignancy

The authors found that a VDT of less than 400 days at 1 year was a very strong predictor of malignancy, with a positive predictive value of 63% in indeterminate nodules, but add that additional features related to the subject and the nodule are needed at baseline screening to improve diagnostic accuracy and avoid repeat studies. They found that because benign lesions may also grow, growth at 3 months had a low diagnostic accuracy, as only 15% of the nodules that exhibited significant growth at 3 months were malignant.

Importantly, the authors found no malignancies in smooth or attached nodules in spite of some of these exhibiting growth. These are indeed common findings on chest CT and comprised 83% of the indeterminate solid nodules found on this study (in fact, all malignant nodules arose from purely intraparencymal nodules). The authors state that 1-year follow up scanning should be sufficient for the follow-up evaluation of these types of nodules. They add that subpleural (or in their words "pleural-based") opacities are typically scars, pleural plaques, or small post infarction fibrotic lesions. With respect to triangular or ovoid nodules adjacent to fissures, they state that these may represent intrapulmonary lymph nodes, which may exhibit rapid growth or decrease in size. Finally, nodules attached to vessels are thought to represent small lymph nodes, small branching vessels or the sequela of small airways disease.

In the case of non-smooth intraparenchymal pulmonary nodules, the authors found that the only predictor of malignancy was baseline volume. They also found that VDT of less than 400 days at 3 months and at 1 year were additional predictors of malignancy. They add that cancer growth is not necessarily linear and that it may exhibit a sigmoid shape characterized by a growth acceleration phase after a period of apparent size stability. Study limitations included the small number of cancers found and the short period of follow-up.

M. L. Rosado de Christenson, MD

Reference

1. Midthun DE, Swensen SJ, Jett JR, et al. Evaluation of nodules detected by screening for lung cancer with low dose spiral computed tomography. *Lung Cancer.* 2003;41(suppl 2):S40.

Differential diagnosis and management of focal ground-glass opacities

Infante M, Lutman RF, Imparato S, et al (IRCCS Istituto Clinico Humanitas, Milan, Italy)

Eur Respir J 33:821-827, 2009

Focal pulmonary ground-glass opacities (GGOs) can be associated with bronchioloalveolar carcinoma. The present retrospective study aimed to test the validity of a multistep approach to discriminate malignant from benign localised (focal) GGOs, identifies useful diagnostic features on computed tomography (CT), and suggests appropriate management guidelines. A stepwise approach, including oral antibiotics, follow-up high-resolution CT (HRCT) 40-60 days later and CT-guided core biopsy, was used. All cases with localised GGOs detected since 2001 were reviewed. CT features were described according to a structured scheme. In total, 40 patients were evaluated. Of these, 11 patients were diagnosed with benign GGOs, 19 patients had lung cancer and 10 were undetermined. Nonpolygonal shape, apparent radial growth and clear-cut margins were associated with a malignant histology. The specificity of CT findings was low. Diagnostic accuracy increased after oral antibiotics, follow-up HRCT and percutaneous core biopsy. Overall, 18 patients underwent surgery for lung cancer. In conclusion, malignant ground-glass opacities have a fairly typical appearance, but some benign lesions closely mimic their malignant counterparts. The stepwise approach adopted in the present study increased the diagnostic specificity and reduced time to definitive diagnosis. Segmentectomy might be the ideal resection volume for such tumours.

▶ The authors retrospectively reviewed 40 cases of what they refer to as focal ground glass opacities (GGO). Their cases included subsolid nodules, that is, those that exhibited pure ground glass opacity (also known as non-solid lesions) and so-called mixed ground glass opacity (part-solid nodules). The latter exhibited GGO with variable amounts of solid components. While the cases were collected at the authors' institution from June 2001 to November 2007, it is not clear how the institution's database was searched to find the 40 patients that formed the basis of the study. The cases were either found during the authors' ongoing lung cancer screening trial (29 cases) or incidentally during the evaluation of unrelated conditions (11 cases). The authors used a stepwise management scheme "on a regular basis" in which patients with focal GGO lesions on chest CT were treated with oral antibiotics (levofloxacin, 500 mg daily for 8 days), followed by repeat chest CT (specifically, high resolution CT of the lesion) 40 days to 60 days later, and finally percutaneous CT guided core biopsy of lesions that showed no resolution or regression. Patients with diffuse GGO and those with GGO associated with large areas of consolidation and/or bronchiolitis were excluded. Other than stating that all the lesions studied were focal GGO lesions, the authors do not report the range of lesion sizes. A synchronous lung cancer was present in 4 patients, 1 patient had a previously resected stage I lung cancer, and 10 patients had 2 or more lesions.

The authors identified absence of a solid component, presence of polygonal shape (roughly linear lesion margins), and ill-defined margins as associated with benign nonevolving GGO lesions. Five lesions (12.5%) exhibited resolution or regression on CT after antibiotic treatment and were presumed benign.

Malignant GGO lesions tended to exhibit nonpolygonal shapes. Radial growth (centrifugal extension of the lesion through adjacent pulmonary lobules) was associated with malignancy, but was also observed in benign lesions. Intrinsic solid components (part-solid lesions) were associated with malignancy in 75% of cases, but the association of solid components with lung cancer was not significant. However, the association of well-defined margins with malignancy was statistically significant.

The authors propose:

- A trial of antibiotic treatment as the first step in the management of non-solid and part-solid lesions
- CT guided core biopsy for non resolving, non responding focal GGO lesions

They state that this management algorithm will allow the noninvasive diagnosis of some infectious lesions while only delaying the diagnosis of neoplastic lesions by about 3 months. They do not recommend the routine use of PET due to its low sensitivity in the diagnosis of GGO lesions.

The authors address the issue of potential overdiagnosis of indolent malignancy, as in some cases GGO lesions grow very slowly and may remain clinically silent for the duration of the patient's life. They recommend that stable pure GGO lesions in patients who are poor operative candidates be followed until signs of progression become evident. They add that segmentectomy may be an acceptable management strategy for some early indolent lung cancers and recommend it for excision of focal GGO lesions with minor solid components.

M. L. Rosado de Christenson, MD

Subsolid pulmonary nodules and the spectrum of peripheral adenocarcinomas of the lung: recommended interim guidelines for assessment and management

Godoy MC, Naidich DP (New York Univ-Langone Med Ctr)

Radiology 253:606-622, 2009

Pulmonary nodule characterization is currently being redefined as new clinical, radiologic, and pathologic data are reported, necessitating a reevaluation of the clinical management, especially of subsolid nodules. These are now known to frequently, although not invariably, fall into the spectrum of peripheral adenocarcinomas of the lung. Strong correlation between the Noguchi histologic classification and computed tomographic (CT) appearances of these lesions, in particular, has been reported. Serial CT findings have further documented that stepwise progression of lesions with ground-glass opacity, manifested as an increase in size or the appearance and/or subsequent increase of solid components, does occur in a select subset of patients. As a consequence, recognition of the potential association between subsolid nodules and peripheral adenocarcinomas requires a review of current guidelines for the management of these lesions, further necessitated by a differential diagnosis that includes benign lesions such as focal inflammation, focal fibrosis, and organizing pneumonia. Specific issues that need to be addressed are the need for consensus regarding an appropriate CT classification, methods for precise measurement of subsolid nodules, including the extent of both ground-glass and solid components, as well as accurate assessment of the growth rates as means for predicting malignancy and prognosis. It is anticipated that interim guidelines may serve to standardize our current management of these lesions, pending further clarification of their natural history.

▶ This article addresses the important and controversial topic of the assessment and management of subsolid lung nodules and provides interim recommendations for the imaging and management of these lesions. The article presents a clear, comprehensive, and beautifully illustrated review of the epidemiology, current pathologic classification, and prognosis of adenocarcinomas of the lung, and clearly summarizes the extensive published literature on the topic of subsolid lung nodules. While guidelines for the assessment and management of small solid lung nodules were published by the Fleischner Society in 2005, the management of subsolid lung nodules remains controversial. It is understood that many subsolid lung nodules fall within the spectrum of primary pulmonary adenocarcinoma (including bronchioloalveolar carcinoma) and the premalignant condition atypical adenomatous hyperplasia, whereas some represent the sequela of infectious/inflammatory processes, focal pulmonary fibrosis, and organizing pneumonia. Furthermore, it is recognized that absence of change over the course of 2 years does not necessarily correlate with a benign histology as some adenocarcinomas exhibit an indolent behavior and long tumor doubling times. In addition, some of these lesions exhibit a weak association with cigarette smoking, and have a higher incidence in women and younger patients.

The authors classify subsolid nodules as non-solid (ground glass nodules) and part-solid (nodules with mixed solid and ground glass attenuation) types. It is recognized that increasing size of solid components within part-solid nodules correlates with increasing likelihood of malignancy. Based on their review of the substantial literature on the subject of subsolid nodules and recognizing the evolving pathologic classification of adenocarcinomas of the lung and their premalignant conditions, the authors provide the following recommendations for assessment and management of these lesions. It should be noted that unlike the Fleischner Society guidelines, patients are not stratified into high- or low-risk categories.

- No imaging follow-up for isolated pure ground glass nodules smaller than 5 mm, particularly when occurring in elderly patients, with the exception of patients enrolled in lung cancer CT screening programs. The authors recognize that some of these lesions may ultimately prove to represent bronchioloalveolar carcinomas.
- Short interval follow-up (3-6 months with consideration of interim antibiotic treatment) of pure ground glass nodules measuring between 5 mm and 10 mm to document resolution of infectious/inflammatory processes.
- If CT surveillance is elected, the follow-up period should be longer than the traditional 2-year follow-up period, with at least 3 consecutive annual follow-up CT studies. The authors recommend such long-term follow-up in lieu of surgery.
- Assessment with thin-section CT and preferably with low dose CT to reduce radiation exposure.
- PET-CT is of questionable diagnostic value in these lesions.
- Transbronchial biopsy is discouraged.
- Resection of solitary ground glass lesions measuring 10 mm or more that persist after 3 months to 6 months. The authors emphasize that PET negativity does not exclude invasive carcinoma in these lesions.
- Resection of persistent part-solid lesions of any size. The authors recommend PET/CT evaluation in these cases, because of the higher likelihood of invasive adenocarcinoma. The authors emphasize the difficulties in distinguishing bronchioloalveolar carcinomas from invasive adenocarcinomas on biopsy, given the inherent sampling errors encountered in transbronchial and transthoracic needle biopsy specimens.
- CT follow-up of multifocal pure ground glass lesions measuring <5 mm for at least 1 year. It is acknowledged that these patients may have a higher risk for developing lung cancer.
- CT follow-up of multifocal 5 mm to 10 mm lesions, as they may represent either atypical adenomatous hyperplasia or respiratory bronchiolitis (in smokers).

- Consideration of surgical excision of dominant lesions (ground glass lesions larger than 10 mm and part-solid lesions) in patients with multifocal lesions, PET-CT evaluation and consideration of lung sparing resections of selected lesions. It is understood that although dominant lesions may be resected, the remaining lesions may continue to grow.

M. L. Rosado de Christenson, MD

Effect of slab thickness on the CT detection of pulmonary nodules: Use of sliding thin-slab maximum intensity projection and volume rendering

Kawel N, Seifert B, Luetolf M, et al (Kantonsspital Graubuenden, Chur, Switzerland)

AJR Am J Roentgenol 192:1324-1329, 2009

Objective.—The objective of this study was to evaluate the detection rates of pulmonary nodules on CT as a function of slab thickness using sliding thin-slab maximum intensity projection (MIP) and volume rendering (VR).

Subjects and Methods.—Eighty-eight oncology patients (33 women, 55 men; mean age, 59 years; age range, 18-81 years) who routinely underwent chest CT examinations were prospectively included. Two radiologists independently evaluated each CT examination for the presence of pulmonary nodules using MIP and VR, with each image reconstructed using three different slab thicknesses (5, 8, 11 mm). The standard of reference was the maximum number of detected nodules, which were classified by localization and size, judged to be true-positives by a consensus panel. Interreader agreement was assessed by kappa value on a nodule-by-nodule basis. Sensitivities for both reconstruction techniques and for the three slab thicknesses were calculated using the proportion procedure for survey data with the patient as the primary sample unit and were compared using the Wilcoxon's signed rank test with Bonferroni correction for both readers separately.

Results.—One thousand fifty-eight true-positive nodules were detected. Interreader agreement was fair to moderate. Sensitivity for pulmonary nodules was superior for 8-mm MIP (reader 1, 84%; reader 2, 81%) and was significantly better than the sensitivities of all other tested techniques for both readers ($p < 0.001$ each) independent of nodule localization and size (except for one reader's analysis of 8-mm MIP versus 11-mm MIP for nodules > 8 mm). A higher sensitivity was achieved using MIP than VR.

Conclusion.—MIP with a slab thickness of 8 mm is superior in the detection of pulmonary nodules to all other tested techniques.

▶ The detection of pulmonary nodules is an important aspect of interpretation of thoracic CT, particularly in the setting of malignancy and in patients with high-risk factors for the development of primary lung cancer. Several tools have been proposed for facilitating the detection of small pulmonary nodules. Among these are postprocessing techniques, including volume rendering (VR) and maximum intensity projection (MIP). The authors explain that VR is a 3D reconstruction technique in which final images contain data from each voxel, whereas in MIP, the highest attenuation voxels are projected through the entire volume to create the final image.

The authors prospectively evaluated 118 consecutive patients with known malignancy imaged between July 9 and August 10, 2007. Patients under the age of 18, those with more than 60 nodules in both lungs or coexisting lung disease, and those who refused to participate were excluded. MIP images were displayed with a window center of −300 HU and a window width of 1600 HU, whereas VR images were reviewed with a center of −500 HU and a window width of 1,500 HU. Only axial postprocessed images were reviewed.

The 1058 true-positive nodules studied exhibited central location (72.7%), pleural contact (19.4%), and hilar location (7.9%). Most nodules (69.5%) were ≤4 mm in size. The mean number of nodules per patient was 22.

The authors found that nodule detection rates were superior using MIP when compared with VR independent of slab thickness. In addition, detection rates were significantly higher using 8 mm MIP than all other tested techniques and slab thicknesses. The superiority of 8 mm MIP was not affected by nodule location or nodule size. The authors stated that slabs larger than 11 mm were not evaluated as they were considered confusing due to the large number of visualized pulmonary vessels. In addition, slabs smaller than 5 mm were felt to be too small to differentiate small nodules from pulmonary vessels seen axially.

The authors list the fair to moderate interreader agreement as a limitation of their study, and state that this may have been related to the fact that the readers were instructed to evaluate the images at the same speed used in clinical practice. In addition, the authors state that while one of the readers was a radiology resident, this individual gained similar sensitivities in nodule detection when compared with the chest radiologist readers. They add that the absence of a true standard of reference constituted another limitation of their study.

M. L. Rosado de Christenson, MD

Lung nodules: CT-guided placement of microcoils to direct video-assisted thoracoscopic surgical resection

Mayo JR, Clifton JC, Powell TI, et al (Univ of British Columbia, Vancouver, Canada)

Radiology 250:576-585, 2009

Purpose.—To prospectively assess the safety and effectiveness of computed tomography (CT)-guided placement of fiber-coated microcoils used to guide video-assisted thoracoscopic surgical (VATS) excision of small peripheral lung nodules, with successful excision as the primary outcome and successful CT-guided microcoil placement and procedural complications as secondary outcomes.

Materials and Methods.—The institutional review board approved the study protocol. Informed consent was obtained from all 69 enrolled patients (30 men, 39 women; mean age, 60.7 years ± 10.1 [standard deviation]) with 75 nodules. At CT, one end of an 80-mm long, 0.018-inch-diameter fiber-coated microcoil was placed deep to the small peripheral lung nodule, and the other end was coiled in the pleural space. VATS excision of lung tissue, nodules, and the microcoil was performed with fluoroscopic guidance.

Results.—Seventy-three (97%) 4-24-mm nodules were successfully removed at fluoroscopically guided VATS excision; two nodules could not be removed. CT-guided microcoil placement was successful in all cases; however, two (3%) of 75 coils were displaced at VATS excision. Pneumothorax requiring chest tube placement occurred in two (3%) patients, and asymptomatic hemothorax occurred in one (1%) patient. The microcoil did not impede intraoperative frozen-section histopathologic analysis, which facilitated accurate clinical management in all patients. For 19 (28%) patients, the preoperative treatment plan based on bronchoscopy, needle biopsy, and positron emission tomography findings changed after VATS excision.

Conclusion.—Microcoil localization of small peripheral lung nodules enabled fluoroscopically guided VATS resection of 97% of the nodules, with a low rate of intervention (3%) for procedural complications.

▶ The authors report the results of fiber coated platinum microcoil localization of 75 pulmonary nodules, performed in 69 consecutive patients who underwent fluoroscopically guided video assisted thoracoscopic surgical (VATS) excision of the nodules with a success rate of 97%. All nodules were smaller than 30 mm in longest dimension and located in the lung periphery within 5 cm of a pleural surface or interlobar fissure and not adjacent to a central pulmonary blood vessel. Based on a previous animal study, the authors determined that 30 mm of the 80 mm microcoil had to be deployed into the lung parenchyma deep to the nodule to anchor the deployed microcoil. The other end of the microcoil was made to form a loop on the visceral pleural surface of the lung overlying the nodule. Post microcoil deployment CT was performed to assess the position of the microcoil and evaluate for resultant hemorrhage or pneumothorax (Fig 3 in the original article). During the VATS procedure the surgeon used the visceral pleural portion of the microcoil to identify the lung tissue to be resected. A histologic diagnosis of benign lesion, completely resected metastasis, or well-differentiated adenocarcinoma marked the end of the procedure. A diagnosis of non-small cell lung cancer led to intraoperative staging and potentially complete resection.

The nodules studied exhibited ground glass opacity (3), semisolid morphology (10), and solid morphology (62). Parenchymal hemorrhage was noted on CT in 5 (7%) of microcoil-localized nodules. Six nodule localizations were accomplished with 2 microcoil placements. In 2 cases, the microcoil was dislodged during the surgical procedure, but the patients still had successful resections as the surgeon was guided by the parenchymal hematoma caused by the dislodged microcoil. Pneumothoraces were noted in 75% of the 71 CT guided procedures, but only 2 patients (3%) required chest tube placement. A small hemothorax was diagnosed during VATS in 1 patient. While 73 of 75 nodules were completely excised during VATS, 2 VATS procedures were changed to thoracotomies because the stapler jaws could not be properly closed around the nodule. Both these lesions required lobectomy, as segmental resection could not be performed due to the deep location of the nodules. The authors found that the presence of the microcoil did not interfere with histopathologic tissue sectioning techniques in any of their cases and add that the hemostatic fiber coating of the microcoil minimizes pulmonary hemorrhage, which could potentially interfere with histologic interpretation.

The microcoil placement technique was markedly successful in the authors' institution. They state that withdrawing the needle and achieving proper placement of the microcoil can be challenging. They report a cost of approximately $180.00 for the set of microcoil, pusher wire, and graduated Chiba needle. The authors add that further experience in other institutions is needed to ensure that the technique and success rates they experienced can be duplicated.

M. L. Rosado de Christenson, MD

Infection

Chest Radiographic and CT Findings in Novel Swine-Origin Influenza A (H1N1) Virus (S-OIV) Infection

Agarwal PP, Cinti S, Kazerooni EA (Univ of Michigan Health Service, Ann Arbor; Univ of Michigan, Ann Arbor)
AJR Am J Roentgenol 193:1488-1493, 2009

Objective.—This article reviews the chest radiographic and CT findings in patients with presumed/laboratory-confirmed novel swine-origin influenza A (H1N1) virus (S-OIV) infection.

Materials and Methods.—Of 222 patients with novel S-OIV (H1N1) infection seen from May 2009 to July 2009, 66 patients (30%) who underwent chest radiographs formed the study population. Group 1 patients ($n = 14$) required ICU admission and advanced mechanical ventilation, and group 2 ($n = 52$) did not. The initial radiographs were evaluated for the pattern (consolidation, ground-glass, nodules, and reticulation), distribution, and extent of abnormality. Chest CT scans ($n = 15$) were reviewed for the same findings and for pulmonary embolism (PE) when performed using IV contrast medium.

Results.—Group 1 patients were predominantly male with a higher mean age (43.5 years versus 22.1 years in group 2; $p < 0.001$). The initial radiograph was abnormal in 28 of 66 (42%) subjects. The predominant radiographic finding was patchy consolidation (14/28; 50%) most commonly in the lower (20/28; 71%) and central lung zones (20/28; 71%). All group 1 patients had abnormal initial radiographs; extensive disease involving ≥ 3 lung zones was seen in 93% (13/14) versus 9.6% (5/52) in group 2 ($p < 0.001$). No group 2 patients had >20% overall lung involvement on initial radiographs compared with 93% of group 1 patients (13/14). PEs were seen on CT in 5/14 (36%) of group 1 patients.

Conclusion.—Chest radiographs are normal in more than half of patients with S-OIV (H1N1) and progress to bilateral extensive air-space disease in severely ill patients, who are at a high risk for PE.

▶ According to the Centers for Disease Control (CDC), the 2009 H1N1 influenza virus, which was previously called novel H1N1 or swine flu is responsible for the first influenza pandemic in over 40 years. Affected patients tend to be young to middle aged. The disease was first reported in Mexico in April of 2009.

This is a retrospective review in which the authors identified all cases positive for influenza A, identified by direct fluorescent antigen-antibody test or viral culture screen for a total of 222 patients. Of these, 31 had further testing with reverse transcription polymerase chain reaction confirming the diagnosis of H1N1. The rest were presumed to have the same virus, as no other viruses were identified in the community with any frequency during the period of study. In fact, the authors include the retrospective design and the absence of definitive pathogen confirmation for all patients among the limitations of their study.

Of the 222 patients, 66 underwent thoracic imaging and formed the basis of the report. The authors divided the patients into 2 groups based on the severity of their disease. Group 1 (14 patients) included patients with severe disease who required advanced mechanical ventilation, and group 2 was comprised of the remaining 52 patients. There were 36 men and 30 women ranging in age from 9 months to 60 years, with a mean age of 26.7 years. It should be noted that the authors reported a high prevalence of obesity among their severely ill patients (group 1).

The authors made the following observations:

- None of the patients in group 1 had a normal initial chest radiograph, whereas 73% of patients in group 2 had a normal initial chest radiograph.
- All patients in group 1 had bilateral radiographic abnormalities.
- The most common radiographic finding in both groups was patchy consolidation, followed by ground glass opacity or a mixture of ground glass opacity and consolidation.
- The abnormalities were most common in the lower lung zones, and airspace disease was more frequent in the central lungs.

• There was a greater extent of disease in patients in group 1, and most patients in group 2 had less than 10% lung involvement.

• Thoracic CT (performed in 10 patients in group 1) most frequently showed a combination of ground glass opacity and consolidation and typically showed diffuse involvement. (Fig 2 in the original article).

• Pulmonary emboli were identified in 36% of patients admitted to the intensive care unit.

In his editorial entitled "Conventional Wisdom: Unconventional Virus",[1] Dr Ketai comments on two series (including this article)[2] and a case report[3] of swine origin influenza A H1N1 virus published in the same issue of the journal in which this article appears. He emphasizes that these articles do not describe imaging findings considered typical of viral pulmonary infection affecting the small airways, which generally include centrilobular nodules, tree-in bud opacities and mosaic perfusion. Instead, the findings described in these articles appear more consistent with bacterial infection. It should be noted that the CDC has documented bacterial superinfection as a complication in patients dying after infection with the H1N1 virus.

M. L. Rosado de Christenson, MD

References

1. Ketai LH. Conventional wisdom: unconventional virus. *AJR Am J Roentgenol.* 2009;193:1486-1487.
2. Ajlan AM, Quiney B, Nicolaou S, "Müller NL. Swine-origin influenza A (H1N1) viral infection: radiographic and CT findings. *AJR Am J Roentgenol.* 2009;193: 1494-1499.
3. Mollura DJ, Asnis DS, Crupi RS, et al. Imaging findings in a fatal case of pandemic swine-origin influenza A (H1N1). *AJR Am J Roentgenol.* 2009;193:1500-1503.

Tuberculosis: Value of Lateral Chest Radiography in Pre-employment Screening of Patients with Positive Purified Protein Derivative Skin Test Results

Eisenberg RL, Romero J, Litmanovich D, et al (Beth Israel Deaconess Med Ctr, Boston, MA)

Radiology 252:882-887, 2009

Purpose.—To test the hypothesis that lateral radiography adds no clinically relevant information to that acquired with posteroanterior (PA) radiography in individuals with positive purified protein derivative (PPD) skin test results.

Materials and Methods.—As part of routine practice, all 875 adults with positive PPD skin test results at pre-employment examinations performed at a medical center during 2007 underwent PA and lateral chest radiography. In this institutional review board-approved HIPAA-compliant study, the patient informed consent requirement was waived. Two radiologists retrospectively and independently interpreted each radiograph for evidence of abnormalities that were indicative of acute or chronic tuberculosis (TB) infection. First, only the PA radiograph was analyzed. Thereafter, both the PA radiograph and the lateral radiograph were analyzed together to determine if any observed finding was identified on only the lateral radiograph. If a finding was seen on both images, a determination was made as to whether the finding on the lateral radiograph changed the radiologist's decision based on the PA radiograph alone.

Results.—The PA radiograph revealed abnormalities in all 91 (10.4%) subjects with positive findings at radiography. The lateral radiograph revealed no abnormality in 75 subjects (83.4%). All abnormalities seen on lateral radiographs were also seen on PA radiographs. When abnormalities were seen on both images, the information on the lateral image never caused the radiologist to change the decision he or she made on the basis of the PA image alone.

Conclusion.—In a pre-employment setting, one PA radiograph is sufficient for TB screening of individuals with positive PPD skin test results. Elimination of the acquisition of lateral radiographs would substantially reduce radiation exposure.

▶ The authors state that according to the guidelines of the Centers for Disease Control and the American Thoracic Society, it is recommended that all individuals in the United States, with a positive purified protein derivative (PPD) skin test, undergo screening chest radiography for the exclusion of radiographic evidence of active pulmonary tuberculosis (TB). In many institutions, such radiographic screening is performed using only posteroanterior (PA) chest radiography, while in others, both PA and lateral chest radiographs are performed. Some authors advocate the use of PA and lateral chest radiography in the screening of children and young adults suspected of having tuberculosis.

The authors retrospectively examined an asymptomatic population of patients who underwent chest radiography because of a positive PPD skin test. The individuals studied were evaluated in the employee health division of the authors' tertiary care hospital. Pregnant women and individuals under 18 years of age were excluded. The study population consisted of 875 adults evaluated between January 1, 2007 and December 31, 2007. There were 483 women (mean age, 37 years; age range, 18-65 years) and 392 men (mean age, 38 years; age range, 18-65 years).

Two chest radiologists blinded to the contents of the original radiographic reports reviewed the images and evaluated them for known radiographic abnormalities of acute or chronic tuberculosis infection. All discrepant findings were rereviewed in an additional reading session, after which a third chest radiologist addressed persistent discrepancies. The 91 chest radiographs that exhibited abnormalities showed a calcified lymph node or granuloma (36, 40%), apical pleural thickening (24, 26%), at least one noncalcified nodule (22, 25%), and fibrous scarring (11, 12%). Although it is not known whether any of the subjects had sputum positive for active tuberculosis, there were no subjects with radiographic findings consistent with active TB disease. Lateral chest radiographs demonstrated abnormalities in 16 (18%) of the 91 subjects, but no findings of fibrous scarring or apical pleural thickening were noted on the lateral radiographs. This is an expected finding as such abnormalities are not optimally evaluated with lateral radiography due to their anatomic location in the apical or apical posterior segments of the upper lobes.

The authors concluded that in this patient population there was no need for obtaining lateral chest radiographs. They cite the work of Meyer et al who surveyed an urban population at high risk for TB and reached similar conclusions. In fact, based on the results of their study, the authors changed the protocol for employment screening radiography in patients with positive PPD at their institution and now obtain only PA chest radiographs. They state that this method should be applicable to the radiographic screening of patients with a positive PPD skin test in a pre-employment setting.

The authors cite the potential benefits of reduction of radiation exposure by approximately 67% per examination, cost savings, and time savings in the interpretation of 1 radiograph instead of 2. They list the following limitations of their study: readers only focused on the issue of potential tuberculosis, readers were not blinded to the results of the PA chest radiograph when interpreting the lateral radiograph, and one of the readers knew the purpose of the study at the time the images were interpreted.

M. L. Rosado de Christenson, MD

Differentiation of bacterial and non-bacterial community-acquired pneumonia by thin-section computed tomography

Ito I, Ishida T, Togashi K, et al (Kurashiki Central Hosp, Japan; Kyoto Univ, Japan; et al)

Eur J Radiol 72:388-395, 2009

Background and Objective.—The management of community-acquired pneumonia (CAP) depends, in part, on the identification of the causative agents. The objective of this study was to determine the potential of thin-section computed tomography (CT) in differentiating bacterial and non-bacterial pneumonia.

Patients and Methods.—Thin-section CT studies were prospectively examined in hospitalized CAP patients within 2 days of admission, followed by retrospective assessment by two pulmonary radiologists. Thin-section CT findings on the pneumonias caused by each pathogen were examined, and two types of pneumonias were compared. Using multivariate logistic regression analyses, receiver operating characteristic (ROC) curves were produced.

Results.—Among 183 CAP episodes (181 patients, 125 men and 56 women, mean age ± S.D.: 61.1 ± 19.7) examined by thin-section CT, the etiologies of 125 were confirmed (94 bacterial pneumonia and 31 non-bacterial pneumonia). Centrilobular nodules were specific for non-bacterial pneumonia and airspace nodules were specific for bacterial pneumonia (specificities of 89% and 94%, respectively) when located in the outer lung areas. When centrilobular nodules were the principal finding, they were specific but lacked sensitivity for non-bacterial pneumonia (specificity 98% and sensitivity 23%). To distinguish the two types of pneumonias, centrilobular nodules, airspace nodules and lobular shadows were found to be important by multivariate analyses. ROC curve analysis discriminated bacterial pneumonia from non-bacterial pneumonia among patients without underlying lung diseases, yielding an optimal point with sensitivity and specificity of 86% and 79%, respectively, but was less effective when all patients were analyzed together (70% and 84%, respectively).

Conclusion.—Thin-section CT examination was applied for the differentiation of bacterial and non-bacterial pneumonias. Though showing some potential, this examination at the present time would not be applicable for patients with underlying lung diseases, severe conditions of pneumonia, or immunocompromised conditions.

▶ The authors review a practical clinical classification of community-acquired pneumonia (CAP) based on sensitivity to antibiotics, with bacterial pneumonias responding to beta-lactam antibiotics and non-bacterial pneumonias failing to respond to such treatment. The authors add that while both types of CAP may respond to new macrolides and quinolones, the emergence of resistant strains is a growing concern.

This was a retrospective review of the imaging studies of patients who formed part of an ongoing long-term prospective study to determine the etiologies of CAP in the authors' institution. The study was performed between January 1998 and June 2001. The authors defined CAP as pulmonary infection in patients who developed their symptoms outside "any hospital or health care institution." They performed conventional and thin section chest CT in those hospitalized patients who gave consent with the exclusion of patients under the age of 15 and those with severe dyspnea, shock, thoracic deformity, disturbance of consciousness, pregnancy, and immunodeficiency or underlying lung disease that could confuse the CT findings. All patients underwent microbiological analysis to determine the infectious organism. The final study population consisted of 125 patients, 94 with bacterial, and 31 with nonbacterial pneumonia (3 patients were excluded because they were infected with both bacterial and nonbacterial organisms).

The authors found that lobular opacities, centrilobular bundles, and thickening of the bronchovascular bundles were more frequent findings in nonbacterial CAP (Fig 2 in the orginal article). Ground glass opacity surrounding airspace consolidation and airspace consolidations were more frequent in patients with bacterial CAP (Fig 5 in the orginal article). These findings correlate with the mode of infection of and the host response to the various organisms. For example, patients with pneumococcal pneumonia, typically, experience a parenchymal inflammatory response to the infection in which edema fills the airspaces and can spread to adjacent airspaces, thus manifesting on CT as airspace disease and ground glass opacities. On the other hand, patients with mycoplasma pneumonia, typically, experience involvement limited to the ciliated airways, thus manifesting radiologically as centrilobular nodules. In fact, the authors found high specificity for airspace nodules and centrilobular nodules with regard to the diagnosis of bacterial and nonbacterial pneumonia respectively.

Study limitations include the fact that only hospitalized patients and then only those with a confirmed microbiological diagnosis were included. In addition, patients with chronic obstructive pulmonary disease were not included. These factors may have contributed to selection bias in spite of the prospective manner in which the patients were accrued into the larger study.

M. L. Rosado de Christenson, MD

Interstitial Lung Disease

Lung Morphology in the Elderly: Comparative CT Study of Subjects over 75 Years Old versus Those under 55 Years Old

Copley SJ, Wells AU, Hawtin KE, et al (Hammersmith Hosp, London, England; Royal Brompton Hosp, London, England; et al)
Radiology 251:566-573, 2009

Purpose.—To describe thin-section pulmonary computed tomographic (CT) features in asymptomatic elderly individuals.

Materials and Methods.—Institutional review board approval was given, and informed consent was obtained. Two study groups (older group, over 75 years of age; younger group, under 55 years) were prospectively identified from outpatient requests for CT of the abdomen or brain. Fifty-six consecutive volunteers (older group: $n = 40$, 18 men, 22 women; younger group: $n = 16$, eight men, eight women) with no known respiratory disease were included. Prone thin-section CT imaging was performed, and two observers independently scored images for the presence and extent of CT features (including reticular pattern, ground glass opacity, and thin-walled cystic air spaces). Group comparisons were made, and

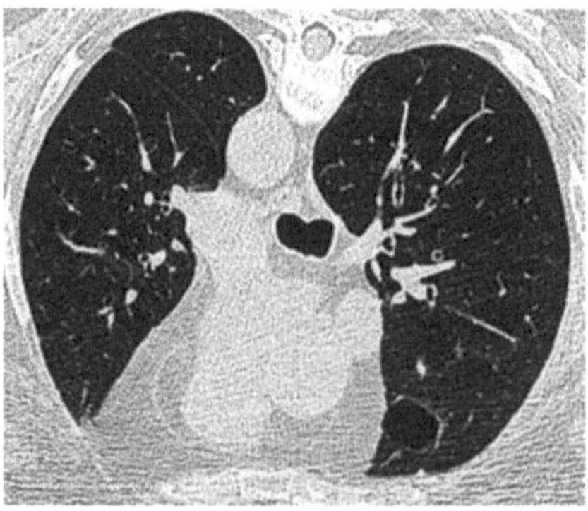

FIGURE 3.—Axial prone thin-section CT image of a 77-year-old woman who was a nonsmoker shows a 2-cm (diameter) thin-walled cystic airspace within the right upper lobe. Bronchial wall thickening was scored as 1 in right and left lower lobes. (Courtesy of Copley SJ, Wells AU, Hawtin KE, et al. Lung morphology in the elderly: comparative CT study of subjects over 75 years old versus those under 55 years old. *Radiology*. 2009;251:566-573. Copyright by the Radiological Society of North America.)

logistic regression analysis was used to assess relationships between CT findings and age and smoking history.

Results.—A limited predominantly subpleural basal reticular pattern was identified in the majority (24 of 40, 60%) of individuals in the older group and was absent (zero of 16) in the younger group ($P < .001$). Cysts were seen in 10 (25%) of the 40 subjects in the older group but were seen in none of the subjects in the younger group ($P = .02$). Bronchial dilation and wall thickening were also seen significantly more frequently ($P < .001$) in the older group (24 [60%] and 22 [55%] of 40, respectively) than in the younger group (both one [6%] of 16). All findings were independent of pack-year smoking history with multiple logistic regression analysis.

Conclusion.—Thin-section CT findings usually associated with interstitial lung disease are frequently seen in asymptomatic elderly individuals and are absent in younger subjects. Therefore, these findings may not necessarily represent clinically relevant disease.

▶ The authors state that the aging lung undergoes physiologic and morphologic changes characterized by loss of alveolar and airway elastic recoil and increasing peripheral lung markings on radiography. Animal models of the aging lung have demonstrated increased collagen deposition and progressive pulmonary fibrosis. The authors prospectively evaluated the thin-section CT features of the lungs of a group of asymptomatic adults to characterize thin-section CT findings in elderly individuals. A series of 56 consecutive outpatient volunteers over the age of 75 years and under the age of 55 years referred for CT of the abdomen or brain between July 2003 and February 2007 comprised the study population. Patients with a history of smoking within the last 5 years, chronic pulmonary disease, recent respiratory symptoms, dyspnea, cardiac failure, malignancy, exposure to carcinogens and/or pneumotoxic drugs, and collagen vascular diseases were excluded. Patients unable to lie in the prone position were also excluded.

To reduce radiation exposure, inspiratory 1 mm CT sections were obtained through the lungs at 20 mm intervals. The images were acquired on an 8-channel multidetector CT scanner. The images were evaluated by 2 experienced thoracic radiologists who were blinded to the clinical information. All but 10 patients underwent pulmonary function tests, and there were no statistically significant differences in the CT features of patients who underwent pulmonary function tests when compared with those who did not.

The authors found an increasing presence of reticular opacities with increasing age, but the difference was not statistically significant. Bronchial dilatation and bronchial wall thickening were significantly more common in the older patient population (Fig 3). There were no statistically significant differences between older ex-smokers and nonsmokers. In addition, basal subpleural cysts were more frequent in elderly patients independent of smoking history (Fig 3). Of note, the above findings were not associated with traction bronchiectasis.

The authors remark that the above-mentioned findings could potentially be mistaken for features of interstitial lung disease. They postulate that the CT abnormalities could be related to increased interlobular septal thickening without inflammation or fibrosis documented in histopathologic studies of the lungs of elderly adults. They also postulate that the thin-walled cysts could result from a check valve phenomenon of partial airway obstruction, and that airway changes in the elderly could relate to age-related impairment of clearance mechanisms and low-grade inflammatory changes.

The authors acknowledge several limitations to their study including the absence of pulmonary function studies in all the study subjects and the inclusion of former smokers. They also acknowledge a possible selection bias based on the inclusion of patients undergoing abdomen or brain CT.

At any rate, the results of this study may prove helpful in the analysis of asymptomatic or minimally symptomatic elderly patients undergoing chest CT.

M. L. Rosado de Christenson, MD

Usual Interstitial Pneumonia and Nonspecific Interstitial Pneumonia with and without Concurrent Emphysema: Thin-Section CT Findings

Akira M, Inoue Y, Kitaichi M, et al (Kinki-Chuo Chest Med Ctr, Osaka, Japan)

Radiology 251:271-279, 2009

Purpose.—To determine whether concurrent emphysema influences the distinction between usual interstitial pneumonia (UIP) and nonspecific interstitial pneumonia (NSIP) at thin-section computed tomography (CT).

Materials and Methods.—Institutional review board approval was obtained for this retrospective study; informed consent was not required. The study included 54 patients with NSIP and 42 patients with UIP (55 men, 41 women; mean age, 60.2 years ± 9.2 [standard deviation]; age range, 33–77 years). Two independent readers assessed the CT images and made a first-choice diagnosis. The appearances of UIP and NSIP at CT were compared with univariate and multivariate analyses. Receiver operating characteristic curves were used to determine how concurrent emphysema influences the distinction of UIP from NSIP at thin-section CT.

Results.—The diagnosis was correct in 136 (71%) of 192 readings. In patients with concurrent emphysema, the diagnosis was correct in 30 (44%) of 68 readings. Sensitivity, specificity, and accuracy for diagnosis were lower in patients with concurrent emphysema than in patients without concurrent emphysema. In patients with concurrent emphysema, there were no significant differences in extent of fibrosis, extent of honeycombing, extent of consolidation, coarseness of fibrosis score, extent of traction bronchiectasis, upper lung irregular lines, peribronchovascular distribution, and nodules between UIP and NSIP. According to multivariate analysis, the CT feature that helped best differentiate UIP from NSIP in patients with emphysema was traction bronchiolectasis.

Conclusion.—Concurrent emphysema influenced the distinction between UIP and NSIP.

▶ Underlying or concurrent emphysema has been shown to have an effect on the imaging manifestations of various pulmonary abnormalities, including lung nodules and pulmonary consolidations. Thus, it may follow that underlying emphysema may also alter or confound the thin section CT findings of common interstitial lung diseases, such as usual interstitial pneumonia (UIP) and non-specific interstitial pneumonia (NSIP). Furthermore, the CT diagnosis of interstitial lung diseases may prove challenging for radiologists without subspecialty training in cardiothoracic radiology, even when it occurs in the absence of underlying emphysema. The diagnosis of NSIP is typically suspected in patients with basilar ground glass and reticular opacities on thin section CT, whereas UIP is characterized by peripheral basilar reticular opacities and honeycomb lung. Subpleural sparing and involvement of more than 15% of the lung parenchyma have been described in patients with NSIP, whereas basilar honeycomb lung and upper lung irregular lines have been described in those with UIP.

The authors performed a retrospective review of the imaging findings of 54 patients with biopsy proven NSIP and 42 patients with biopsy proven UIP between the years 1993 and 2005. Thirty-one patients with UIP and 29 patients with NSIP were former smokers. The authors found that underlying emphysema influenced the distinction of NSIP from UIP or idiopathic pulmonary fibrosis. They state that bullae or emphysematous changes could be potentially mistaken for honeycomb lung. In fact, they observed that the misdiagnosis of UIP in patients with NSIP and concurrent emphysema was associated with the presence of coarser fibrosis, emphysema in areas of ground glass opacity that mimicked honeycomb lung, and the absence of other useful differentiating CT features in these cases. They found that in patients without emphysema, the CT findings that best differentiated UIP from NSIP were traction bronchiolectasis and reticular opacity-to-ground glass opacity ratio, whereas in patients with concurrent emphysema the feature that best helped in differentiation of UIP from NSIP were traction bronchiolectasis and extent of ground glass attenuation.

The authors concede that their study is limited by selection bias. Because all their cases were biopsy proven, the cases of UIP may have exhibited atypical imaging features, as in many instances typical CT findings of UIP provide a presumptive diagnosis and biopsy is not obtained.

At any rate, the evaluation of interstitial lung disease should include a review of the patient's smoking history and an attempt to account for concurrent imaging features of emphysema characterized by centrilobular lucencies without perceptible walls in the case of centrilobular emphysema. Concurrent paraseptal emphysema with small subpleural cystic spaces may potentially be more difficult to differentiate from honeycomb lung.

M. L. Rosado de Christenson, MD

Pulmonary Thromboembolic Disease

Do emergency physicians use serum D-dimer effectively to determine the need for CT when evaluating patients for pulmonary embolism? Review of 5,344 consecutive patients

Corwin MT, Donohoo JH, Partridge R, et al (Warren Alpert Med School of Brown Univ and Rhode Island Hosp, Providence, RI)

AJR Am J Roentgenol 192:1319-1323, 2009

Objective.—The purpose of our study was to investigate whether D-dimer screening is being used effectively to determine the need for MDCT in diagnosing acute pulmonary embolism (PE) in emergency department patients.

Materials and Methods.—We performed a retrospective review of all patients who underwent D-dimer testing or MDCT in the emergency department from January 1, 2003, through October 31, 2005. A D-dimer value of >0.43 microg/mL was considered positive. Diagnosis of PE was made on the basis of the MDCT. Clinical algorithms for diagnosing PE mandate that patients with a low clinical suspicion for PE undergo D-dimer testing, then MDCT if positive. For patients with a high clinical suspicion for PE, MDCT should be performed without D-dimer testing.

Results.—Of 3,716 D-dimer tests, 1,431 (39%) were positive and 2,285 (61%) were negative. MDCT was performed in 166 (7%) patients with negative D-dimer results and in 826 (58%) patients with positive D-dimer results. The prevalence of PE in patients with a high clinical suspicion and no D-dimer testing was 9% (139/1,628), which was higher than the rate of PE in the positive D-dimer group at 2% (19/826) ($p < 0.0001$). There was no significant difference in the prevalence of PE in the positive and negative D-dimer groups (2% vs 0.6%, respectively) ($p = 0.23$). The sensitivity and negative predictive value of D-dimer for PE were 95% (95% CI, 73.1-99.7%) and 99% (95% CI, 96.2-99.9%), respectively.

Conclusion.—D-dimer screening is not used according to established diagnostic algorithms to determine the need for MDCT in diagnosing acute pulmonary embolism in our emergency department.

▶ The authors state that serum D-dimer is a fibrin degradation product, which is elevated in patients with thromboembolic disease. D-dimer has been shown to have high specificity, sensitivity, and negative predictive value for acute pulmonary embolus (PE), it is considered, especially, useful in patients with low clinical probability for PE, and when negative in this population, effectively excludes the diagnosis. When D-dimer is positive, further evaluation is undertaken, typically with CT pulmonary angiography. On the other hand, if the patient has a high clinical probability for pulmonary embolus, then CT angiography is performed without the need for an intervening D-dimer test.

As is the case in many other institutions, the authors observed an increase in the use of CT pulmonary angiography that caused them to question whether emergency department physicians were using the D-dimer test according to published recommendations. To this end, they performed a retrospective review of all patients who had D-dimer testing or CT pulmonary angiography from January 1, 2003 to October 31, 2005. They assumed that according to published algorithms patients undergoing MDCT had a high clinical suspicion for pulmonary embolus, whereas those undergoing D-dimer testing had a low clinical suspicion, an assumption that may or may not be valid in this patient population. Patients with nondiagnostic or inconclusive CT pulmonary angiograms and those with chronic pulmonary embolism were excluded as well as those with D-dimer tests followed by ventilation-perfusion scintigraphy.

Interestingly, CT pulmonary angiography was not performed in 605 patients with a positive D-dimer, and these patients were discharged with a variety of diagnoses, the most common being "other" and "chest pain or dyspnea not otherwise specified." Thus 496 patients were discharged from the emergency department with diagnoses unrelated to their cardiopulmonary symptoms or without an explanation for their complaints (Table 1 in the original article).

The data raises an interesting question given the fact that 42% of patients with a positive D-dimer who should have had CT pulmonary angiography did not, and 7% of those with negative D-dimer who should not have had a CT pulmonary angiogram did. The authors state that 166 CT pulmonary angiograms performed in patients with negative D-dimers would have been deemed unnecessary had the clinical algorithm been followed. They acknowledge that in some of these cases, the clinical presentation may have changed during the patient's stay in the emergency department or that the CT pulmonary angiogram may have been performed in error when a contrast-enhanced chest CT was the examination ordered.

The authors comment on the absence of a significant difference between the population of patients with PE in the positive and negative D-dimer patient groups, and on the low prevalence of PE in both patients with high and low clinical suspicion (9% and 2% respectively). They cite other articles that describe a trend of decreasing rates of PE seen on CT pulmonary angiograms performed in emergency departments. This together with published literature stating that only 23% of surveyed physicians use published guidelines in assessing the pretest probability for pulmonary embolism may help explain the increasing use of CT for evaluation of pulmonary embolism and the decreasing rates of PE.

In an era of great pressures toward cost containment, and of increasing efforts toward the protection of the population from ionizing radiation, this study reveals the role played by practice patterns in the authors' emergency department. It would be interesting to see the results of a similarly designed multi-institutional study to determine the prevalence of such practice patterns in our nation. The authors close by stating that the decreasing threshold for evaluation for PE with CT pulmonary angiography decreases the prevalence of PE in the emergency department. They also state that the high number of false-positive results in this study limits the utility of the D-dimer test, in spite of its high negative predictive value.

M. L. Rosado de Christenson, MD

Radiation dose savings for adult pulmonary embolus 64-MDCT using bismuth breast shields, lower peak kilovoltage, and automatic tube current modulation

Hurwitz LM, Yoshizumi TT, Goodman PC, et al (Duke Univ Med Ctr, Durham, NC)

AJR Am J Roentgenol 192:244-253, 2009

Objective.—The purpose of this study was to assess whether radiation dose savings using a lower peak kilovoltage (kVp) setting, bismuth breast shields, and automatic tube current modulation could be achieved while preserving the image quality of MDCT scans obtained to assess for pulmonary embolus (PE).

Materials and Methods.—CT angiography (CTA) examinations were performed to assess for the presence or absence of pulmonary artery emboli using a 64-MDCT scanner with automatic tube current modulation (noise level = 10 HU), two kVp settings (120 and 140 kVp), and bismuth breast shields. Absorbed organ doses were measured using anthropomorphic phantoms and metal oxide semiconductor field effect transistor (MOSFET) detectors. Image quality was assessed quantitatively as well as qualitatively in various anatomic sites of the thorax.

Results.—Using a lower kVp (120 vs 140 kVp) and automatic tube current modulation resulted in a dose savings of 27% to the breast and 47% to the lungs. The use of a lower kVp (120 kVp), automatic tube current modulation, and bismuth shields placed directly on the anterior chest wall reduced absorbed breast and lung doses by 55% and 45%, respectively. Qualitative assessment of the images showed no change in image quality of the lungs and mediastinum when using a lower kVp, bismuth shields, or both.

Conclusion.—The use of bismuth breast shields together with a lower kVp and automatic tube current modulation will reduce the absorbed radiation dose to the breast and lungs without degradation of image quality to the organs of the thorax for CTA detection of PE.

▶ The authors state that whole body effective radiation dose using 16-array and higher multidetector CT can range between 14 mSv and 20 mSv, which may in turn result in a 0.2% to 2.2% increased lifetime relative risk for either breast or lung cancer. However, they state that lowering the kVp may increase image noise and absorbed radiation in the breasts. Bismuth breast shields have been shown to reduce radiation to the ventral thorax, but may increase image noise. The authors used 2 anthropomorphic female phantoms (a small and a large phantom) to mimic different body habituses and studied the effects of using Bismuth breast shields, lower peak kilovoltage, and automatic tube current modulation on dose and image quality.

The authors state that lowering of the mA results in a linear decrease in dose, while lowering the kVp reduces overall dose by approximately the power of 2. They add that Bismuth shields (available for the reduction of dose to superficial organs including the eye, thyroid, and breast) allow selective radiation reduction by filtering out lower energy photons from the polychromatic X-ray beam.

The study yielded the following findings:

(1) There are dose savings when bismuth shields are used together with tube current modulation.

(2) The dose to the breast and internal organs can be further reduced by using bismuth breast shields and a lower kVp.

(3) The breast shields provided a greater dose reduction in the anterior thorax. The authors suggest that using shields around the entire thorax may reduce radiation further and suggest that manufacturers should explore the development of such shields or of shields that filter out the lower energy photons before impacting the patient.

(4) The use of bismuth shields maintains an acceptable noise level (< 25 Hounsfield units in their study).

The authors believe that while their study was designed with the pulmonary embolus CT protocol in mind, the results could be extrapolated to other thoracic vascular imaging protocols that use similar parameters. While they evaluated 2 different body types and documented dose savings, they did not evaluate the effect of the modifications in patients with significant subcutaneous fat nor did they assess the effects of noise on the diagnosis of pulmonary thromboembolic disease. However, they advocate the use of Bismuth shields and automatic tube current modulation to decrease absorbed doses indicating that these maneuvers may be especially valuable for large patients who require higher radiation doses to maintain image quality.

M. L. Rosado de Christenson, MD

Vascular enhancement and image quality of MDCT pulmonary angiography in 400 cases: comparison of standard and low kilovoltage settings

Matsuoka S, Hunsaker AR, Gill RR, et al (Brigham and Women's Hosp and Harvard Med School, Boston, MA)

AJR Am J Roentgenol 192:1651-1656, 2009

Objective.—The purpose of this study was to investigate the vascular enhancement and image quality of pulmonary CT angiography performed with lower peak kilovoltage settings in a large patient sample.

Materials and Methods.—This retrospective study was approved by our institutional review board, which waived the requirement for informed consent. Four hundred patients believed to have a pulmonary embolism were studied. All patients underwent 16- or 64-MDCT with automatic tube current modulation. The 200 patients in the standard peak kilovoltage group (mean age, 57 years; range, 22-95 years) underwent MDCT at 130 or 120 kVp. The 200 patients in the low peak kilovoltage group (mean age, 56 years; range, 21-92 years) underwent MDCT at 110 or 100 kVp. Vascular enhancement was evaluated by measurement of the attenuation value in the main pulmonary artery and segmental and subsegmental arteries. Image noise was quantified by measurement of the SD of the attenuation value in the main pulmonary artery. One blinded radiologist assessed image quality using visual scores. Wilcoxon's rank test was used to evaluate differences between the groups.

Results.—Mean vascular enhancement in the main pulmonary artery had significantly higher attenuation values in the low peak kilovoltage group (376.1 ± 102.9 HU) than in the standard peak kilovoltage group (309.2 ± 94.8 HU) ($p < 0.0001$). Mean attenuation values in all measured segmental and subsegmental arteries were significantly higher in the low peak kilovoltage group than in the standard peak kilovoltage group ($p < 0.0001$). Image noise in the low peak kilovoltage group was significantly higher than in the standard peak kilovoltage group ($p < 0.0001$). There was no significant difference in the image quality scores of the two groups ($p = 0.116$).

Conclusion.—Lowering kilovoltage improved vascular enhancement without deterioration of image quality. The results of our study confirm previously reported preliminary findings.

▶ The authors state that the prevalence of pulmonary embolism among patients undergoing CT pulmonary angiography is less than 10%. The high number of negative results in patients undergoing CT angiography for exclusion of pulmonary thromboembolic disease highlights the exposure of this population to ionizing radiation in the absence of a positive diagnosis. The authors suggest that lowering the peak kilovoltage setting for CT is an effective method for reducing the radiation dose received by patients. They cite reports that quote a 44% reduction of estimated effective dose. They also cite reports that state that using 80 kVp to 100 kVp instead of 120 kVp to 140 kVp can improve vascular contrast enhancement. This added to concurrent automatic tube current modulation techniques that may result in significant dose reductions for patients undergoing these imaging studies.

The authors performed a retrospective evaluation of MDCT studies in which the reduction of peak kilovoltage had been used to improve contrast enhancement and reduce radiation dose. The images were evaluated with a window level of 100 HU and a window width of 450 HU. There were no statistically significant differences in the populations or in the prevalence of pulmonary emboli in the 2 groups. The authors found that vascular enhancement was greater when using the low peak kilovoltage settings without concurrent deterioration of image quality. They explain that this is due to the increase in the attenuation of iodine-based contrast agents with reduced X-ray energy distribution due to the high relative atomic number of iodine and the k-edge in such energy levels.

Most practicing radiologists have at one time or another found themselves describing a CT pulmonary angiogram as a limited study, due to poor timing of the contrast bolus, poor vascular enhancement, or patient body habitus. This study documents greater vascular enhancement at low peak kilovoltage in central, segmental and subsegmental pulmonary arteries. Further, the authors estimate a 40% to 50% reduction in the radiation dose with a lowering of the peak kilovoltage by 20 kVp.

The authors list a series of limitations to their study:

- They did not calculate the dose with the use of a phantom.
- They did not evaluate the detectability of pulmonary emboli as they did not perform conventional angiography (which may not still be the gold standard for the evaluation of pulmonary embolus).
- They only evaluated a single apical and basilar subsegmental pulmonary artery branch in each patient.
- They used 2 different CT scanners and noted a higher image noise in the studies performed with the 64-channel multidetector CT.

Based on their results, the authors recommend a CT protocol of 100 kVp to 110 kVp for the evaluation of pulmonary embolism in clinical practice. However, it should be noted that the authors do not mention whether the size and weight of their patient population was representative of the wide range of body habitus encountered in many clinical practices.

M. L. Rosado de Christenson, MD

Pulmonary embolism in pregnancy: comparison of pulmonary CT angiography and lung scintigraphy

Ridge CA, McDermott S, Freyne BJ, et al (St Vincent's Univ Hosp, Dublin, Ireland)

AJR Am J Roentgenol 193:1223-1227, 2009

Objective.—The purpose of this study was to retrospectively compare the diagnostic adequacy of lung scintigraphy with that of pulmonary CT angiography (CTA) in the care of pregnant patients with suspected pulmonary embolism.

Materials and Methods.—Patient characteristics, radiology report content, additional imaging performed, final diagnosis, and diagnostic adequacy were recorded for pregnant patients consecutively referred for lung scintigraphy or pulmonary CTA according to physician preference. Measurements of pulmonary arterial enhancement were performed on all pulmonary CTA images of pregnant patients. Lung scintigraphy and pulmonary CTA studies deemed inadequate for diagnosis at the time of image acquisition were further assessed, and the cause of diagnostic inadequacy was determined. The relative contribution of the inferior vena cava to the right side of the heart was measured on nondiagnostic CTA images and compared with that on CTA images of age-matched nonpregnant women, who were the controls.

Results.—Twenty-eight pulmonary CTA examinations were performed on 25 pregnant patients, and 25 lung scintigraphic studies were performed on 25 pregnant patients. Lung scintigraphy was more frequently adequate for diagnosis than was pulmonary CTA (4% vs 35.7%) ($p = 0.0058$).

Pulmonary CTA had a higher diagnostic inadequacy rate among pregnant than nonpregnant women (35.7% vs 2.1%) ($p < 0.001$). Transient interruption of contrast material by unopacified blood from the inferior vena cava was identified in eight of 10 nondiagnostic pulmonary CTA studies.

Conclusion.—We found that lung scintigraphy was more reliable than pulmonary CTA in pregnant patients. Transient interruption of contrast material by unopacified blood from the inferior vena cava is a common finding at pulmonary CTA of pregnant patients.

▶ The authors state that pulmonary thromboembolic disease is the leading cause of maternal death during pregnancy. Although there are no published guidelines for the evaluation of pulmonary embolus in this patient population, CT pulmonary angiography is considered the test of choice in the general population. The authors state that CT may be a less reliable diagnostic test in the pregnant population when compared with the nonpregnant population based on the hyperdynamic state of pregnancy, hemodilution, and interruption of contrast secondary to unopacified blood from the inferior vena cava (IVC) resulting in poor opacification of the pulmonary arteries.

To compare CT pulmonary angiography and lung scintigraphy in their ability to diagnose pulmonary emboli in the pregnant population, the authors performed a retrospective review of consecutive pregnant patients referred for either CT pulmonary angiography or scintigraphy between July 1, 2006 and April 1, 2008. While the studies were ordered by referral physicians according to their practice preferences, all patients with pulmonary abnormalities on chest radiography were studied with CT angiography. The authors performed a keyword search for the words "inadequate", "indeterminate", "poor opacification", "suboptimal", and "nondiagnostic".

The authors then classified CT pulmonary angiograms as adequate or inadequate for the diagnosis of PE. The studies on pregnant patients were evaluated for attenuation of the pulmonary trunk with adequate opacification determined if it was greater than 250 HU. The images were also assessed for the presence of transient interruption of contrast by unopacified blood from the IVC.

Based on their results (presented in the abstract), the authors concluded that in pregnant patients lung scintigraphy produces diagnostic images more frequently than CT angiography and found that most nondiagnostic CT angiograms showed transient interruption of contrast material by unopacified blood from the IVC. The authors review the literature and document various explanations for the transient contrast interruption effect in pregnancy, which is thought to arise from the increase in venous return to the right heart that occurs after deep inspiration maneuvers that are typically used for acquiring CT pulmonary angiograms. It has been shown that venous return to the heart increases by approximately 50% during deep inspiration, with most of the returning blood contributed by the IVC. Deep inspiration is thought to produce a decrease in intrathoracic pressure and a pressure gradient known as the thoracoabdominal pump with a reported 3% incidence in nonpregnant patients. Pregnant patients have been documented to have a 6-fold increase in IVC pressure in the supine position, which likely contributes to this artifact. Other explanations for this phenomenon may involve vertebral and azygos collaterals, which may help counteract the mechanical obstruction of the IVC caused by the pregnant uterus. In two of the 10 non-diagnostic studies the artifact was not present, and the authors postulate that increases in cardiac output, total vascular resistance, heart rate, and plasma volume may have contributed to dilution of contrast material. The authors add that it has been shown that cardiac output is inversely related to peak arterial enhancement and time of arrival of contrast in the thoracic aorta. Thus, the elevated cardiac output of pregnancy may further compromise contrast enhancement in the pulmonary arteries in these patients.

Based on their experience and the results of their study, the authors suggest acquisition of images in pregnant patients during shallow inspiration or held expiration to reduce transient interruption of contrast and the resultant dilution of contrast-opacified blood in the right atrium. They add that lung scintigraphy should be considered the study of choice for evaluating pregnant patients suspected of having a pulmonary embolus unless CT angiography image quality can be optimized with a modification of the typical breathing maneuvers.

The authors list several limitations of their study, including the small number of patients and the absence of echocardiographic data to exclude the possibility of an occult patent foramen ovale or atrial septal defect in the pregnant patients studied with CT angiography, which may have contributed to the nondiagnostic studies. In addition, the fact that all patients with pulmonary abnormalities on chest radiography were evaluated with CT angiography may have introduced selection bias.

M. L. Rosado de Christenson, MD

Technique

Retrodiaphragmatic portion of the lung: how deep is the posterior costophrenic sulcus on posteroanterior chest radiography?

Oh JK, Ahn MI, Kim HL, et al (The Catholic Univ of Korea, Seoul)
Clin Radiol 64:786-791, 2009

Aim.—To determine the depth of the posterior costophrenic sulcus (CPS) on posteroanterior (PA) chest radiography in relation to the diaphragmatic dome and lateral CPS.

Materials and Methods.—Five hundred and forty consecutive PA chest radiographs that were performed for general health screenings were retrospectively reviewed. Among them 282 radiographs were selected that met the following criteria: visualization of the inferior boundary of the posterior CPS behind the right hemidiaphragm; and no abnormal findings that affected the shape and level of the diaphragm. The selected chest radiographs were from 155 men and 127 women with a mean age of 40.7 ± 8.4 years. On 282 PA chest radiographs, the distances between the right diaphragmatic dome and posterior CPS (total diaphragmatic height), the diaphragmatic dome and lateral CPS (diaphragmatic dome height), and the lateral and posterior CPS (posterior CPS depth) were

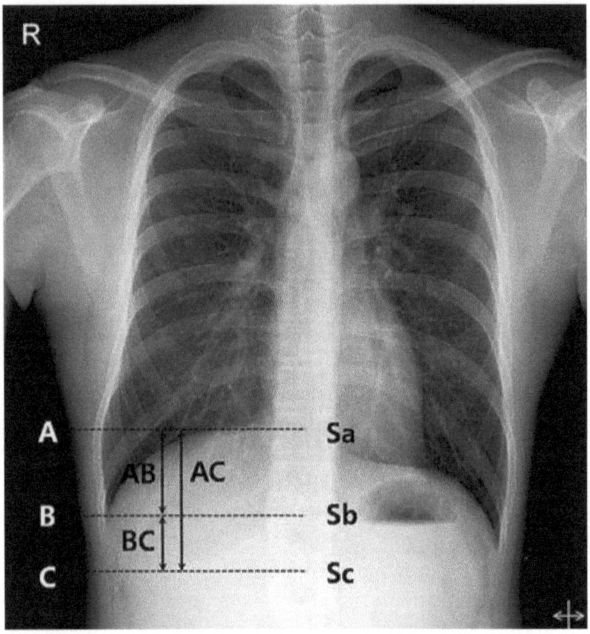

FIGURE 1.—PA chest radiograph showing radiographic measurements. A = horizontal line tangent to the top of the right diaphragmatic dome; B = horizontal line tangent to the lowest point of the right lateral CPS; C = horizontal line tangent the lowermost level of the right posterior CPS; AC = diaphragm height; BC = posterior CPS depth; Sa = spine level of diaphragmatic dome; Sb = spine level of lateral CPS; Sc = spine level of posterior CPS. In this individual, AC was 87.7 mm; AB, 51.2 mm; BC, 36.5 mm, and the spine level scores were 11, 13, 14 for Sa, Sb, and Sc, respectively. (Reprinted from Oh JK, Ahn MI, Kim HL, et al. Retrodiaphragmatic portion of the lung: how deep is the posterior costophrenic sulcus on posteroanterior chest radiography? *Clin Radiol.* 2009;64:786-791, with permission from The Royal College of Radiologists.)

measured. In addition levels of the right lateral and posterior CPS were scored in relation to levels of the thoracic and lumbar spines. The relationships between the posterior CPS depth and demographic and physical data and other radiographic measurements were analysed.

Results.—The mean right posterior CPS depth was 29.2 ± 15.6 mm. The average level of the posterior CPS in relation to the spine was 13.5 ± 0.6, i.e., the level of lower half of the L1 vertebral body. The posterior CPS depth had a tendency to be deeper in those participants who were taller ($r = 0.17$, $p < 0.01$), had a higher body mass index (BMI; $r = 0.25$, $p < 0.01$), longer total diaphragmatic height ($r = 0.55$, $p < 0.01$), and shorter diaphragmatic dome height ($r = -0.18$, $p < 0.01$).

Conclusion.—As the posterior CPS is deeper than the lateral CPS by approximately 3 cm, and reaches, on average, to L1, the standard chest PA radiograph must include >3 cm below the level of lateral CPS, or should include the L1 spine.

▶ Radiography is the most frequently performed imaging study in the initial assessment of patients presenting with thoracic complaints. With the increasing demand for advanced thoracic imaging with CT, MR, positron emission tomography (PET), and ultrasound, radiologists in busy practices could be tempted to neglect quality assurance in the performance of chest radiography. However, keeping in mind the large numbers of studies performed in all institutions, radiologists must continue to implement quality initiatives in the performance of chest radiography. This article reminds us that a significant amount of lung is imaged through the diaphragm and can be assessed on well exposed chest radiographs. The authors recommend inclusion of the posterior costophrenic sulcus (CPS) in its entirety on all frontal chest radiographs.

The American College of Radiology (ACR) Practice Guidelines for the Performance of Pediatric and Adult Chest Radiography revised in 2006 state that: the chest radiograph should include both of the lung apices and costophrenic angles.[1] The midthoracic vertebral bodies and the left retrocardiac pulmonary vessels should be appropriately defined. Although quality assurance efforts in chest radiography universally mandate the inclusion of the bilateral costophrenic angles on the frontal radiograph, this practice may result in exclusion of a portion of the retrodiaphragmatic lung parenchyma.

The authors based their observations on a retrospective review of 282 posteroanterior (PA) chest radiographs of asymptomatic subjects with normal chest radiographic findings or minor abnormalities that did not affect the level or shape of the right hemidiaphragm. All the studies included allowed visualization of the inferior border of the right posterior CPS and were obtained at full inspiration with the subject in the upright position. Fig 1 shows the different measurements obtained with respect to the right hemidiaphragm height and the location of the lateral and posterior CPS. The figure also shows the parameters measured, including diaphragm height and posterior CPS depth. The authors correlated the location of the different right hemidiaphragm landmarks with the corresponding portion of the vertebral body located at the same level.

The authors highlight the importance of inclusion of the posterior CPS on frontal radiographs and the careful evaluation of this portion of the lung for significant pulmonary pathology, including lung cancer. They further state that the CPS depth forms approximately one third of the total right diaphragmatic height. They found that the posterior CPS was deeper in individuals with a higher body mass index (BMI) and in thin tall individuals. They conclude that the right posterior CPS is on average 3 cm deeper than the lateral CPS and that in 95% of cases it was situated between the inferior half of T12 and the superior half of L2. They also point out that in some cases the posterior CPS may be situated approximately 6 cm inferior to the lateral CPS.

Thus, quality chest radiography should not be limited to the inclusion of the lateral CPS but should strive to include the entirety of the lung bases bordered by the posterior CPS. Radiologists must assess frontal radiographs for the presence of disease in this area and work with their technical staff to improve the quality of chest radiography in their institutions.

M. L. Rosado de Christenson, MD

Reference

1. ACR Practice Guideline for the Performance of Pediatric and Adult Chest Radiography. www.acr.org/SecondaryMainMenuCategories/quality_safety/guidelines/dx/chest/chest_radiography.aspx. Accessed October 7, 2009.

Miscellaneous

Pleural Effusion: Characterization with CT Attenuation Values and CT Appearance

Abramowitz Y, Simanovsky N, Goldstein MS, et al (Hadassah-Hebrew Univ Med Ctr, Jerusalem, Israel)
AJR Am J Roentgenol 192:618-623, 2009

Objective.—The purpose of this study was to assess the utility of CT in characterizing pleural effusions on the basis of attenuation values and CT appearance.

Materials and Methods.—We retrospectively analyzed 100 pleural effusions in patients who underwent chest CT and diagnostic thoracentesis within 48 hours of each other. On the basis of Light's criteria, effusions were classified as exudates or transudates using laboratory biochemistry markers. The mean value in Hounsfield units of an effusion was determined using a region of interest on the three slices with the greatest quantity of fluid. All CT scans also were reviewed for the presence of additional pleural features such as fluid loculation, pleural thickening, and pleural nodules.

Results.—Twenty-two of the 100 pleural effusions were transudates and 78 were exudates. The mean attenuation of the exudates (7.2 HU; [SD] 9.4 HU; range, 21–28 HU) was not significantly lower than the mean attenuation of the transudates (10.1 HU; 6.9 HU; range, 0.3–32 HU), ($p = 0.24$). None of the additional CT features accurately differentiated exudates from transudates ($p > 0.1$). Fluid loculation was found in 58% of exudates and in 36% of transudates. Pleural thickening was found in 59% of exudates and in 36% of transudates.

Conclusion.—The clinical use of CT attenuation values to characterize pleural fluid is not accurate. Although fluid loculation, pleural thickness, and pleural nodules were more commonly found in patients with exudative effusions, the presence of these features does not accurately differentiate between exudates and transudates.

▶ This is a retrospective review of 100 pleural effusions evaluated with both CT and thoracentesis (performed within 48 hours of each other) with the object of finding differentiating CT features between transudates and exudates. The pleural effusions consisted of 78 exudates and 22 transudates (diagnosed based on Light's criteria) affecting 62 men and 38 women ranging in age from 17 to 94 years (mean age; 62.4 years). The authors state that as previous CT studies that identified differentiating features between exudates and transudates were performed over a decade ago, they wanted to analyze the contribution of modern CT scanners to the diagnosis of pleural effusion. The scans were performed on one of two 16-channel multidetector CT scanners or on a 64-channel multidetector CT scanner. Pleural effusions were evaluated by 2 experienced radiologists noting CT attenuation, loculation, pleural thickening (thickness >4 mm), pleural nodules, and increased attenuation of subpleural fat.

The authors concluded that pleural fluid attenuation values were not useful in classifying pleural effusions as exudates or transudates with considerable overlap between the 2 types of effusions. Although it is expected that exudative pleural effusions will contain higher levels of protein, lactate dehydrogenase (LDH), and bilirubin, which should result in increased attenuation, the authors did not find significant differences between the attenuation values of exudates and transudates. In fact, all 13 pleural effusions with negative attenuation values were diagnosed as exudates. The authors state that exudates contain a larger amount of cholesterol when compared with transudates and that cholesterol content may negate the elevated attenuation expected from increased pleural fluid protein, thus contributing to decreased attenuation. In addition, both pleural thickening and loculation were seen in more than one third of patients with documented transudates, and pleural nodules, thought to be highly specific for malignant (typically exudative) pleural effusions, were found in one of the transudates (5%).

The authors acknowledge several limitations of their study, including the retrospective nature of the review and the fact that the thoracentesis and CT were not performed at the same time. In fact, intervening treatment (including diuresis) has been shown to alter the biochemistry of pleural fluid. The study population included a large number of exudates. The authors postulate that selection bias for complicated effusions related to malignancy or infection may have resulted in a much larger number of patients with exudative effusions undergoing both thoracentesis and chest CT, as these diagnostic tests are rarely used in evaluating patients with pleural effusions secondary to heart failure (typically transudates). In addition, 3 different CT scanners were used for evaluating the pleural effusions and both contrast enhanced and unenhanced studies were included.

Studies have shown that CT attenuation measurements (in the evaluation of emphysema) are influenced by equipment calibration[1] and reconstruction algorithms.[2] The authors do not mention whether they assessed the reconstruction algorithm used or whether the CT scanners were calibrated for water. Streak artifacts from adjacent structures (such as the patient's upper extremities) may also influence CT measurements. The authors do not mention whether they took the above issues into consideration. At any rate, the retrospective nature of the investigation and the different scanners used would make it difficult to address these issues.

Nevertheless, it is important to understand that absolute attenuation measurements are not diagnostic of pleural fluid type and that thoracentesis remains the gold standard in the differentiation between exudates and transudates and is critical in the evaluation of complicated pleural effusions.

M. L. Rosado de Christenson, MD

References

1. Parr DG, Stoel BC, Stolk J, Nightingale PG, Stockley RA. Influence of calibration on densitometric studies of emphysema progression using computed tomography. *Am J Respir Crit Care Med.* 2004;170:883-890.
2. Boedeker KL, McNitt-Gray MF, Rogers SR, et al. Emphysema: effect of reconstruction algorithm on CT imaging measures. *Radiology.* 2004;232:295-301.

2 Breast Imaging

Introduction

I have had a wonderful 11 years serving in the capacity of breast imaging editor for the YEAR BOOK OF DIAGNOSTIC RADIOLOGY. Anne Osborn's invitation to be a part of this comprehensive publication allowed me an exposure to a vast array of information published throughout the world. I learned much more than I taught, and for that I am grateful.

This edition begins with an article on the risk/benefit ratio of obtaining screening mammograms on asymptomatic women 80 years and older. The controversial recommendations issued in late 2009 by the US Preventative Services Task Force was an unwelcome surprise to the medical community, particularly the portion regarding the reversal of support for screening women in their 40s. This was an unfortunate conclusion drawn by the task force, and much has been written in contradiction of their findings. However, I feel that a portion of the task force report has merit. Doing less, when appropriate, is a good message and for asymptomatic women over 80 (and even younger according to the task force), finding benign or in situ disease may indeed cause more harm than good.

The second article focuses on situations where, rather than doing less, we need to consider doing more. Regarding specific ethnic groups/races, this important study outlines the need to consider initiating screening at ages younger than 40, and possibly more frequently than yearly, because cancer appears to develop at younger ages.

I chose to focus the second portion of this edition's breast articles on the management of palpable breast masses assessed with ultrasound. High quality US imaging should be practiced, and when solid masses with benign characteristics are found, the palpable nature of the mass should not, a priori, demand a biopsy recommendation. Far too much time, money, and intervention are directed toward lesions with a very low likelihood of breast cancer. With excellent technique, high-quality equipment, and experience, breast imagers should be able to triage lesions into those truly suspicious for cancer and those that can be followed clinically and/or with imaging.

Also included in this edition is information regarding breast MR, both technical advances/challenges and impact on patient management; diffusion-weighted imaging techniques; nonsurgical intervention procedures; and an overview of molecular imaging and its application to breast imaging.

I look forward to next year's YEAR BOOK edition to see what interests and excites Dr Osborn's newly appointed breast imaging editor, Dr Tanya W. Stephens. I had the pleasure of meeting Dr Stephens recently, and I am confident the readership interested in breast imaging is in great hands.

Robyn L. Birdwell, MD

Weighing the Benefits and Burdens of Mammography Screening Among Women Age 80 Years or Older

Schonberg MA, Silliman RA, Marcantonio ER (Beth Israel Deaconess Med Ctr, Brookline, MA; Boston Univ Med Ctr, MA)

J Clin Oncol 27:1774-1780, 2009

Purpose.—To examine outcomes of mammography screening among women ≥ 80 years to inform decision making.

Patients and Methods.—We conducted a cohort study of 2,011 women without a history of breast cancer who were age ≥ 80 years between 1994 and 2004 and who received care at one academic primary care clinic or two community health centers in Boston, MA. Medical record data were abstracted on all screening and diagnostic mammograms, breast ultrasounds and biopsies performed, all breast cancers diagnosed through December 31, 2006, and on sociodemographics. Date and cause of death were confirmed using the National Death Index.

Results.—The majority of patients (78.6%) were non-Hispanic white and 51.4% (n = 1,034) had been screened with mammography since age 80 years. Among women who were screened, eight were diagnosed with ductal carcinoma in situ, 16 with early stage disease (1.5%), two with late stage disease, and one died as a result of breast cancer. Many (110; 11%) experienced a false-positive screening mammogram that led to 19 benign breast biopsies, eight refused work-up, and three experienced a false-negative screening mammogram; 97 were screened within 2 years of their death from other causes. There were no significant differences in the rate, stage, recurrence rate, or deaths due to breast cancer between women who were screened and those who were not screened.

Conclusion.—The majority of women ≥ 80 years are screened with mammography yet few benefit. Meanwhile, 12.5% experience a burden from screening. The data from this study can be used to inform elderly women's decision making and potentially lead to more rational use of screening (Table 3).

▶ The topic of age (when to start—when to end mammographic screening) is very important to consider. Very high risk very young women (maybe even <25) may benefit from the sensitivity of MRI surveillance. On the other end of the spectrum are the many women who are 80 years and older. In these women, if their clinical examination is negative, having a mammogram may

TABLE 3.—Follow-Up Data on Women Who Were Screened After Age 80 Years Compared With Those Who Were Not Screened

Characteristic	Mammography Screening Since Age 80 Years (n = 1,034) No.	%	No Mammography Screening Since Age 80 Years (n = 977) No.	%
Developed breast cancer including DCIS	28		20	
Mean Charlson Comorbidity Score for those who were diagnosed with breast cancer	1.5		1.9	
Alive after December 31, 2006	16; 4 with DCIS	57	7	35
Alive but unconfirmed	0		1*	
Total deaths	12; 4 with DCIS	43	12	60
Death as a result of breast cancer	1		2	
Death as a result of other causes†	11		10	
Cardiovascular disease	5		2	
Cancer (non-breast)	0		3	
Infections	1		0	
CNS disorders (including dementia)	2		0	
Pulmonary disease	0		1	
Stroke			1	
Gastrointestinal disease	1		1	
Metabolic (including diabetes)	1		0	
Unknown	0		0	
Did not develop breast cancer or DCIS	1,006		957	
Alive after December 31, 2006	450	45	279	29
Alive but unconfirmed	135	13	159	17
Death as a result of cause other than breast cancer	421	42	519	54
Cause of death	421		519	
Cardiovascular disease	147	35	171	33
Cancer other than breast	67	16	88	17
Infections	47	11	53	10
CNS disorders (including dementia)	48	11	54	10
Pulmonary disease	28	7	38	7
Stroke	24	6	44	8
Gastrointestinal disease	16	4	24	5
Renal disease	17	4	16	3
Metabolic (including diabetes)	12	3	17	3
Other	13	3	12	2
Unknown	2	0.5	2	0.4

Abbreviation: DCIS, ductal carcinoma in situ.

*The last note in the medical record was from May 2006 from a hospitalization when the patient had pneumonia and there was no mention of breast cancer recurrence.

†For women diagnosed with breast cancer we confirmed cause of death listed in the National Death Index data with medical record data. Because more women in the non-screened group died as a results of cancer other than breast than did women in the screened group, we present their data here: none had breast cancer recurrence; one died as a result of bladder tumor, one as a result of colon cancer, and one as a result of non-Hodgkin's lymphoma (National Death Index and medical record data agreed in all three cases).

not make a difference in their outcome and, in fact, may cause unnecessary intervention and treatment. Dr Schonberg and her group looked very closely at the benefit/burden question in women 80 and over and found, in fact, the benefits to be few while 12.5% experienced a frank burden from the screening study. Does an 80-year-old asymptomatic woman (with or without comorbid medical problems) benefit with a diagnosis of ductal carcinoma in situ (DCIS), or maybe even with a diagnosis of a small tubular carcinoma? Although it is a sensitive topic, with attempts to limit spending, rethinking how best to provide benefit without overburdening an increasingly large population of women is one that I do not think we can afford to ignore.

R. L. Birdwell, MD

African-American and Low–Socioeconomic Status Patients Have a Worse Prognosis for Invasive Ductal and Lobular Breast Carcinoma: Do Screening Criteria Need to Change?

Yang R, Cheung MC, Franceschi D, et al (Univ of Miami, FL; et al)
J Am Coll Surg 208:853-870, 2009

Background.—Determine the effect of race, socioeconomic status (SES) and other demographic variables on outcomes of patients with invasive ductal and lobular breast cancer.

Study Design.—Florida cancer registry and inpatient hospital data were queried for patients diagnosed with invasive breast cancer from 1998 to 2002.

Results.—A total of 63,472 patients with breast cancer were identified. Overall, 90.5% of patients were Caucasian, 7.6% African American, and 8.7% Hispanic. African-American patients presented at a younger age and with more-advanced disease, 10.5% presented with breast cancer before the age of 40 years, and 22.4% before 45 years of age. African-American patients were less likely to undergo operations. Similarly, low-SES patients were less likely to have operations and presented more often with larger tumors. Stepwise multivariate analysis revealed a substantial drop in the hazard ratio for African-American patients once correction for stage of presentation was made, suggesting that disparities in breast cancer outcomes are, in part, a result of advanced stage at presentation. Race and low SES were independent predictors of worse prognosis when controlling for patient comorbidities and treatment.

Conclusions.—Dramatic disparities by patient race and SES exist in breast cancer. Our study integrates previous smaller studies, providing comprehensive insight into African-American patients and their outcomes for breast cancer. Earlier screening programs and greater access to cancer care for the poor and African Americans are needed. Successful institution of such programs will not completely erase disparities in outcomes for breast cancer in African-American patients (Figs 1-3).

▶ This article was chosen as a representative of those publications emphasizing the differences in age at presentation and prognosis for those women who are either African-American or have a low socioeconomic status. While the issues are complex, there are some absolute facts reiterated by these authors: (1) While breast cancer incidence among African-American women is lower than

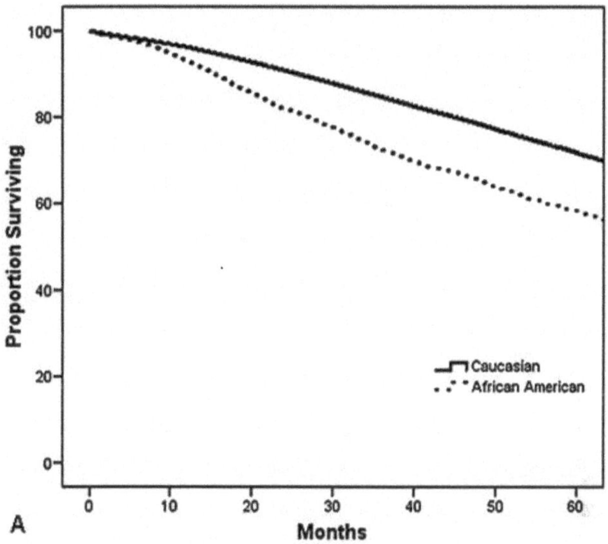

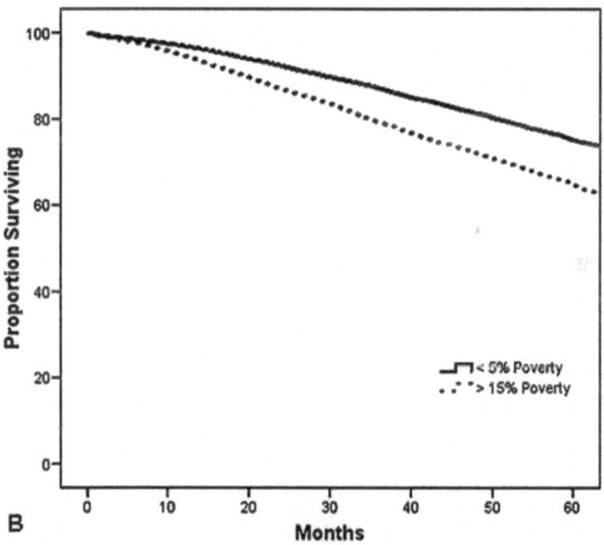

FIGURE 1.—Kaplan-Meier survival curves for patients with invasive breast carcinoma according to (A) race and (B) poverty level. (Reprinted from Yang R, Cheung MC, Franceschi D, et al. African-American and low–socioeconomic status patients have a worse prognosis for invasive ductal and lobular breast carcinoma: do screening criteria need to change? *J Am Coll Surg.* 2009;208: 853-870, with permission from the American College of Surgeons.)

Caucasian counterparts, African-American breast cancer patients suffer a higher mortality rate once diagnosed; (2) African-American women are diagnosed at a younger age with more advanced disease than Caucasian women; (3) African-American race is a predictor of poor outcomes with invasive breast

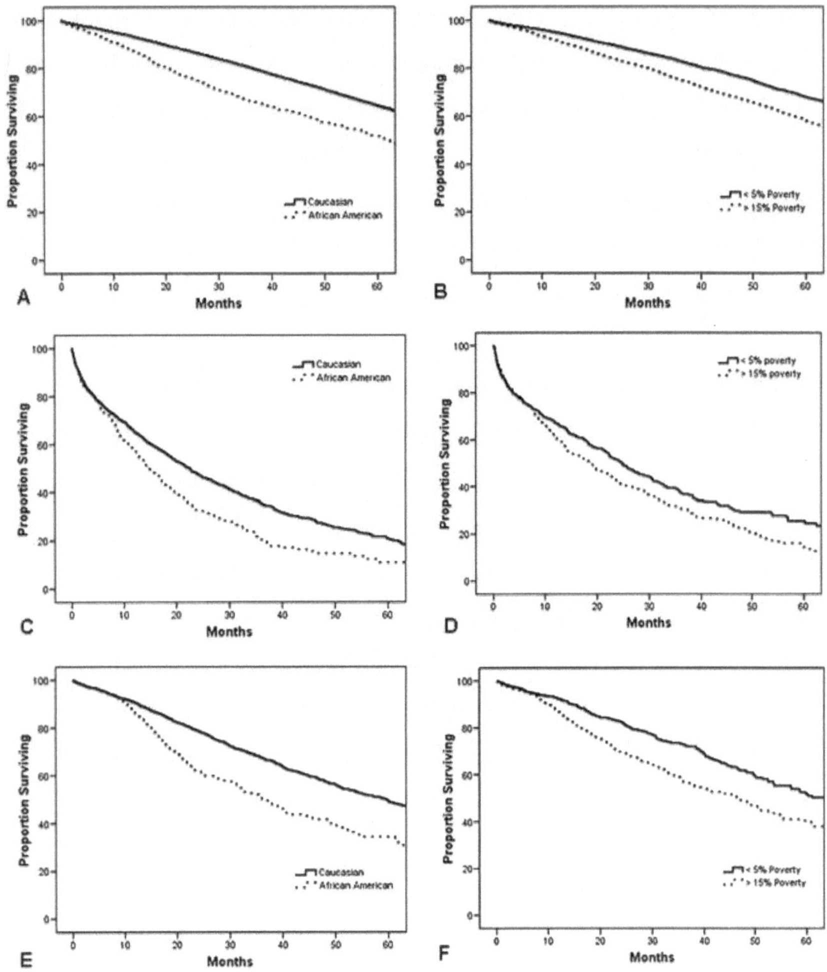

FIGURE 2.—Kaplan-Meier survival curves for patients with invasive breast carcinoma according to race and (A) patient age older than 65 years, (C) distant disease, and (E) tumors > 5 cm. Patient poverty level and (B) patient age > 65 years old, (D) distant disease and (F) tumors > 5 cm are also shown. (Reprinted from Yang R, Cheung MC, Franceschi D, et al. African-American and low–socioeconomic status patients have a worse prognosis for invasive ductal and lobular breast carcinoma: do screening criteria need to change? *J Am Coll Surg*. 2009;208:853-870, with permission from the American College of Surgeons.)

cancer independent of demographics, socioeconomic status, and tumor biology; (4) Women in the lowest socioeconomic group are diagnosed more often with larger tumors, high-grade tumors, and more advanced stage; they are also treated less frequently with surgical therapy, and survival rates decrease as poverty level increases. Some of the aspects of these ethnic and socio-economic differences may in fact have no obvious solution, but consideration should be given to the initiation of screening at a younger age (such as 33 in

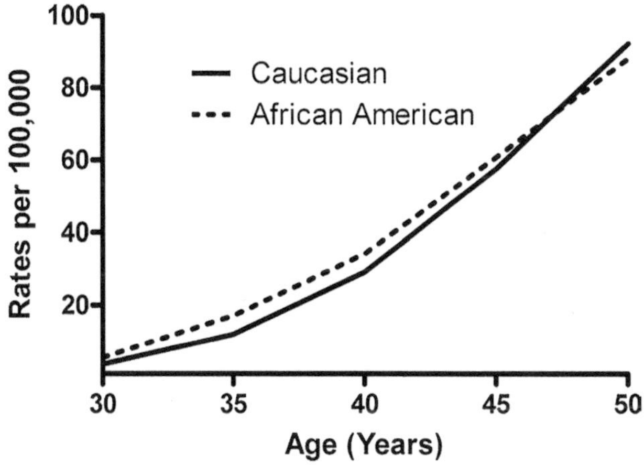

FIGURE 3.—Age-adjusted incidence rate of invasive breast carcinoma according to race. (Data from the National Cancer Institute's Surveillance Epidemiology and End Results database, adjusted to the Standard US Population year 2000.) (Reprinted from Yang R, Cheung MC, Franceschi D, et al. African-American and low–socioeconomic status patients have a worse prognosis for invasive ductal and lobular breast carcinoma: do screening criteria need to change? *J Am Coll Surg*. 2009;208:853-870, with permission from the American College of Surgeons.)

African-American women—an age that parallels the occurrence of breast cancer in Caucasian women recommended to commence screening at age 40) and to narrowing the gap in time between diagnosis and treatment (greater in African-American than Caucasian women).

R. L. Birdwell, MD

Short-Term Hormone Therapy Suspension and Mammography Recall: A Randomized Trial

Buist DSM, Anderson ML, Reed SD, et al (Univ of Washington, Seattle)
Ann Intern Med 150:752-765, 2009

Background.—Without population-based evidence, some clinicians recommend short-term suspension of hormone therapy to improve the performance of mammography. Hormone therapy increases breast density, and abnormal screening mammograms are more common among women with denser breasts and among women using hormone therapy.

Objective.—To test whether 1 to 2 months of hormone therapy suspension before screening mammography decreases additional mammographic imaging (recall) in women age 45 to 80 years.

Design.—3-group randomized, controlled trial.

Setting.—Integrated health plan in western Washington from 2004 to 2007.

Patients.—1704 women age 45 to 80 years who used hormone therapy at their most recent screening (index) mammography, were due for screening (study) mammography, and were still using hormone therapy.

Intervention.—Block random assignment (by breast density and hormone therapy type) to no hormone therapy suspension ($n = 567$) or suspension for 1 month ($n = 570$) or 2 months ($n = 567$) before study mammography. One blinded expert radiologist interpreted all mammograms.

Measurements.—Recall was the primary outcome, and change in mammographic breast density (percentage and dense area) between the index and study mammograms was the secondary outcome.

Results.—Mammography recall rates were 11.3% (61 of 542 women in the no-suspension group), 12.3% (50 of 478 women in the 1-month suspension group), and 9.8% (44 of 451 women in the 2-month suspension group). No subgroups were identified in which brief suspension of hormone therapy resulted in decreased mammography recall. With suspension, decreases in percentage of breast density were orderly and statistically significant: 0.1% (no-suspension group), −0.9% (1-month suspension group), and −1.5% (2-month suspension group). Similar ordered decreases were observed for dense area. Women in the suspension groups experienced increased menopause symptoms.

Limitations.—Results can only be generalized to women age 45 to 80 years who have used hormone therapy for at least 1 year and will consider short-term suspension; most eligible women (61%) declined participation. Mammography recall was determined by 1 expert radiologist.

Conclusion.—Brief hormone therapy suspension was associated with small changes in breast density and did not affect recall rates. No evidence supports short-term hormone therapy suspension before mammography.

▶ An interesting suggestion was presented by an Italian investigator at the recent Fifth International Congress on MR-Mammography held in Jena, Germany in September 2009. Regarding the decreased sensitivity of breast MR in the setting of marked background parenchymal enhancement, particularly prominent in some young premenopausal women for whom breast MR is used for screening (high-risk BRCA genetic mutations and Hodgkin disease survivors), would a short exposure to tamoxifen help subdue the enhancement thus improving sensitivity? While very likely not applicable to the clinical setting, the information in this article from a group in Seattle, Washington, suggests that short-lived alterations in hormonal milieu are unlikely to alter breast density substantially, and by inference, might have little effect on background enhancement. Of course, there may be no direct correlation between background enhancement and breast tissue volume, but at least for mammography recall reduction, it appears that the practice many of us have followed (the suggestion of cessation of hormone replacement therapy (HRT) followed by a short interval repeat mammogram) may not be effective in the evaluation of mammographic findings.

R. L. Birdwell, MD

Comparison of Digital Mammography and Screen-Film Mammography in Breast Cancer Screening: A Review in the Irish Breast Screening Program
Hambly NM, McNicholas MM, Phelan N, et al (Eccles Screening Unit, Dublin, Ireland; et al)
AJR Am J Roentgenol 193:1010-1018, 2009

Objective.—Clinical trials to date into the use of full-field digital mammography (FFDM) for breast cancer screening have shown variable results. The aim of this study was to review the use of FFDM in a population-based breast cancer screening program and to compare the results with screen-film mammography.

Materials and Methods.—The study included 188,823 screening examinations of women between 50 and 64 years old; 35,204 (18.6%) mammograms were obtained using FFDM. All films were double read using a 5-point rating scale to indicate the probability of cancer. Patients with positive scores were recalled for further workup. The recall rate, cancer detection rate, and positive predictive value (PPV) of FFDM were compared with screen-film mammography.

Results.—The cancer detection rate was significantly higher for FFDM than screen-film mammography (6.3 vs 5.2 per 1,000, respectively; $p = 0.01$). The cancer detection rate for FFDM was higher than screen-film mammography for initial screening and subsequent screening, for invasive cancer and ductal carcinoma in situ, and across all age groups. The cancer detection rate for cancers presenting as microcalcifications was significantly higher for FFDM than for screen-film mammography (1.9 vs 1.3 per 1,000, $p = 0.01$). The recall rate was significantly higher for FFDM than screen-film mammography (4.0% vs 3.1%, $p < 0.001$). There was no significant difference in the PPVs of recall to assessment for FFDM and screen-film mammography (15.7% and 16.7%, $p = 0.383$).

Conclusion.—FFDM resulted in significantly higher cancer detection and recall rates than screen-film mammography in women 50-64 years old. The PPVs of FFDM and screen-film mammography were comparable. The results of this study suggest that FFDM can be safely implemented in breast cancer screening programs (Tables 1 and 2).

▶ Change is difficult, whether it is to a new job or to a new imaging modality. Much is lost in leaving the comfort zone inhabited for maybe decades and learning new skills and inheriting new problems. However, the mammography digital "train" is very unlikely to reverse its course and with more than 50% of United States practicing having at least one digital acquisition unit, the fact that this modality performs well and may in fact have the potential to find additional screen-detected cancers as well as be poised to enable future technology such as tomosynthesis, is good news indeed. This particular article is reviewed because it reports on the performance of a large and very well-organized screening program outside the United States where digital equipment replaced analog throughout an entire country during the time period of 2005 to April 2008, at which time the Irish screening program became the first national

TABLE 1.—Previous Studies Comparing Full-Field Digital Mammography (FFDM) to Screen-Film Mammography (SFM)

Study [Reference no.]	Study Design: Retrospective or Prospective	Study Design: Population	Imaging Technique(s)	Patient Age (y)	Study Size (No. of Examinations)	Results: Recall Rate (p)	Results: Cancer Detection Rate (p)	Results: PPV_1 (p)
INBSP Study [current study]	Retrospective review	Population-based screening program; initial and subsequent screenings	FFDM or SFM	50–64	188,823 total: 35,204 FFDM + 153,619 SFM	FFDM > SFM: 4.0% vs 3.1% (< 0.001)	FFDM > SFM: 0.63% vs 0.52% (0.01)	FFDM = 15.7%, SFM = 16.7% (0.383)
Oslo I [10]	Prospective case control study	Population-based screening program; subsequent screenings only	FFDM and SFM	50–69	3,683 paired examinations: 3,683 FFDM and 3,683 SFM	FFDM > SFM: 4.6% vs 3.5% (NS)	SFM > FFDM: 0.76% vs 0.62% (NS)	SFM = 20%, FFDM = 12% (NA)
Oslo II [13]	Prospective randomized control trial	Population-based screening program; initial and subsequent screenings	FFDM or SFM	45–69	23,929 total: 16,985 SFM + 6,944 FFDM	FFDM > SFM: 4.2% vs 2.5% (< 0.001)	FFDM > SFM: 0.59% vs 0.38% (0.02)	SFM = 15.1%, FFDM = 13.9% (0.68)
DMIST [11]	Prospective study	Population recruited over 2 years at 33 sites in the United States and Canada	FFDM and SFM		42,760 paired examinations: 42,760 FFDM and 42,760 SFM	8.4% for both FFDM and SFM	Accuracy of FFDM was significantly higher in pre- or perimenopausal women < 50 y with dense breasts[a]	
Lewin et al. [9]	Prospective study	Women presenting for screening at two institutions	FFDM and SFM	≥ 40	6,736 paired examinations: 6,736 FFDM and 6,736 SFM	SFM > FFDM: 14.9% vs 11.8% (< 0.001)	SFM > FFDM: 0.49% vs 0.4% (NS)	FFDM = 3.4%, SFM = 3.3% (NS)
Heddson et al. [17]	Retrospective review	Population-based screening program; subsequent screening only	PCDR, CR, or SFM	40–74	52,172 total: 25,901 SFM + 9,841 PC-DR + 16,430 CR	SFM > PCDR: 1.4% vs 1.0% (< 0.001) SFM > CR: 1.4% vs 1% (< 0.001)	PCDR > SFM 0.49% vs 0.31% (0.01) CR > SFM 0.38% vs 0.31% (0.22)	PCDR = 47%, CR = 39%, SFM = 22% (< 0.001)

Del Turco et al. [16]	Retrospective review	Population-based screening program; initial and subsequent screenings	FFDM or SFM	50–69	28,770 total: 14,385 FFDM + 14,385 SFM	FFDM > SFM: 4.56% vs 3.96% (0.01)	FFDM > SFM (NS) 0.72% vs 0.58% (0.14)	FFDM = 15.9%, SFM = 14.7% (0.65)
Vigeland et al. [15]	Retrospective review	Population-based screening program; initial screenings only	FFDM only; compared with SFM data from previous 9 y	50–69	343,002 total: 18,239 FFDM + 324,763 SFM	FFDM = SFM: 4.09% vs 4.16% (NS)	FFDM > SFM for all cancers: 0.77% vs 0.65% (0.058) DCIS: 0.21% vs 0.11% (< 0.001)	FFDM = 16.6%, SFM = 13.5% (0.014)

Note—PPV_1 = positive predictive value of recall to assessment, NS = nonsignificant, NA = not available, PCDR = photon-counting direct radiography, CR = computed radiography, DM = digital mammography, DMIST = Digital Mammographic Imaging Screening Trial, INBSP = Irish National Breast Screening Programme, DCIS = ductal carcinoma in situ.

[a]The end point of the study was diagnostic accuracy.

TABLE 2.—Age Distribution and Screening Round Compared for Screen-Film Mammography (SFM) and Full-Field Digital Mammography (FFDM)

Characteristic	SFM ($n = 153{,}619$) No.	%	FFDM ($n = 35{,}204$) No.	%
Age (y)				
50–54	53,229	34.7	12,188	34.6
55–59	56,209	36.6	12,493	35.5
60–64	44,181	28.8	10,523	29.9
Screening round				
Initial	44,156	28.7	9,546	27.1
Subsequent	109,463	71.3	25,658	72.9

screening center in Europe to be fully digitized. There are differences in the practice patterns of the Irish screening program as compared with that of the United States, including the fact that all screening mammograms are double read, the minimum number of screening mammograms interpreted/radiologist/year is > 5000 (and as high as 20 000), the invitation to screening is for women aged 50 to 64 years, and the interval for all of these screened women is every 2 years. These sorts of practice differences always raise the question as to whether the results of one group (here an entire country's screening program) may be applied to another, but the most recent study from Norway (so-called Oslo II)[1] also found overall improved full-field digital mammography (FFDM) performance as compared with analog, and it is likely that with increasing experience with this new modality, at least the detection of asymptomatic cancer within the denser breast mammograms may be aided with the use of digital mammography.

R. L. Birdwell, MD

Reference

1. Skaane P, Hofvind S, Skjennald A. Randomized trial of screen-film versus full-field digital mammography with soft-copy reading in population-based screening program: follow-up and final results of Oslo II study. *Radiology*. 2007;244: 708-717.

Best Ways to Provide Feedback to Radiologists on Mammography Performance

Aiello Bowles EJ, Geller BM (Group Health Ctr for Health Studies, Seattle, WA; Univ of Vermont, Burlington)
AJR Am J Roentgenol 193:157-164, 2009

Objective.—The Mammography Quality Standards Act of 1992 required a minimum performance audit of radiologists performing mammography. Since then, no studies have evaluated radiologists' perceptions of their audit reports, such as which performance measures are the most or least useful, or what the best formats are to present performance data.

Materials and Methods.—We conducted a qualitative study with focus groups and interviews of 25 radiologists currently practicing mammography. All radiologists practiced at one of three sites in the Breast Cancer Surveillance Consortium (BCSC). The discussion guide included open-ended questions to elicit opinions on the following subjects: the most useful performance outcome measures, examples of reports and formats that are easiest to understand (e.g., graphs or tables), thoughts about comparisons between individual-level and aggregate data, and ideas about additional performance measures they would find useful. All discussions were tape-recorded and transcribed. We developed a set of themes and used ethnographic software to qualitatively analyze and extract quotes from transcripts.

Results.—Radiologists thought that almost all performance measures were useful. They particularly liked seeing individual data presented in graphic form with a national benchmark or guideline for each performance measure clearly marked on the graph. They appreciated comparisons between their individual data and their peers' data (within their facility or state) and requested comparisons with national data (such as the BCSC). Many thought customizable, Web-based reports would be useful.

Conclusion.—Radiologists think that most audit statistics are useful; however, presenting performance data graphically with clear benchmarks may make them easier to understand.

▶ I am including this article as a way to focus thought on the positive associations of understanding required and other audit information with the fact that feedback as to these numbers has been shown to improve performance. The Mammography Quality Standards Act (MQSA) requires a minimal audit but does not offer either standard audits or feedback systems to aid in the individual radiologist's self-evaluation. The United States differs from other countries with breast screening programs in a number of ways, including the required number of mammograms to be interpreted per year (eg, > 5000 in the United Kingdom) and the methods used to evaluate the performance. Perry[1] outlines a 2-week training program in the United Kingdom that includes working with a specialist for 3 sessions per week interpreting screening mammograms. Each radiologist also receives personal audit reports. Learning the level of radiologists' understanding of audit reports and enlightening the radiologists as to the fact that this sort of information can indeed improve performance[2] are 2 of the goals of this chosen article. The interviewed community radiologists were not only enthusiastic to learn the benefits of understanding and using audit information but also had definite ideas as to how best to "see" the data (Fig 1 in the original article). I found this information very informative and look forward to working with our connectivity group toward the creation of a robust web-based tool, allowing a better understanding of and flexibility for the display of our audit data accompanied by benchmarks established by the Breast Cancer Surveillance Consortium (BCSC).[3,4]

R. L. Birdwell, MD

References

1. Perry NM. Interpretive skills in the National Health Service Breast Screening Programme: performance indicators and remedial measures. *Semin Breast Disease.* 2003;6:108-113.
2. Jamtvedt G, Young JM, Kristoffersen DT, O'Brien MA, Oxman AD. Does telling people what they have been doing change what they do? A systematic review of the effects of audit and feedback. *Qual Saf Health Care.* 2006;15:433-436.
3. Rosenberg RD, Yankaskas BC, Abraham LA, et al. Performance benchmarks for screening mammography. *Radiology.* 2006;241:55-66.
4. Sickles EA, Miglioretti DL, Ballard-Barbash R, et al. Performance benchmarks for diagnostic mammography. *Radiology.* 2005;235:775-790.

Digital Breast Tomosynthesis: Observer Performance Study

Gur D, Abrams GS, Chough DM, et al (Univ of Pittsburgh, PA; Magee-Womens Hosp, Pittsburgh, PA)

AJR Am J Roentgenol 193:586-591, 2009

Objective.—The purpose of this study was to compare in a retrospective observer study the diagnostic performance of full-field digital mammography (FFDM) with that of digital breast tomosynthesis.

Materials and Methods.—Eight experienced radiologists interpreted images from 125 selected examinations, 35 with verified findings of cancer and 90 with no finding of cancer. The four display conditions included FFDM alone, 11 low-dose projections, reconstructed digital breast tomosynthesis images, and a combined display mode of FFDM and digital breast tomosynthesis images. Observers rated examinations using the screening BIRADS rating scale and the free-response receiver operating characteristic paradigm. Observer performance levels were measured as the proportion of examinations prompting recall of patients for further diagnostic evaluation. The results were presented in terms of true-positive fraction and false-positive fraction. Performance levels were compared among the acquisitions and reading modes. Time to view and interpret an examination also was evaluated.

Results.—Use of the combination of digital breast tomosynthesis and FFDM was associated with 30% reduction in recall rate for cancer-free examinations that would have led to recall if FFDM had been used alone ($p < 0.0001$ for the participating radiologists, $p = 0.047$ in the context of a generalized population of radiologists). Use of digital breast tomosynthesis alone also tended to reduce recall rates, an average of 10%, although the observed decrease was not statistically significant ($p = 0.09$ for the participating radiologists). There was no convincing evidence that use of digital breast tomosynthesis alone or in combination with FFDM results in a substantial improvement in sensitivity.

Conclusion.—Use of digital breast tomosynthesis for breast imaging may result in a substantial decrease in recall rate.

▶ Mammography is a good tool and has been shown to find small clinically occult breast cancers. Since the initiation of screening mammography, the mortality of breast cancer has gone down. This is great news, but all of us in the field know that this examination, like so many, is far from perfect. Tomosynthesis, with 3D reconstruction, is a technical method designed to help reduce the difficulty of overlapping breast tissue as seen in 2D mammography. This article's conclusion that screening recall rates could be decreased is what would be expected if the imager really had a better chance to "see through" the busy background of the normal tissue and not necessarily have to bring the woman back for additional focused spot compression views. Unfortunately, the development of tomosynthesis for clinical use has been slow and as of 2009, I am curious as to what role this technique (and how it should best be performed) may actually play in day-to-day clinical practice.

R. L. Birdwell, MD

Breast Imaging Reporting and Data System Lexicon for US: Interobserver Agreement for Assessment of Breast Masses

Abdullah N, Mesurolle B, El-Khoury M, et al (McGill Univ Health Ctr, Quebec, Canada)

Radiology 252:665-672, 2009

Purpose.—To retrospectively evaluate the interobserver agreement of radiologists who used the Breast Imaging Reporting and Data System (BI-RADS) lexicon to characterize and categorize ultrasonographic (US) features of breast masses.

Materials and Methods.—No institutional review board approval or patient consent was required. Five breast radiologists retrospectively independently evaluated 267 breast masses (113 benign and 154 malignant masses in 267 patients) by using the BIRADS US lexicon. Reviewers were blinded to mammographic images, medical history, and pathologic findings. Interobserver agreement was assessed with the Aickin revised κ statistic.

Results.—Interobserver agreement varied from fair for evaluation of mass margins ($\kappa = 0.36$) to moderate for evaluation of lesion boundary ($\kappa = 0.48$), echo pattern ($\kappa = 0.58$), and posterior acoustic features ($\kappa = 0.47$) to substantial for evaluation of mass orientation ($\kappa = 0.70$) and shape ($\kappa = 0.64$). For small (≤ 0.7 cm; $n = 49$) or malignant ($n = 154$) masses, low concordance was noted for margin descriptors ($\kappa = 0.30$ and 0.28, respectively) and BI-RADS category ($\kappa = 0.21$ and 0.26, respectively). Overall, only fair agreement was obtained for BI-RADS category ($\kappa = 0.30$). Agreement for subdivisions 4a, 4b, and 4c of BI-RADS category 4 was fair ($\kappa = 0.33$), fair ($\kappa = 0.32$), and poor ($\kappa = 0.17$), respectively.

Conclusion.—Reproducibility of US BI-RADS terminology is good except for margin evaluation. A trend toward lower concordance was noted for the evaluation of small masses and malignant lesions. Classification into subdivisions 4a, 4b, and 4c was poorly reproducible.

▶ As a reality rebuttal to my belief that we should be able to limit the number of ultrasound (US)-guided biopsies of solid masses (both nonpalpable and palpable) that have no suspicious US characteristics and are oval and smooth,[1] I chose this report of a retrospective evaluation of interobserver agreement (more disagreement, unfortunately) for the assessment of ultrasound characteristics of breast masses. Regarding margins, the group found only a fair agreement level ($k = 0.36$), but if the mass was small or malignant, the margin assessment agreement levels decreased to as low as 0.28. Of importance is the fact that even when the reviewers disagreed as to the margin status or level of concern categorization (as in Fig 1 in the original article), biopsy was recommended and the lesion was not placed into either an imaging follow-up or benign category. Rather than throwing the baby out with the bath water, I would hope we could take the information from these investigators as a challenge to further refine our US image characteristic evaluations with the goal of limiting unnecessary biopsies for those lesions very likely to be benign while working harder to limit misclassification of those lesions with suspicious (and in particular non-smooth margins) imaging findings.[1]

R. L. Birdwell, MD

Reference

1. Shin JH, Han BK, Ko EY, Choe YH, Nam SJ. Probably benign breast masses diagnosed by sonography: is there a difference in the cancer rate according to palpability? *AJR Am J Roentgenol.* 2009;192:187-191.

Short-Term Follow-Up of Palpable Breast Lesions With Benign Imaging Features: Evaluation of 375 Lesions in 320 Women

Harvey JA, Nicholson BT, LoRusso AP, et al (Univ of Virginia, Charlottesville; et al)

AJR Am J Roentgenol 193:1723-1730, 2009

Objective.—The purpose of this study was to evaluate the feasibility of short-term follow-up of palpable masses that have benign imaging features.

Materials and Methods.—The cases of all women with round, oval, or lobular palpable masses with circumscribed margins and homogeneous ultrasound echotexture for which short-term follow-up was recommended from July 1997 through December 2003 were retrospectively identified. Evaluation was by ultrasound and/or mammography and focused clinical

TABLE 2.—Age Distribution of Masses Assessed as Probable Fibroadenoma

Age (y)	No. of Women ($n = 262$)[a]	No. of Lesions ($n = 304$)[a]	No. of Women With Multiple Lesions[b]
12–19	29 (11.1)	37 (12.2)	8 (27.6)
20–29	88 (33.6)	109 (35.9)	11 (12.5)
30–39	81 (30.9)	88 (28.9)	6 (7.4)
40–49	43 (16.4)	47 (15.5)	4 (9.3)
50 and greater	21 (8.0)	23 (7.6)	1 (4.8)

[a]Values in parentheses are percentages.
[b]Values in parentheses are percentage of women in age group with probable fibroadenoma.

TABLE 3.—Outcome of 375 Palpable Lesions in 320 Women

Lesion Outcome	No.	No. of Cases of Breast Cancer
Sampled because of patient or health care provider preference or other reason	85 (22.7)	1
Resolved (BI-RADS 1)	86 (22.9)	0
Decreased in size or level of suspicion (BI-RADS 2)	65 (17.3)	0
Stable (BI-RADS 2) (mean follow-up, 30 mo; range, 12–94 mo)	113 (30.1)	0
Increase in size or level of suspicion at follow-up	26 (6.9)	0 (24/26 sampled)

Note—Values in parentheses are percentages.

TABLE 4.—Comparison of Palpable and Nonpalpable Breast Masses With Recommendation of Short-Term Follow-Up

Nature of Lesion	Time Period	Cancer Incidence	95% CI (%)
Palpable	7/97–12/03	1/375 (0.3%)	0.0001–0.015
Nonpalpable	1/98–12/00	10/641 (1.6%)	0.8–2.9

examination. Outcome was assessed with imaging or clinical follow-up lasting at least 12 months. The cancer incidence for palpable lesions was compared with that for nonpalpable lesions recommended for short-term follow-up.

Results.—In 379 women, 443 palpable masses with benign features for which shortterm follow-up was recommended were identified. Outcome data were available on 375 masses in 320 women. Lesions were evaluated with mammography and ultrasound ($n = 186$) or ultrasound alone ($n = 189$). Masses were typically identified only with ultrasound ($n = 258$, 68.8%); were oval ($n = 275$, 73.3%), of equal density to normal breast tissue on mammograms ($n = 95$ on 117 mammograms, 81.2%), and hypoechoic ($n = 336$ in 372 ultrasound examinations, 90.3%); and were prospectively believed to be fibroadenoma ($n = 304$, 81.1%). Eighty-five lesions (22.7%) were biopsied soon after evaluation, and one 1.5-mm ductal carcinoma in situ was diagnosed. At follow-up (mean, 2.7 years), 26 lesions (6.9%) had grown. Twenty-four of the 26 lesions were biopsied, and no cancer was diagnosed. The overall cancer prevalence was similar for palpable (0.3%) and nonpalpable (1.6%) masses. The cost of short-term follow-up was less than that of biopsy.

Conclusion.—Short-term follow-up is a reasonable alternative to biopsy of palpable breast lesions with benign imaging features, particularly for young women with probable fibroadenoma (Tables 2–4).

▶ This is a very important article attacking the issue of whether nonsuspicious solid palpable breast masses can safely be followed with imaging alone. The underlying important assumption is that breast imagers should have equipment and training adequate enough to provide them with the knowledge of what constitutes a nonsuspicious solid breast mass. Attention to detail, particularly of the margin characteristics of small lesions is imperative. However, assuming that accredited imagers have the ability to characterize lesions using the American College of Radiology Breast Imaging Reporting and Data System (ACR BI-RADS) lexicon as a guide, the mere fact that a lesion is palpable does not in and of itself necessitate the need for a biopsy recommendation. It is very important that any suspicious finding be acted on, but this article underscores what many of us know from our own practice—there are many breast lesions and in particular those in young women that can indeed be followed with imaging. Although the cost of imaging follow-up is not zero, it is less than that of biopsy.

R. L. Birdwell, MD

Probably Benign Breast Masses Diagnosed by Sonography: Is There a Difference in the Cancer Rate According to Palpability?

Shin JH, Han B-K, Ko EY, et al (Sungkyunkwan Univ School of Medicine, Seoul, South Korea)

AJR Am J Roentgenol 192:W187-W191, 2009

Objective.—The study was designed to assess retrospectively whether there is a difference in the cancer rates between palpable and nonpalpable probably benign breast nodules detected by sonography. We further investigated the clinicopathologic results of sonographically false-negative cases.

Materials and Methods.—This study included 352 women who had undergone sonographically guided core biopsies for 374 BI-RADS category 3 masses between March 2004 and February 2005. No masses were diagnosed with definite malignant findings on mammography. The cancer rates of nonpalpable and palpable masses were compared using

Fisher's exact test. The clinicopathologic results of sonographically false-negative cases were investigated.

Results.—Among the 374 masses, 86 masses (23%) that were lost to follow-up were excluded. Of the 288 masses with follow-up or excision, the cancer rate was 2.4% (7/288). The cancer rate of the nonpalpable masses was 2.1% (4/194) (95% CI, 0.6–5.2%), and the cancer rate of the palpable masses was 3.2% (3/94) (95% CI, 0.6–9.0%), with no statistically significant difference ($p = 0.6864$). Of the seven cancers, five were diagnosed by a sonographically guided core biopsy and two were diagnosed by surgical excision after a benign biopsy. The seven sonographically false-negative masses in seven patients were identified as three invasive ductal carcinomas, two ductal carcinomas in situ, one mucinous carcinoma, and one papillary carcinoma on the basis of pathology results.

Conclusion.—There is no statistically significant difference between the cancer rates of palpable and nonpalpable BI-RADS category 3 masses seen on sonography.

▶ This is an important addition to the literature of breast imaging because it focuses on one of the areas in which we can affect change regarding cost. Yes, there are differences in breast ultrasound expertise, but with strict well-educated attention to good technique using improved imaging technology, we should raise the bar as to when we recommend biopsies for solid masses and follow with imaging the appropriate oval, smooth, nonsuspicious masses that are very apt to be benign. Further, this paradigm needs to include both the palpable and mammographically detected masses (that have benign characteristics by ultrasound). This study looked only at those lesions for which a category 3 (short interval imaging follow-up) had been assessed but then went on to biopsy, and therefore was an excellent cadre of exactly the sort of lesions for which we want to know their outcome. At 2.4%, the authors' overall results of cancer detection are just slightly greater than the < 2% cancer rate recommended for those mammographic lesions placed into category 3. Of importance as to palpable versus nonpalpable lesions there was no statistical difference (3.2% vs 2.1%). (As an important aside, the original article included 2 figures of nonpalpable lesions that I would critique based on the appearance of their margins as not appropriate to have placed into the Breast Imaging Reporting and Data System [BI-RADS] 3 category. However, perfection is not attainable, and the very good overall statistics as to missed cancers is the important take-home point in this audit.)

Other investigators[1,2] have reported similar reassurances regarding letting morphologic imaging characteristics rather than palpability (alone) determine biopsy recommendations. I recently attended the Royal College of Radiology annual 2009 meeting held in Belfast, Northern Ireland, and was asked what I might suggest to help contain the costs associated with breast imaging/biopsy. My answer was that we can make a difference in the arena of limiting biopsies of those oval, smooth, nonsuspicious solid breast masses, both nonpalpable and palpable, and use imaging follow-up with assurance that with strict attention to imaging quality and mass detail, we may look toward achieving acceptable very low missed cancer rates while saving untold thousands of dollars and unnecessary interventions.

R. L. Birdwell, MD

References

1. Graf O, Helbich TH, Fuchsjaeger MH, et al. Follow-up of palpable circumscribed noncalcified solid breast masses at mammography and US: can biopsy be averted? *Radiology.* 2004;233:850-856.
2. Raza S, Chikarmane SA, Neilsen SS, Zorn LM, Birdwell RL. BI-RADS 3, 4, and 5 lesions: value of US in management—follow-up and outcome. *Radiology.* 2008; 248:773-781.

Sonography of Axillary Masses: What Should be Considered Other Than the Lymph Nodes?

Kim EY, Ko EY, Han B-K, et al (Samsung Med Ctr, Seoul, Korea; et al)

J Ultrasound Med 28:923-939, 2009

Objective.—The purpose of this study was to review the sonographic findings of various axillary masses other than lymph nodes in correlation with other imaging and pathologic findings.

Methods.—From a sonographic database, we collected interesting cases of axillary masses with pathologic or other imaging corroboration from the last 10 years.

Results.—Images of various soft tissue masses were reviewed. They included masses associated with accessory breasts (fibroadenomas, hamartomas, fat necrosis, and cancer arising from axillary breasts), other soft tissue masses (lipomas, schwannomas, hemangiomas, fibromatosis, epidermoid cysts, and malignant fibrous histiocytomas), and complications presenting as masses after axillary lymph node dissection (seromas, hematomas, suture granulomas, pseudoaneurysms, and lymphangiectasia).

Conclusions.—Awareness of the characteristic sonographic findings of various disease entities that cause axillary masses will help in the correct diagnosis of axillary masses.

▶ Readers are encouraged to read this article in part because it demonstrates a number of uncommon axillary masses and in part because the images are beautiful (Figs 3, 4, 6, 10, and 17). The many entities that may present within the axilla of both men and women cover a broad spectrum of conditions. However, one condition not common to the axilla is that of simple cysts. I want to caution us as imagers to be very wary of what may appear initially as anechoic axillary masses and be mistakenly assessed as cysts. Much more likely with such findings is the presence of tumor-filled markedly hypoechoic axillary nodes. Using color Doppler to assess what may initially be mistaken as cysts may be very important in making the correct diagnosis of both lymph node and vascular abnormalities (Fig 16).

R. L. Birdwell, MD

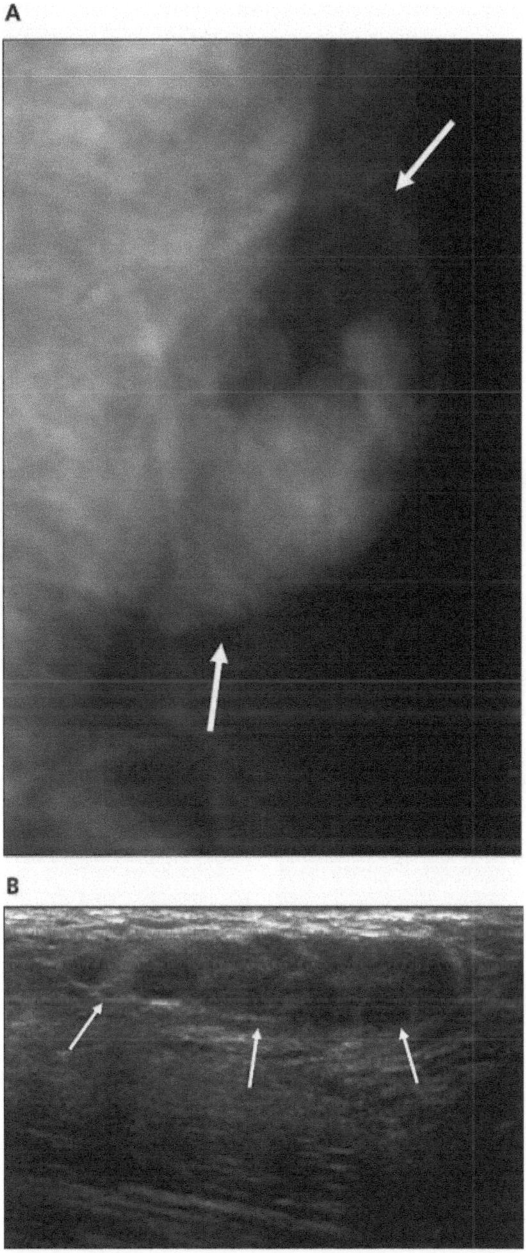

FIGURE 3.—Images from a 38-year-old female patient with a hamartoma that developed in an accessory breast. **A**, Medio-lateral oblique mammography shows an encapsulated circumscribed mass (arrows) containing fat density located within the axillary breast. **B**, Sonography shows an elongated well-circumscribed mixed echoic mass (arrows) originating from accessory breast tissue. (Courtesy of Kim EY, Ko EY, Han B-K, et al. Sonography of axillary masses: what should be considered other than the lymph nodes? *J Ultrasound Med.* 2009;28:923-939.)

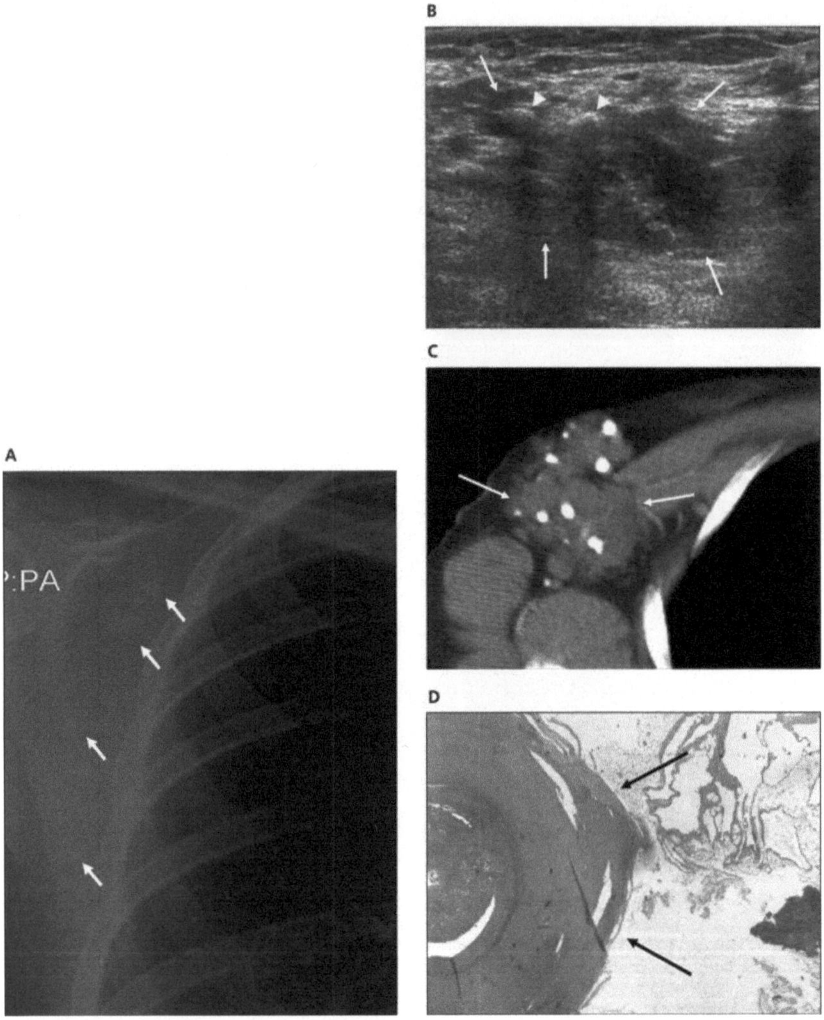

FIGURE 4.—Images from a 32-year-old female patient with massforming fat necrosis in the right axilla. **A**, Chest radiography shows multiple small round calcifications (arrows) in the right axilla. **B**, Sonography shows poorly defined hypoechoic lesions in the right axilla (arrows). There are curvilinear echogenic foci with posterior shadowing (arrowheads) within the mass. **C**, Contrast-enhanced axial CT shows a lobulated low-density mass (arrows) in the right axilla. Within the mass, dense round calcifications are scattered. **D**, Photomicrograph shows a round fibrocalcific nodule (arrows) in the background of fat necrosis (hematoxylin-eosin, original magnification ×40). (Courtesy of Kim EY, Ko EY, Han B-K, et al. Sonography of axillary masses: what should be considered other than the lymph nodes? *J Ultrasound Med.* 2009;28:923-939.)

A

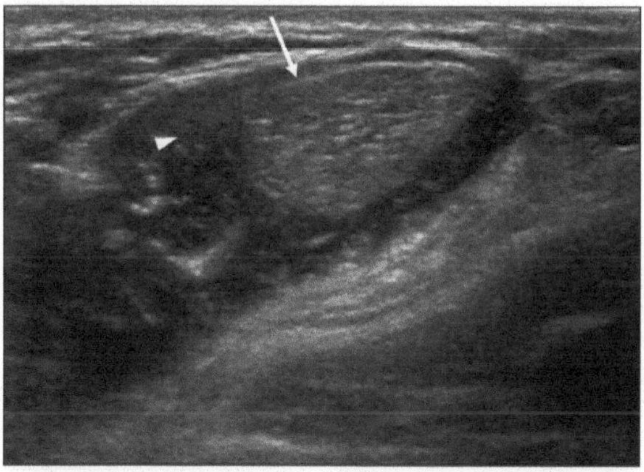

B

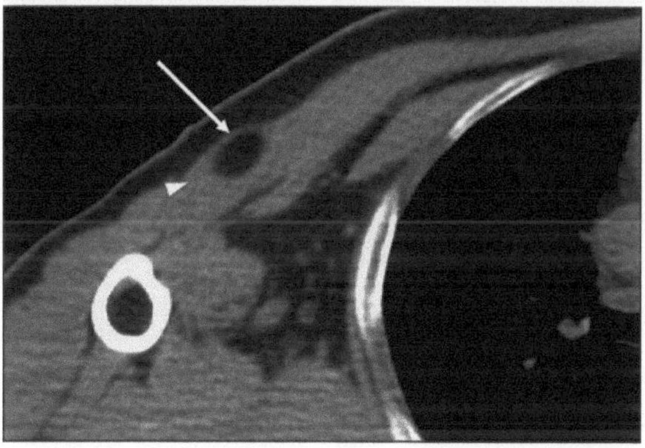

FIGURE 6.—Images from a 69-year-old male patient with an intramuscular lipoma of the right axilla. A, Sonography shows a small well-circumscribed hyperechoic mass (arrow) in the right axilla. The mass is surrounded by the hypoechoic pectoralis major muscle (arrowhead). B, Noncontrast CT shows a homogeneous low-density mass (arrow) within the pectoralis major muscle (arrowhead). The CT attenuation was measured as −60 Hounsfield units, suggesting fat. (Courtesy of Kim EY, Ko EY, Han B-K, et al. Sonography of axillary masses: what should be considered other than the lymph nodes? *J Ultrasound Med.* 2009;28:923-939.)

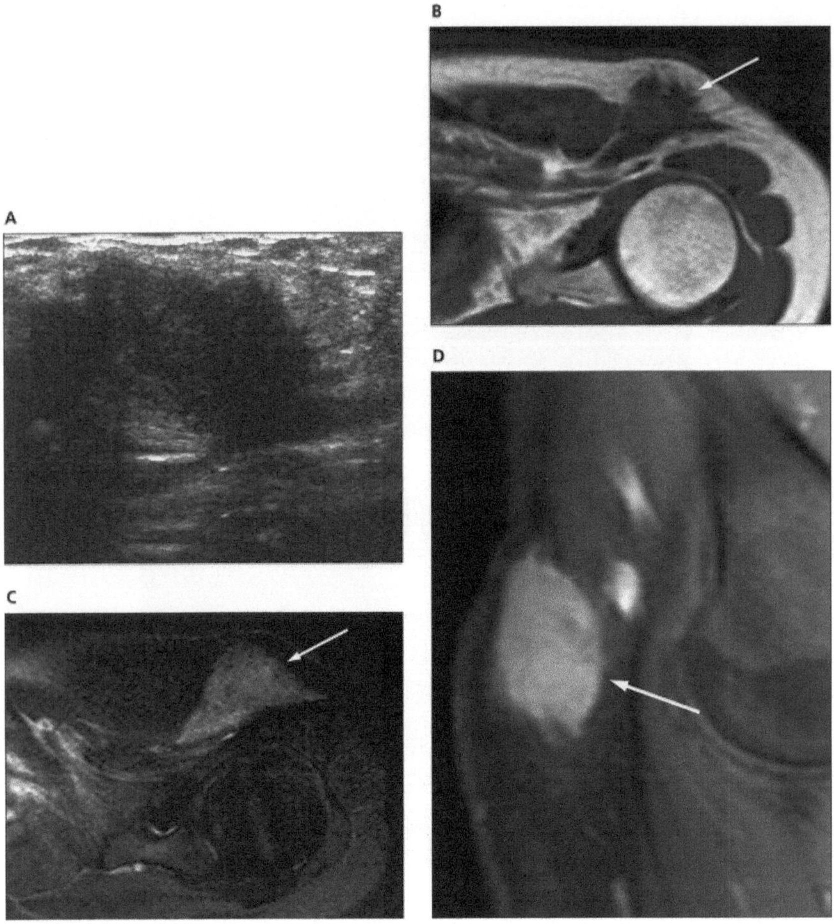

FIGURE 10.—Images from a 15-year-old female patient with fibromatosis in the left axilla. **A**, Sonography shows an irregular hypoechoic mass with posterior acoustic shadowing. These features mimic breast cancer. **B**, Axial T1-weighted MRI shows a triangular isointense mass (arrow) in the left axilla. **C**, Axial fatsaturated T2-weighted MRI shows hyperintensity of the mass (arrow). **D**, Postenhancement sagittal fat-saturated T1-weighted MRI shows strong enhancement of the mass (arrow). The surface of the mass has a shaggy appearance, indicating the possibility of fibromatosis. (Courtesy of Kim EY, Ko EY, Han B-K, et al. Sonography of axillary masses: what should be considered other than the lymph nodes? *J Ultrasound Med.* 2009;28:923-939.)

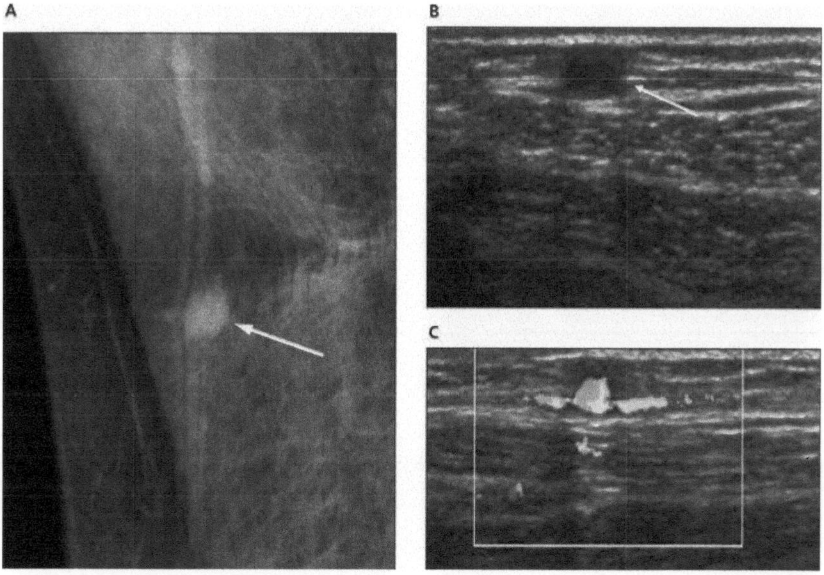

FIGURE 16.—Images from a 42-year-old female patient with a palpable nodule in the right axilla. The patient underwent mass excision, and a pathologic examination revealed dilated vessels with an organizing thrombus and fibrosis, compatible with a pseudoaneurysm. One year previously, the patient underwent partial mastectomy and sentinel lymph node biopsy for breast carcinoma. **A**, Mediolateral oblique mammography of the right breast shows a small area of nodular density (arrow) in the axilla, which is located close to the axillary vessels. **B**, Gray scale sonography shows a small hypoechoic nodular lesion (arrow) in the subcutaneous fatty layer of the axilla. C, On color Doppler imaging, the nodular lesion shows intense vascularity and a connection with the superficial axillary vessel. These findings are compatible with a postoperative pseudoaneurysm. (Courtesy of Kim EY, Ko EY, Han B-K, et al. Sonography of axillary masses: what should be considered other than the lymph nodes? *J Ultrasound Med*. 2009;28:923-939.)

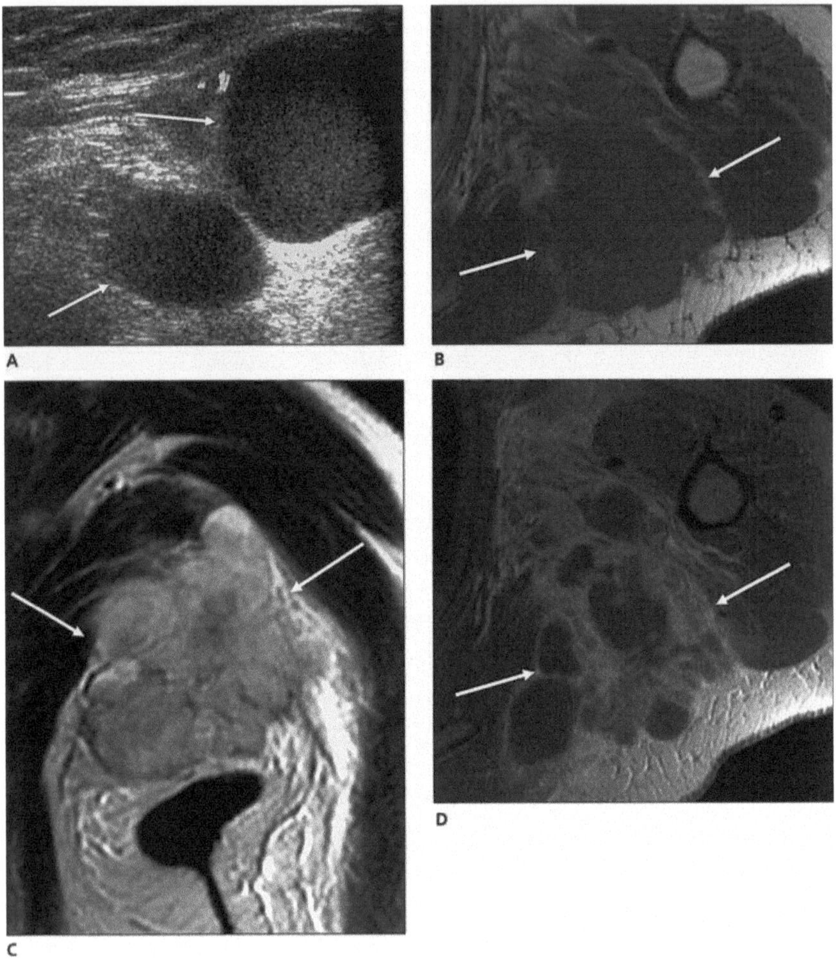

FIGURE 17.—Images from a 69-year-old female patient with an enlarging palpable mass in the left axilla. After mass excision, the final diagnosis was cystic lymphangiectasia with chronic inflammation. Several years previously, the patient underwent ipsilateral modified radical mastectomy and axillary lymph node dissection for Paget disease. **A**, Sonography shows conglomerated variably sized cystic masses (arrows) in the left axilla. Some of the lesions have echogenic debris. On color Doppler imaging, the periphery of the cystic lesions shows mild vascularity. **B**, Axial T1-weighted MRI shows an irregularly shaped isointense mass (arrows) in the left axilla. **C**, Coronal T2-weighted MRI shows a multiseptated hyperintense cystic mass (arrows) in the left axilla. **D**, Gadolinium-enhanced axial T1-weighted MRI shows peripheral and septal enhancement of the multiseptated cystic mass (arrows) after contrast agent administration. (Courtesy of Kim EY, Ko EY, Han B-K, et al. Sonography of axillary masses: what should be considered other than the lymph nodes? *J Ultrasound Med.* 2009;28:923-939.)

Utility of Targeted Sonography for Breast Lesions That Were Suspicious on MRI

DeMartini WB, Eby PR, Peacock S, et al (Univ of Washington, Seattle)
AJR Am J Roentgenol 192:1128-1134, 2009

Objective.—Data remain sparse to guide the use of targeted sonography for suspicious breast lesions found with MRI. The purpose of our study was to determine the usefulness of sonography in this setting.

Materials and Methods.—Retrospective review of our breast MRI database was performed to identify all nonpalpable, mammographically occult MRI-detected suspicious lesions from January 1, 2003, to December 30, 2004, that underwent targeted sonography followed by imaging-guided biopsy. Frequency of sonography detection was determined, and MRI features (BI-RADS lesion type and size), histopathology outcomes, and positive biopsy rates of sonographically detected and occult findings were compared using chi-square analysis.

Results.—Targeted sonography was performed for 167 of 201 MRI lesions. Outcomes were benign in 120 (72%) and malignant in 47 (28%) findings. Lesions assessed on sonography were 84 (50%) masses, 53 (32%) nonmasslike enhancements, and 30 (18%) foci, with mean size of 16 mm. Overall, 76 of 167 (46%) lesions were detected with sonography. Depiction was significantly more frequent for masses (58%) than for nonmasslike enhancement (30%) or foci (37%) ($p = 0.003$). Lesion size did not affect the frequency of detection. Sonographically depicted findings had a significantly higher positive biopsy rate (36%) than sonographically occult lesions (22%) ($p = 0.05$).

Conclusion.—Targeted sonography detected nearly half (46%) of MRI lesions evaluated, with depiction being most frequent for masses. Sonographically occult lesions had a probability of malignancy (22%), which warrants biopsy despite lack of sonographic detection (Table 1).

► I was quite interested at the report of the survey of the Society of Breast Imaging[1] where of those who responded 54% reported "always" performing targeted ultrasound for suspicious MRI findings while 14% said "never," "rarely," or "sometimes." Why might a targeted ultrasound not be performed? The answer to that question has a great deal to do with the fact that many MRI lesions do not have detectable ultrasound correlates (the DeMartini group in this article report 46% of their MRI lesions had ultrasound correlates, which is less than 1/2 but almost double the 23% reported by LaTrenta et al[2]), and there is concern about possible delay in diagnosis/scheduling/performance/reporting of the ultrasound study that may not be beneficial. The primary reason I chose this article, however, was that the authors asked the question as to the impact of lesion type on the likelihood of ultrasound detection. Is MRI lesion type (focus, mass, nonmass-like enhancement, or NMLE) predictive of success rate in finding an ultrasound correlate? Yes, in fact, MRI lesion type does make a significant difference in ultrasound correlate: masses (58%), foci

TABLE 1.—MRI and Histopathology Characteristics of Lesions Evaluated with Targeted Sonography Versus Those Not Evaluated with Targeted Sonography Before Biopsy

	Targeted Sonography Before Biopsy		
Lesion Characteristics	Yes ($n = 167$)	No[a] ($n = 34$)	p
Mean size (mm)	15.8	22.3	0.07
MRI type			0.0001
Focus or foci	30 (18.0)	4 (11.8)	
Mass	84 (50.3)	6 (17.6)	
Nonmass	53 (31.7)	24 (70.6)	
Outcome			0.24
Benign	120 (71.9)	21 (61.8)	
Malignant	47 (28.1)	13 (38.2)	

Note—Except where indicated, data are numbers (%) of lesions.
[a]Lesion went directly to MRI-guided biopsy without prior targeted sonography.

(37%), NMLE (30%). At our institution we will omit the recommendation for targeted ultrasound in cases of very small (< 3 mm) foci/masses and many of the NMLE lesions and proceed directly to MR-guided biopsy.

R. L. Birdwell, MD

References

1. Bassett LW, Dhaliwal SG, Eradat J, et al. National trends and practices in breast MRI. *AJR Am J Roentgenol.* 2008;191:332-339.
2. LaTrenta LR, Menell JH, Morris EA, Abramson AF, Dershaw DD, Liberman L. Breast lesions detected with MR imaging: utlitiy and histopathologic importance of identification with US. *Radiology.* 2003;227:856-861.

Patients' Perceptions of Breast MRI: A Single-Center Study

Zakaria S, Brandt KR, Degnim AC, et al (The Chrysalis, Rancho Mirage, CA; Mayo Clinic, Rochester, MN)

AJR Am J Roentgenol 192:1149-1154, 2009

Objective.—The high sensitivity and lower specificity of breast MRI for detecting breast cancer are well documented in the literature. As the use of breast MRI continues to grow, medical controversy persists over the appropriate use of this test. However, patients' perceptions of breast MRI are unknown. The aim of this study was to determine patients' perception of the impact of breast MRI on their care and to compare patients' perception with objective assessments.

Materials and Methods.—Patients undergoing breast MRI between January 2004 and December 2005 for cancer staging or high-risk surveillance were surveyed regarding comfort during, satisfaction with, and perception of MRI's impact on their clinical care. The respondents' perceptions of the impact of MRI on their care were compared with objective assessments based on retrospective chart review for breast cancer patients.

Results.—Of 300 patients, the survey was completed by 227 (76%). Ninety-one percent (203/224) reported mild or no discomfort during the examination, and 91% (206/226) perceived the breast MRI examination as having a positive impact on their care. Breast cancer patients were significantly more likely to perceive the impact of MRI as beneficial in comparison with the objective assessment (91% vs 35%, respectively; $p < 0.001$).

Conclusion.—Most women undergoing breast MRI for cancer staging and screening found it to be a comfortable test and perceived it to have a positive impact on their care. Breast cancer patients perceived the clinical impact of their breast MRI examination to be significantly more positive than when assessed objectively.

▶ Much of the controversy surrounding the use of the breast MRI is directed toward those women who have a recent diagnosis of breast cancer where MRI is recommended for staging of the ipsilateral breast and assessment for occult disease in the contralateral breast. Both this group of women and asymptomatic women who were assessed at high risk for the development of breast cancer were surveyed regarding comfort during, satisfaction with, and perception of the MRIs impact on their clinical care. Overall the perception was positive and I was intrigued to see that, at least in the state of Minnesota, only 5.4% of the woman reported that they had to pay for the examination out of pocket. I agree with caution as to the use of MRI to stage recently diagnosed breast cancer and that we must be careful as to the climbing number of mastectomies in settings where breast conservation treatment would be appropriate. However, to assess the contralateral breast and for those asymptomatic high-risk women who have MRI for surveillance (among other evidence-based uses of breast MRI), I believe this is a good examination.

R. L. Birdwell, MD

Diagnostic Breast Magnetic Resonance Imaging and Contralateral Prophylactic Mastectomy

Sorbero MES, Dick AW, Beckjord EB, et al (RAND Corporation, Pittsburgh, PA; Magee-Womens Hosp of UPMC, Pittsburgh, PA)

Ann Surg Oncol 16:1597-1605, 2009

Background.—Preoperative use of breast magnetic resonance imaging (MRI) in women with breast cancer may increase rates of mastectomy. This study investigated relationships between breast MRI and therapeutic and contralateral prophylactic mastectomy (CPM) in women with breast cancer.

Methods.—A total of 3606 women diagnosed with stage 0–III breast cancer from 1998 through 2000 ($n = 1743$; early period) or from 2003 through 2005 ($n = 1863$; late period) were retrospectively identified. Patient demographic and clinical characteristics were obtained from our institution's tumor registry. MRI use in the diagnostic evaluation was obtained from a prospective radiology database. Rates of therapeutic mastectomy, CPM, and associations with breast MRI were compared between the two time periods by multiple logistic regressions controlling for disease stage, age, family history, and calendar year of diagnosis.

Results.—A total of 14.2% of women underwent MRI, 29.0% had mastectomy, and 5.3% had CPM. Use of breast MRI increased substantially between the two time periods (4.1% to 23.7%, $P < 0.001$). Mastectomy rates increased from 28% to 30% ($P > 0.05$). The rate of CPM increased by >50% from the early to late period (4.1% to 6.4%, $P < 0.002$). Women who underwent MRI were nearly twice as likely to have CPM (9.2 vs. 4.7%, $P < 0.001$). Multivariate models found MRI was associated with increased rates of CPM for women with stage I or II disease (odds ratio 2.04, $P = 0.001$).

Conclusions.—MRI changes the surgical treatment of breast cancer among subsets of women diagnosed with breast cancer, suggesting there are hidden monetary and nonmonetary costs associated with its use.

▶ The controversy as to whether or not MRI should be used in all/any women with newly diagnosed breast cancer includes concerns about changes in surgical management, reported increase in contralateral prophylactic mastectomies, costs, and whether or not there is any impact on survival. Dr Monica Morrow et al[1] again argues for caution as reports (such as this one chosen to act as an introduction to this controversial topic) demonstrate an increase in mastectomy rates, both ipsilateral and prophylactic contralateral. Awareness of genetic mutations such as breast cancer (*BRCA*) account for only a small number of cases where bilateral mastectomy may be an appropriate choice based on life expectancy of invasive breast cancer; improvement in surgical reconstruction postmastectomy procedures also may play a role in patient-driven decision-making. Other changes in individualized patient treatment regimens such as the addition of breast MRI and the manner in which information is conveyed/available to the patient must be evaluated. To date there is no compelling evidence that preoperative MRI in newly diagnosed breast cancer affects survival rates.[2,3] Also, there is evidence that patient-involved decision-making may lead to somewhat unexpected results. In a sample of Caucasian women without contraindications to breast conserving therapy (BCT), mastectomy rates differed based on how the patient identified the "director" of her care. In the case where the surgeon was the primary decision maker, the rate was 5.3% as compared with when the patient identified herself as the decision maker where the mastectomy rate was reported to be 27%.[4] As Dr Morrow points out despite the advances in "molecular imaging" where treatment choices are and will increasingly be made/changed on the basis of molecular-level testing, the present setting suggests that to the patient, "being safe"; carries with it responses to treatment options that may have little basis in fact.

R. L. Birdwell, MD

References

1. Morrow M, Harris JR. More Mastectomies: is this what patients really want? *J Clin Oncol.* 2009;27:4038-4040.
2. Katipamula R, Degnim AC, Hoskin T, et al. Trends in mastectomy rates at the Mayo Clinic Rochester: effect of surgical year and preoperative magnetic resonance imaging. *J Clin Oncol.* 2009;27:4082-4088.
3. Fischer U, Zachariae O, Baum F, von Heyden D, Funke M, Liersch T. The influence of preoperative MRI of the breasts on recurrence rate in patients with breast cancer. *Eur Radiol.* 2004;14:1725-1731.
4. Solin LJ, Orel SG, Hwang WT, Harris EE, Schnall MD. Relationship of breast magnetic resonance imaging to outcome after breast-conservation treatment with radiation for women with early-stage invasive breast carcinoma or ductal carcinoma in situ. *J Clin Oncol.* 2008;26:386-391.

Kinetic Curves of Malignant Lesions Are Not Consistent Across MRI Systems: Need for Improved Standardization of Breast Dynamic Contrast-Enhanced MRI Acquisition

Jansen SA, Shimauchi A, Zak L, et al (Univ of Chicago, IL)
Am J Roentgenol 193:832-839, 2009

Objective.—The purpose of this study was to compare MRI kinetic curve data acquired with three systems in the evaluation of malignant lesions of the breast.

Materials and Methods.—The cases of 601 patients with 682 breast lesions (185 benign, 497 malignant) were selected for review. The malignant lesions were classified as ductal carcinoma in situ (DCIS), invasive ductal carcinoma (IDC), and other. The dynamic MRI protocol consisted of one unenhanced and three to seven contrast-enhanced images acquired with one of three imaging protocols and systems. An experienced radiologist analyzed the shapes of the kinetic curves according to the BI-RADS lexicon. Several quantitative kinetic parameters were calculated, and the kinetic parameters of malignant lesions were compared across the three systems.

Results.—Imaging protocol and system 1 were used to image 304 malig nant lesions (185 IDC, 62 DCIS); imaging protocol and system 2, 107 lesions (72 IDC, 21 DCIS); and imaging protocol and system 3, 86 lesions (64 IDC, 17 DCIS). Compared with those visualized with imaging protocols and systems 1 and 2, IDC lesions visualized with imaging protocol and system 3 had significantly less initial enhancement, longer time to peak enhancement, and a slower washout rate ($p < 0.004$). Only 47% of IDC lesions imaged with imaging protocol and system 3 exhibited washout type curves, compared with 75% and 74% of those imaged with imaging protocols and systems 2 and 1, respectively. The diagnostic accuracy of kinetic analysis was lowest for imaging protocol and system 3, but the difference was not statistically significant.

Conclusion.—The kinetic curve data on malignant lesions acquired with one system showed significantly lower initial contrast uptake and a different curve shape in comparison with data acquired with the other two systems. Differences in k-space sampling, T1 weighting, and magnetization transfer effects may be explanations for the difference.

▶ This work is included because it highlights a very important and concerning issue regarding the inconsistency of kinetic curve data acquired across different MRI systems. The authors focus on the 2 principles expected by the interpreting imager: 1. Malignant lesions tend to exhibit rapid uptake and washout, whereas benign lesions tend to have slower uptake and persistent contrast uptake over time; and 2. The kinetic characteristics of DCIS show a considerable overlap with those of benign lesions, often showing slower contrast uptake and washout than invasive ductal carcinoma. The field of breast MRI does not include widespread standardization of image acquisition, and as the authors point out, "There are no universally applied quality assurance procedures to ensure that as newer MRI systems are used clinically, the measured kinetic curves continue to satisfy the two general principles." The findings of this article emphasize the need for the clinical community to be aware that inconsistency in kinetic curve information among MRI systems exist, and it is important that those of us working within this imaging field make efforts to demand standardization of contrast-enhanced breast MRI acquisition protocols as newer technology is implemented to assure that what we are shown (a conspicuously enhancing malignancy) is what we have been taught to expect.

R. L. Birdwell, MD

Increasing Accuracy of Detection of Breast Cancer with 3-T MRI

Elsamaloty H, Elzawawi MS, Mohammad S, et al (Univ of Toledo Med Ctr, OH; et al)

AJR Am J Roentgenol 192:1142-1148, 2009

Objective.—The purposes of this study were to assess the sensitivity and specificity of 3-T MRI compared with those of mammography and sonography in the evaluation of breast cancer and to evaluate the results in light of previously published data obtained with 1-T and 1.5-T systems.

Materials and Methods.—A retrospective review was conducted with data on 434 women (mean age, 53 years) who underwent MRI evaluation of the breasts ($n = 868$). Results obtained from 3-T MRI, mammographic, and sonographic examinations of all patients were compared. Sensitivity, specificity, and predictive values were calculated, and statistics were analyzed with the chi-square test.

Results.—The sensitivities of MRI, mammography, and sonography in the detection of malignancy were 100%, 81.8%, and 86.4%, and the specificities were 93.9%, 99%, and 98.1%. The specificity of MRI increased from 92.8% to 94.5% over the course of the study (2006–2007). Compared with mammography ($p < 0.001$) and sonography ($p = 0.001$), MRI depicted a significantly higher number of malignant tumors of the breast. There was no significant difference between mammography and sonography ($p = 0.095$). Results were compared with those in earlier reports in the literature.

Conclusion.—MRI at 3 T is more sensitive than mammography and sonography in the detection of breast cancer and the characterization of small lesions (reaching 4 mm) but has lower specificity. Compared with the results of 1-T and 1.5-T MRI in the literature, 3-T MRI has higher sensitivity in the detection of breast cancer with no significant difference in specificity.

▶ This study from Toledo, Ohio, had as its objective to assess the sensitivity and specificity of breast MRI at 3 T as compared with mammography and ultrasound and evaluate the results in light of previously published data using 1 T and 1.5 T. Not surprising were the findings of higher sensitivity than mammography and ultrasound, but of some disappointment was the fact that there was no significant difference in specificity when comparing the higher strength magnet with the literature using 1 T and 1.5 T. Improving specificity was one of the hoped for advances offered by the higher field magnet. Do we really need 3 T for clinical breast MR? Many feel that for diagnostic purposes alone, the answer is no. A recently published abstract by Brian Hargreaves[1] outlined benefits and challenges of higher strength magnets. His conclusion regarding the differences between 1.5 and 3 T includes emphasis on subtle effects such as variation of contrast enhancement due to RF inhomogeneity or the effect of flip angle echo trains that may require have an overall impact on diagnostic accuracy. Further he states that "many of the gains of using higher field strength can be made by improving receiver coil arrays, an option that should be explored." However, more advanced breast imaging techniques including spectroscopy and multinuclear imaging do benefit from higher field strength. It is prudent for those designing breast imaging centers to be clear as to clinical practice needs and expectations when deciding whether on 1.5 T or beyond.

R. L. Birdwell, MD

Reference

1. Hargreaves BA. Breast MR at 3T and beyond: is it necessary? *Eur Rad.* 2009;19: S831-S834.

Diffusion-Weighted Magnetic Resonance Imaging of Breast Lesions: First Experiences at 3 T

Lo GG, Ai V, Chan JKF, et al (Hong Kong Sanatorium and Hosp, Happy Valley, Hong Kong)

J Comput Assist Tomogr 33:63-69, 2009

Purpose.—Dynamic contrast-enhanced magnetic resonance imaging (MRI) has been established as a valuable tool for the detection of breast cancer. There is evidence suggesting that diffusion-weighted imaging (DWI) may be useful to distinguish between malignant and benign breast lesions. We seek to evaluate the ability of DWI to differentiate between malignant and benign breast lesions at 3 T.

Methods.—Dynamic contrast-enhanced MRI and DWI of the breasts were performed in 31 female patients (age: mean, 46 years; range, 34–69 years) with suspected breast lesions on mammography and ultrasound using a 3-T scanner (MAGNETOM Tim Trio; Siemens Medical Solutions, Erlangen, Germany). Each lesion was assigned as either malignant or benign, blinded to the results of mammography and ultrasound, according to their imaging characteristics on contrast-enhanced MRI, DWI, and apparent diffusion coefficient (ADC) measurements. Tissue samples were obtained from all lesions by either needle or excision biopsy. Using histological results as the gold standard, the diagnostic accuracies of the dynamic contrast-enhanced MRI, DWI, and ADC were calculated and compared.

Results.—All breast lesions (n = 31) were identified on both the dynamic contrast-enhanced MRI and DWI scans. The threshold ADC value was determined to be 0.00121 mm^2/s, below which a lesion was considered malignant. The sensitivities/specificities of the dynamic contrast-enhanced MRI, qualitative DWI, and quantitative ADC were 95%/91%, 95%/63.6%, and 90%/91%, respectively. The differences in sensitivities, specificities, positive and negative predictive values, and diagnostic accuracies between the 3 examinations were statistically insignificant.

Conclusions.—Diffusion-weighted imaging at 3 T is highly sensitive in the detection of malignant breast lesions even with qualitative assessment alone, whereas ADC measurement offers quantitative assessment and increases the specificity to more than 90%. Further studies involving a larger cohort size and a wider spectrum of breast lesions are indicated.

▶ Scientific interest in diffusion-weighted imaging (DWI) of the breast as an adjunct to contrast-enhanced magnetic resonance imaging (MRI) to improve specificity is high, and a fair amount of work outside the United States has been reported.[1,2] DWI is an MRI sequence based on the normal diffusion of water molecules and how this movement is affected by cell swelling, cell membrane permeability, and other molecular and cellular changes that differ within the normal and tumor environments. If indeed this molecular imaging tool is perfected such that both the specificity and sensitivity of differentiation between benign and malignant breast tumors are high, it is predicted that at some future time this imaging could replace the need for contrast injection.

R. L. Birdwell, MD

References

1. Pereira F, Martins G, Figueiredo E, et al. Assessment of breast lesions with diffusion-weighted MRI: comparing the use of different B values. *AJR Am J Roentgenol.* 2009;193:1030-1035.

2. Woodhams R, Kakita S, Hata H, et al. Diffusion-weighted imaging of mucinous carcinoma of the breast: evaluation of apparent diffusion coefficient and signal intensity in correlation with histologic findings. *AJR Am J Roentgenol.* 2009; 193:260-266.

¹H MR Spectroscopy and Diffusion-Weighted Imaging of the Breast: Are They Useful Tools for Characterizing Breast Lesions Before Biopsy?

Tozaki M, Fukuma E (Kameda Med Ctr, Chiba, Japan)

AJR Am J Roentgenol 193:840-849, 2009

Objective.—The objective of our study was to determine whether proton (^{1}H) MR spectroscopy (MRS) and diffusion-weighted imaging might be useful tools for characterizing breast lesions before biopsy.

Materials and Methods.—Single-voxel ^{1}H MRS and diffusion-weighted imaging were performed in 171 suspicious or highly suspicious lesions. Using the residual water signal as a reference (4.7 ppm), a choline peak at 3.22–3.23 ppm was defined as malignant. If a high-signal-intensity lesion was detected in high-b-value ($b = 1{,}500$ s/mm^2) images, that lesion was defined as positive for malignancy. Among the patients with positive results on diffusion-weighted imaging, the apparent diffusion coefficient (ADC) values of the mass or focus were calculated from two different gradient factors ($b_1 = 500$ s/mm^2 and $b_2 = 1{,}500$ s/mm^2).

Results.—The diagnostic sensitivity and specificity of ^{1}H MRS were 44% (40/91) and 85% (68/80), respectively ($p < 0.001$). If ^{1}H MRS was applied for mass lesions larger than 15 mm, the diagnostic sensitivity and specificity were 82% (28/34) and 69% (11/16), respectively. Of the high-b-value images, 24 benign lesions and eight nonmass ductal carcinoma in situ were visually negative. With the use of a cutoff ADC value of 1.13×10^{-3} mm^2/s, a specificity of 67% (43/64) and sensitivity of 97% (61/63) was obtained on diffusion-weighted imaging.

Conclusion.—^{1}H MRS was useful for characterizing breast lesions measuring 15 mm or larger, and diffusion-weighted imaging was useful for characterizing lesions of any size. However, these two techniques still have potential pitfalls in relation to the diagnosis of non-mass breast lesions.

▶ MR spectroscopy (MRS) is an exciting and promising tool. It has the potential to improve the specificity of breast MR and, in future, may be designed in such a way as to to examine the entire breast for cancer metabolites and therefore be useful as a screening tool. The technical challenges of MRS are real and magnet strength, location, careful shimming, coil characteristics, lesion size, the proximity of percutaneously placed metallic clips and, possibly, even the type of contrast injected all play a part in the final spectra.[1,2] Diffusion-weighted imaging may prove even of greater use as there appear to be fewer potential obstacles.[3] The work done by these 2 authors is very good and of great interest.

R. L. Birdwell, MD

References

1. Mountford CE, Stanwell P, Ramadan S. Breast MR imaging at 3T. *Radiology*. 2008;248:319-320.
2. Stanwell P, Mountford C. In vivo proton MR spectroscopy of the breast. *Radiographics*. 2007;27:S253-S266.
3. Woodhams R, Kakita S, Hata H, et al. Diffusion-weighted imaging of mucinous carcinoma of the breast: evaluation of apparent diffusion coefficient and signal intensity in correlation with histologic findings. *AJR Am J Roentgenol*. 2009; 193:260-266.

Cryotherapy for Breast Cancer: A Feasibility Study without Excision

Littrup PJ, Jallad B, Chandiwala-Mody P, et al (Karmanos Cancer Inst, MI; Wayne State Univ, Detroit, MI)

J Vasc Interv Radiol 20:1329-1341, 2009

Purpose.—To assess the feasibility of percutaneous multiprobe breast cryoablation (BC) for diverse presentations of cancers that remained in situ after BC.

Materials and Methods.—After breast magnetic resonance (MR) imaging and thorough consultation, patients underwent BC after giving informed consent. This study was approved by the institutional review board. In 12 BC sessions, 22 breast cancer foci (stages I–IV) were treated in 11 patients who refused surgery by using multiple 2.4-mm cryoprobes. Five patients had recurrent disease and six had new diagnoses. With use of only local anesthesia, six patients were treated with ultrasonographic (US) guidance and five were treated with both computed tomographic (CT) and US guidance. Saline injections and warming bags were used to protect the skin. Procedure success was defined as 1 cm visible ice beyond all tumor margins. MR imaging and/or clinical follow-up were available for up to 72 months after BC.

Results.—US produced sufficient ice visualization for small tumors, whereas CT helped confirm overall ice extent. The mean pretreatment breast tumor diameter was 1.7 cm ± 1.2 (range, 0.5–5.8 cm), and an average of 3.1 cryoprobes produced 100% procedural success with mean ice diameters of 5.1 cm ± 2.2 (range, 2.0–10.0 cm). No significant complications, retraction, or scarring were noted. Biopsies at the margins of the cryoablation site immediately after BC and at follow-up were all negative. No local recurrences have been noted at an average imaging follow-up of 18 months.

Conclusions.—In conjunction with thorough pre- and postablation MR imaging, CT/US-guided multiprobe BC safely achieved 1 cm visible ice beyond tumor margins with minimal discomfort, good cosmesis, and no short-term local tumor recurrences (Fig 2).

▶ I included this article specifically to focus the readership on what may be a part of the future of the beneficial uses of breast MRI. In 2004, the ACR

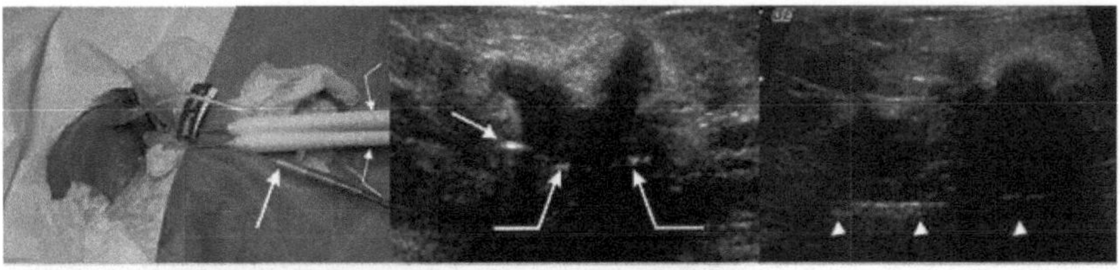

a.

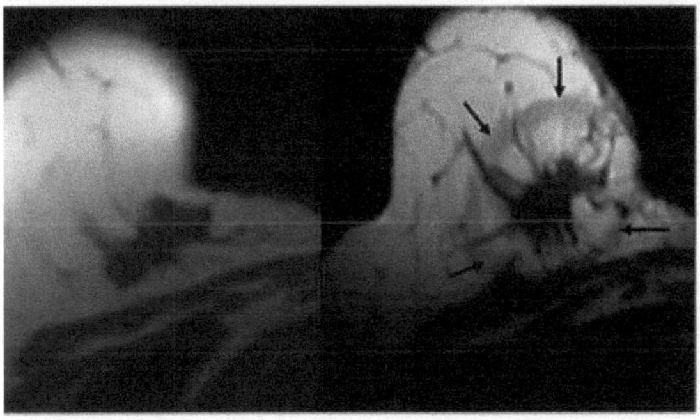

b.

FIGURE 2.—(a) US-guided cryotherapy for local control. Left image, BC performed with two cryoprobes (curved arrow) and a thermocouple (straight arrow) for an irregular 2.0-cm mass using technique planning noted in **Figure 1b** in the original article for tumors closer to the chest wall. As this was our second patient, this procedure also highlighted the need for better circumferential posterior visualization of the iceball near the chest wall, which was better provided with CT guidance. Middle image, Axial US image through the largest portion of the irregular mass shows more posterior placements of cryotherapy probes (curved arrows) and cross-section of a thermocouple en route to the superior tumor (straight arrow). Despite the cancer being near the chest wall, its medial location gave it a relatively superficial position for good US visualization. Right image, sagittal US image shows the posterior course of a thermocouple, which measured −57°C after ~8 minutes of the second freeze cycle at the superior tumor margin. The mobile nature of this tumor allowed substantial retraction off the chest wall in addition to saline injection. (**b**) MR images obtained before and four weeks after cryotherapy for local control. Left image, axial T1-weighted MR image obtained before cryotherapy at a similar tumor level as in **Figure 2a**. The tumor is causing retraction but does not involve the skin. Right image, MR image obtained four weeks after cryotherapy shows a low-signal-intensity rim of edema (black arrows) surrounding an ovoid zone of avascular necrosis that extended approximately 1 cm beyond all visible tumor margins but without extension into the chest wall. The wheelchair-bound patient, who was not a candidate for surgery, continued to do well with tumor involution of the primary and no further progression of distant metastases on anastrozole (Arimidex; AstraZeneca Pharmaceuticals, Delaware City, Delaware) alone until her death two years later. (Courtesy of Littrup PJ, Jallad B, Chandiwala-Mody P, et al. Cryotherapy for breast cancer: a feasibility study without excision. *J Vasc Interv Radiol.* 2009;20:1329-1341.)

guidelines for breast MRI had screening at the bottom of the list while in 2009, based on published data, this recommendation moved to the top of the list. What might be seen, then, on this guideline list in 2014 or 2019? Forward thinkers include percutaneous treatment of breast cancer where MRI may serve not only as a diagnostic tool and follow-up tool, but also may aid in the determination of margin status. This particular feasibility study (with very specific patient inclusion criteria, including agreement for additional nonsurgical systemic and local treatment regimens) used US and CT to guide and confirm treatment success as defined by 1 cm visible ice formation beyond the tumor margins. They included biopsies at the margins of the ablated site immediately after and at follow-up following the cryoablation and report no local recurrences as determined by MRI for an average of 18 months posttreatment. In such cases as those women for whom surgery is ill-advised or who refuse surgery, percutaneous tumor ablation may prove to be a very viable option, including the possibility of cure in some small tumors.

R. L. Birdwell, MD

Molecular Imaging of the Breast

Birdwell RL, Mountford CE, Iglehart JD (Brigham and Women's Hosp and Harvard Med School, Boston, MA; et al)

AJR Am J Roentgenol 193:367-376, 2009

Objective.—Personalized health care centers around the concept that each tumor and its host environment is unique; optimal treatment and expected response for any given woman presenting with a newly diagnosed breast cancer differ from the care and response of other women.

Conclusion.—As more is understood about the molecular subtypes of breast cancer and as development of targeted therapies progresses, the

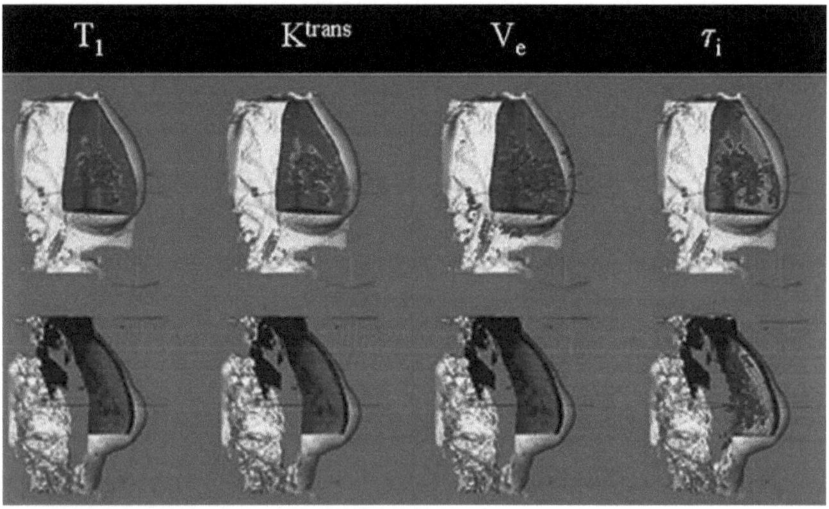

FIGURE 6.—Pretreatment (*top row*) and posttreatment (*bottom row*) 3D volume renderings of each parameter's assessment of tumor extent of representative patient are presented. Each volume is rendered at 50% of each respective parameter's maximum value obtained from pretreatment study—that is, only those voxels with values > 50% of maximum pretreatment value are displayed. Comparing top row with bottom row, quantitative map (T_1), volume transfer constant (K^{trans}), and extravascular extracellular volume fraction (v_e) maps indicate tremendous drop in extent of disease, whereas τ_i map presents some residual areas of elevated τ_i. It is important to note that T_1, K^{trans}, and v_e maps still have voxel values that are elevated above healthy values, but they are no longer about 50% of maximum value obtained from pretreatment study. Posttreatment apparent diffusion coefficient (ADC) rendering (not shown) revealed increase in overall ADC compared with pretreatment rendering. (Reprinted with permission from [45]). Editor's note: Please check the original article for full reference. (Reprinted from Birdwell RL, Mountford CE, Iglehart JD. Molecular imaging of the breast. *AJR Am J Roentgenol.* 2009;193:367-376, with permission from the American Roentgen Ray Society.)

possibility of earlier treatment response assessment and even improved detection will be realized (Fig 6).

▶ I have included this article because it presents both an overview of molecular imaging as it is used now and may be used in the future in the imaging of the breast and as an introduction to the field of the molecular subtypes of breast cancer and their association with treatment options and prognosis. As we summarize in the concluding paragraph, "In both screening and treatment, the role of the breast imager is expanding, not only encompassing expertise in identifying anatomic imaging changes but also necessitating the development of new skills focused on the fact that with molecular imaging techniques, the actual molecular signatures of many diseases can be visualized."

R. L. Birdwell, MD

3 Musculoskeletal System

Introduction

This year, the sports medicine/trauma literature seemed to exhibit continued interest in femoral acetabular impingement of the hip and consequent early development of osteoarthritis. Interestingly, the musculoskeletal community is finally recognizing that the measurements (specifically, the alpha angle) are not the primary abnormality of interest. Rather, identification of early cartilage damage is the abnormality that must be sought in order to guide treatment; this is a much more difficult finding to identify. I expect that we will continue to see articles seeking to refine the methodology for identification of subtle cartilage loss in the hip. Similarly, the shoulder trauma articles chosen seek to identify not just the fairly obvious lesions (for example, SLAP or dislocations), but also associated injuries that thoughtful surgeons may find will alter treatment.

With continued development and testing of new disease-modifying drug therapy, we will see more articles regarding early diagnosis and modified scales for evaluation of treatment, based in large part on imaging. Much of the literature at this point is suboptimal, with very small patient numbers, inadequate length of follow-up, or poorly controlled imaging parameters. As more of this work is performed by radiologists in conjunction with rheumatologists in a prospective fashion, the value of the information should improve. There were several laudable articles chosen for inclusion this year, some of which may serve as a model for further work. The general radiologist should gain valuable insights from the offerings, particularly involving detection of various arthridities in the axial skeleton.

Many of the selections this year involved sarcoma work. Some evaluated and questioned old signs of tumor malignancy, some showed promising new work with molecular imaging, and some serve as a guide to following sites of tumor treatment. Since patients are surviving primary bone and soft tissue tumors longer, some of the old "rules" are changing regarding metastases; the radiologist must understand these trends in order to provide thoughtful and accurate analysis to the referring clinician.

B. J. Manaster, MD, PhD

Foot and Ankle

Correlation of Arthrodesis Stability with Degree of Joint Fusion on MDCT

Dorsey ML, Liu PT, Roberts CC, et al (Mayo Clinic Arizona, Scottsdale; et al)
AJR Am J Roentgenol 192:496-499, 2009

Objective.—The purpose of our study was to correlate clinically determined joint stability with the degree of bone fusion in the ankle or subtalar joint on MDCT examinations after arthrodesis.

Materials and Methods.—We performed a retrospective review of 42 consecutive MDCT examinations from 29 patients. All patients had previously undergone arthrodesis of their ankle or subtalar joints and had persistent or recurrent hindfoot or ankle pain. Two musculoskeletal radiologists examined in consensus sagittal 2-mm-thick reformatted slices, measuring on each image the length of the joint surface and the length of the fused portion of the joint space. The sum of the lengths of the fused segments on all slices was then divided by the sum of the lengths of the joint surfaces to calculate the fusion ratio. For the standard of reference, the medical records were reviewed and operative reports, diagnostic injections, and physical examinations were used to classify the joints as stable or unstable.

Results.—Twelve clinically unstable joints had fusion ratios of 0–32.8%, whereas 30 clinically stable joints had fusion ratios of 33.2–100%. Using receiver operating characteristic analysis, we selected the cutoff level that maximized Youden's index (the sum of sensitivity and specificity). Using a 33% fusion ratio as the lower limit cutoff for joint stability, the sensitivity was 100%; specificity, 100%; and accuracy, 100%.

Conclusion.—After arthrodesis of the ankle or subtalar joint, MDCT scans can be used to determine whether that joint is likely to be stable if >33% of the joint has visible bone fusion on sagittal MDCT images.

▶ Arthrodesis continues to be one of the most effective methods of providing pain-free function of the tibiotalar, subtalar, and other joints of the ankle and foot. However, arthrodesis is not uniformly successful. It may be particularly difficult in patients with poor bone stock, including patients with rheumatoid arthritis, osteoporosis, and those using steroids. Radiologists are often asked to evaluate the degree of fusion, as patients are anxious to start bearing weight or even having hardware removed. Radiographs often yield incomplete information, because of overlying hardware or an oblique or curved morphology of the joint. MDCT is now routinely performed for these evaluations. It is important that radiologists know the techniques required to produce a useful CT, including thin sections with reformatting. This article provides another tool, giving a rational method of measuring fusion ratios (Fig 1 in the original article), which in turn can be given a likelihood of stability. Using this method, radiologists can provide a more helpful evaluation of the joint undergoing arthrodesis.

B. J. Manaster, MD, PhD

Prediction of Midfoot Instability in the Subtle Lisfranc Injury: Comparison of Magnetic Resonance Imaging with Intraoperative Findings

Raikin SM, Elias I, Dheer S, et al (Thomas Jefferson Univ Hosp, Philadelphia, PA)

J Bone Joint Surg Am 91:892-899, 2009

Background.—The objective of the present study was to assess the utility of magnetic resonance imaging for the diagnosis of an injury to the Lisfranc and adjacent ligaments and to determine whether conventional magnetic resonance imaging is a reliable diagnostic tool, with manual stress radiographic evaluation with the patient under anesthesia and surgical findings being used as a reference standard.

Methods.—Magnetic resonance images of twenty-one feet in twenty patients (ten women and ten men with a mean age of 33.6 years [range, twenty to fifty-six years]) were evaluated with regard to the integrity of the dorsal and plantar bundles of the Lisfranc ligament, the plantar tarsal-metatarsal ligaments, and the medial-middle cuneiform ligament. Furthermore, the presence of fluid along the first metatarsal base and the presence of fractures also were evaluated. Radiographic observations were compared with intraoperative findings with respect to the stability of the Lisfranc joint, and logistic regression was used to find the best predictors of Lisfranc joint instability.

Results.—Intraoperatively, seventeen unstable and four stable Lisfranc joints were identified. The strongest predictor of instability was disruption of the plantar ligament between the first cuneiform and the bases of the second and third metatarsals (the pC1-M2M3 ligament), with a sensitivity, specificity, and positive predictive value of 94%, 75%, and 94%, respectively. Nineteen (90%) of the twenty-one Lisfranc joint complexes were correctly classified on magnetic resonance imaging; in one case an intraoperatively stable Lisfranc joint complex was interpreted as unstable onmagnetic resonance imaging, and in another case an intraoperatively unstable Lisfranc joint complex was interpreted as stable on magnetic resonance imaging. The majority (eighteen) of the twenty-one feet demonstrated disruption of the second plantar tarsal-metatarsal ligament, which had little clinical correlation with instability.

Conclusions.—Magnetic resonance imaging is accurate for detecting traumatic injury of the Lisfranc ligament and for predicting Lisfranc joint complex instability when the plantar Lisfranc ligament bundle is used as a predictor. Rupture or grade-2 sprain of the plantar ligament between the first cuneiform and the bases of the second and third metatarsals is highly suggestive of an unstable midfoot, for which surgical stabilization has been recommended. The appearance of a normal ligament is suggestive of a stable midfoot, and documentation of its integrity may obviate the need for a manual stress radiographic evaluation under anesthesia for a patient with equivocal clinical and radiographic examinations.

▶ This article serves 2 purposes. First, the graphic illustrations and MR images provide a fine review of the Lisfranc ligament complex. More importantly, it serves as an initial evidence-based investigation of how the various ligaments relate to clinical instability of the foot following injury. It has long been established that the subtle radiographic findings of Lisfranc injury can be missed and that the clinical findings may be underestimated due to patient pain with weight-bearing imaging or testing of stability on clinical examination. This has resulted in the requirement of examination for instability under anesthesia in cases where the clinical suspicion for this injury is high. A relatively high number of these patients show stability under anesthesia (10%-20%), which is worrisome, given the risks of anesthesia. The article demonstrates a high correlation between rupture and grade-2 sprain of the plantar Lisfranc ligament (extending from medial cuneiform to base of metatarsals 2 & 3), or avulsion fracture at the second metatarsal medial base, and unstable midfoot. However, although disruption of the other regional ligaments is often seen, this does not appear to correlate with midfoot instability. Therefore, it seems logical to base a surgical decision on precise MR findings. The radiologist, by understanding the anatomy and imaging appearance of the midfoot, and knowing the implications for stability suggested by this study, may help obviate the need for examination for stability of the foot under anesthesia in most cases, and direct patients toward appropriate treatment.

B. J. Manaster, MD, PhD

Hip

Acetabular Cartilage Delamination in Femoroacetabular Impingement: Risk Factors and Magnetic Resonance Imaging Diagnosis

Anderson LA, Peters CL, Park BB, et al (Univ of Utah School of Medicine, Salt Lake City)

J Bone Joint Surg Am 91:305-313, 2009

Background.—Delamination of acetabular articular cartilage is a common abnormality in hips with femoroacetabular impingement. The purpose of the present study was to identify clinical and radiographic factors predisposing to delamination and to assess the diagnostic accuracy of magnetic resonance arthrography for the detection of these lesions.

Methods.—Following a retrospective review of records, we determined that acetabular cartilage delamination had been present in twenty-eight of sixty-four hips that had undergone a surgical dislocation procedure for the treatment of femoroacetabular impingement. Multivariable logistic regression was performed to assess the correlation of radiographic findings (i.e., magnetic resonance imaging and computerized tomography findings) with the status of delamination. The preoperative interpretations of the magnetic resonance arthrograms for twenty-seven hips that underwent surgical dislocation were reviewed to assess the accuracy of detecting delamination. At the time of surgery, nine of these twenty-seven hips were found to have delamination. Magnetic resonance arthrography interpretations that did not correlate with operative findings were subjected to conspicuity assessment and error analysis.

Results.—The rate of delamination of the acetabular cartilage as noted at the time of surgical dislocation was 44% (twenty-eight of sixty-four). Delamination was strongly associated with male sex and femoral sided signs of impingement; however, it was not associated with acetabular overcoverage (center-edge angle, >40°) (odds ratio = 0.16; $p < 0.05$). While there was no significant difference in the prevalence of labral lesions between groups, whenever labral and delamination lesions were found in the same hip, they were directly adjacent to one another. Preoperative magnetic resonance arthrography had a low sensitivity for delamination (22%) but had a high specificity (100%). Two-thirds of the delamination lesions were visible on retrospective review of these images when the reader was unblinded to the surgical findings. Delamination could most often be identified on the sagittal T1-weighted image and on the proton-density sequences with fat saturation.

Discussion.—There should be a high level of suspicion for articular cartilage delamination in men and in patients with primarily cam-type femoroacetabular impingement. Acetabular overcoverage may be protective against delamination. Preoperative high-quality magnetic resonance arthrograms should be carefully analyzed for evidence of delamination in this patient population.

▶ Radiologists and orthopedic surgeons specializing in hip disease are well aware of the importance of morphologic abnormalities of the femoral head-neck junction as well as acetabular overcoverage in predisposing the patient to development of impingement and early osteoarthritis. We have also become fairly adept at diagnosing labral tears. However, cartilage damage remains difficult to evaluate, even with a well-performed MR arthrogram. Yet the extent of cartilage damage should be prognostic for the success of hip surgery and likelihood of delaying the onset of significant osteoarthritis. Therefore, evaluation for cartilage damage should be the ultimate goal of imaging studies (Fig 1). The article provides a retrospective review of MR diagnostic accuracy and of risk factors for acetabular cartilage delamination in femoroacetabular impingement (FAI). It provides new insights, which may assist the imager in interpretation of MR arthrograms and detection of cartilage injury. The reader will find that cartilage delamination occurs far more frequently in cam-type than pincer-type of FAI. Furthermore, cartilage delamination tends to occur adjacent to the labral injury. Retrospectively, delaminated cartilage could be seen on MR arthrogram in two-third of the surgically proven cases. Therefore, it seems that the radiologist might be able to improve his/her sensitivity for cartilage

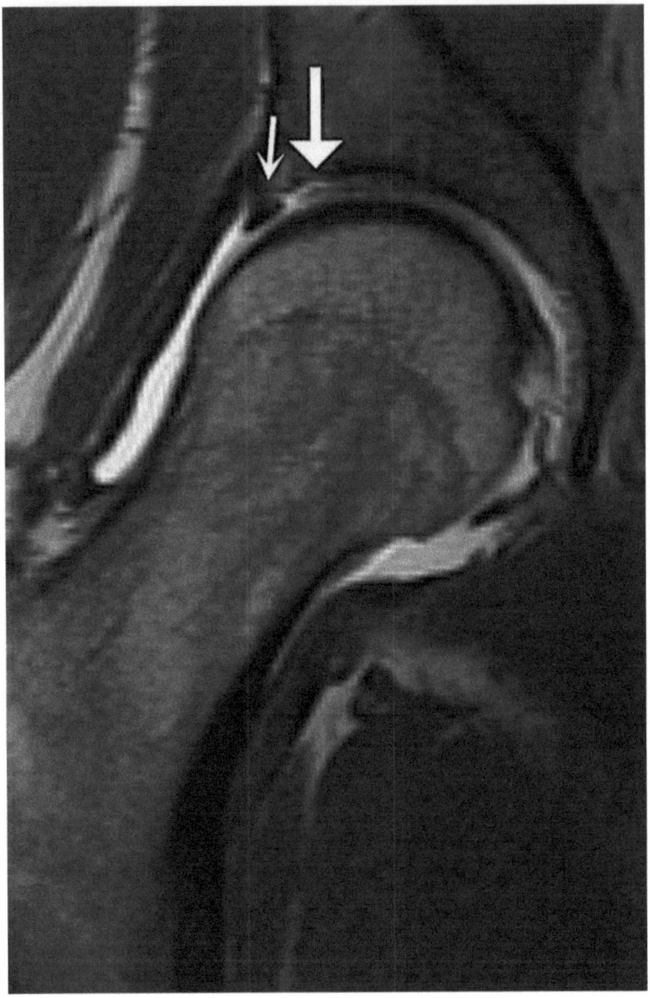

FIGURE 1.—Coronal T1-weighted magnetic resonance arthrogram with fat saturation. Delamination is less often visible in this plane but is obvious in this particular image. The large arrow points to a delaminated area adjacent to a partially detached labrum (small arrow), forming a "labral-chondral transition zone disorder"-type combination lesion. Note the contrast medium undermining the partially detached labrum (small arrow). (Reprinted from Anderson LA, Peters CL, Park BB, et al. Acetabular cartilage delamination in femoroacetabular impingement. Risk factors and magnetic resonance imaging diagnosis. *J Bone Joint Surg Am*. 2009;91:305-313, with permission from The Journal Of Bone And Joint Surgery, Inc.)

damage by paying special attention to the cam-type FAI, particularly in the vicinity of the labral tear. Note also that the sagittal image was found to be particularly sensitive, a finding with which this author concurs based on personal experience.

B. J. Manaster, MD, PhD

Cam-type femoral-acetabular impingement: Is the alpha angle the best MR arthrography has to offer?

Lohan DG, Seeger LL, Motamedi K, et al (David Geffen School of Medicine at UCLA, Los Angeles, CA; et al)

Skeletal Radiol 38:855-862, 2009

Introduction.—In our institutional experience, determination of the alpha (α) angle at MR arthrography as an indicator of the likelihood of cam-type femoroacetabular impingement (FAI) is fraught with inconsistency. The aims of this study were to quantify the degree of variability in and calculate the diagnostic accuracy of the alpha angle in suggesting a diagnosis of cam impingement, to determine the accuracy of a positive clinical impingement test, and to suggest alternative MR arthrographic measures of femoral head-neck overgrowth and determine their diagnostic utilities.

Materials and Methods.—We carried out a retrospective analysis of MR arthrographic studies performed during a 4-year period, combined with chart analysis, which allowed identification of 78 patients in whom surgical correlation was also available. The status of a preoperative clinical impingement test was also noted. Patients were designated as having cam-type FAI (Group A, $n = 39$) if intra-operative femoral head-neck junction bony osteochondroplasty/arthoscopic femoral debridement was performed. Group B ($n = 39$) acted as controls. Three radiologists independently and blindly performed a series of measurements (alpha angle and two newly proposed measurements) in each patient on two separate occasions. An alpha angle of greater than 55° was considered indicative of the presence of cam-type FAI.

Results.—Performance values for alpha angle measurement were poor for each observer. There was considerable (up to 30% of the mean value) intra-observer variability between the first and second alpha angle measurements for each subject. Binary logistic regression analysis confirmed that the alpha angle is of no value in predicting the presence or absence of cam-FAI. A statistically significant difference existed between Groups A and B with regard to the newly proposed anterior femoral distance (AFD; $p = 0.004$). Using an AFD value of 3.60 mm or greater as being indicative of the presence of cam-FAI yields a 0.67 performance measure (95% confidence interval 0.55–0.79). The second proposed parameter (femoral neck ratio) was of no value in suggesting the presence or absence of this condition. The sensitivity, specificity, and positive and negative predictive values of the clinical impingement test were 76.9%, 87.2%, 85.7% and 79.1% respectively.

Conclusions.—Femoral α angle measurement is associated with considerable variability. This index performed poorly in our patient population and was statistically of no value in suggesting the presence or absence of cam-FAI. One of our proposed measures, the AFD, outperformed the alpha angle, though to an insufficient degree to suggest its routine incorporation into clinical practice. Our experience suggests that the clinical impingement test remains the most reliable predictor of the presence of this condition.

▶ There has been a recent flurry of research articles and presentations discussing the accuracy of the alpha angle with respect to the accuracy of the clinical exam in predicting cam-type femoral acetabular impingement (FAI). This article proves no exception showing significant intraobserver variability (up to 30%) in measurement of the alpha angle. It also introduces 2 other potential measurements but concludes that neither exceeds the accuracy of the alpha angle. The conclusion, as with other recent articles, is that the alpha angle is a flawed measurement in predicting femoral acetabular impingement and that the clinical exam should be strongly relied upon. I view the issue somewhat differently. It is not the alpha angle per se that is of interest. Clinicians do not treat the alpha angle but rather the impinging structures themselves. The alpha angle, with a sensitivity of 85% for cam-type femoral acetabular impingement, is just one of several morphologic abnormalities which are sought on radiograph to suggest to the clinician that FAI should be a consideration. Even though the specificity is poor, it is reasonable to use this sign in conjunction with other observations to direct the clinician to perform a good clinical provocative test for FAI. If this is positive, the magnetic resonance (MR) arthrogram should be performed. Once the MR arthrogram is obtained, I believe the alpha angle is of much less interest. The more pertinent observation to be made on the arthrogram is the presence and extent of labral and chondral lesions. This is the information needed to direct prognosis and therapeutic decisions. Chondral injury is particularly difficult to evaluate accurately. Direct observation of fluid-filling chondral defects may be seen infrequently. The radiologist should pay close attention to low signal foci within the cartilage on T1 or intermediate weighted MR sequences to detect subtle injury. Additionally, the radiologist should pay close attention to the higher correlation of cam-type FAI and male gender with chondral injury.

B. J. Manaster, MD, PhD

MR Arthrography of Acetabular Cartilage Delamination in Femoroacetabular Cam Impingement

Pfirrmann CWA, Duc SR, Zanetti M, et al (Univ Hosp Balgrist, Zurich, Switzerland)

Radiology 249:236-241, 2008

Purpose.—To retrospectively assess the frequency and performance of magnetic resonance (MR) arthrography to help diagnose acetabular cartilage delamination in femoroacetabular impingement (FAI).

Materials and Methods.—Institutional review board approval and informed consent were waived for this retrospective study. Forty-four consecutive patients with FAI of the cam type were included (mean age, 30.7 years; range, 16–49 years), including 30 men (mean age, 30.5

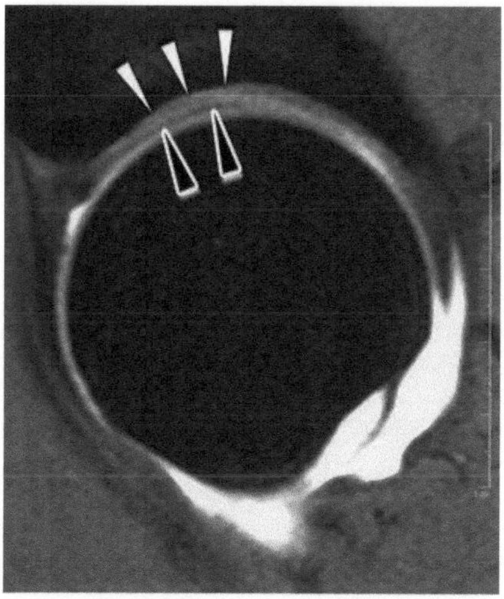

FIGURE 5.—Sagittal water-excitation three-dimensional double-echo steady-state MR image (24/6.5; flip angle, 25°) shows line in acetabular cartilage (white arrowheads) parallel to subchondral bone plate. Interface between acetabular and femoral cartilage (black arrowheads) may represent truncation artifact. (Reprinted from Pfirrmann CWA, Duc SR, Zanetti M, et al. MR arthrography of acetabular cartilage delamination in femoroacetabular cam impingement. *Radiology.* 2008;249:236-241. Copyright by the Radiological Society of North America.)

years; range, 16–49 years) and 14 women (mean age, 31.4 years; range, 18–48 years). The inclusion criteria were no previous surgery, surgery within 3 months after MR imaging, and availability of a detailed surgical report with acetabular cartilage findings. MR arthrographic findings were assessed independently by two blinded readers. Findings at surgery served as reference standard. Sensitivity, specificity, accuracy, and κ statistics for interobserver agreement were calculated.

Results.—At surgery, acetabular cartilage delamination was seen in 23 (52%) of 44 patients (mean size of cartilage flap from acetabular rim, 7.6 mm; range, 2–30 mm). At MR, patients with fluid signal intensity under the cartilage delamination had a respective sensitivity, specificity, and accuracy of 22%, 95%, and 57% for reader 1 and 30%, 95%, and 61% for reader 2. A hypointense line in the acetabular cartilage on sagittal three-dimensional double-echo steady-state images with water excitation demonstrated moderate diagnostic performance (respective sensitivity, specificity, and accuracy were 70%, 57%, and 64% for reader 1 and 70%, 62%, and 66% for reader 2). Hypointense areas in the acetabular cartilage were quite specific on both coronal intermediate-weighted fat-saturated images (respective sensitivity, specificity, and accuracy were 52%, 90%, and 70% for reader 1 and 74%, 90%, and 82% for reader 2) and coronal T1-weighted images (respective sensitivity, specificity, and accuracy were 35%, 90%, and 61% for reader 1 and 61%, 95%, and 77% for reader 2).

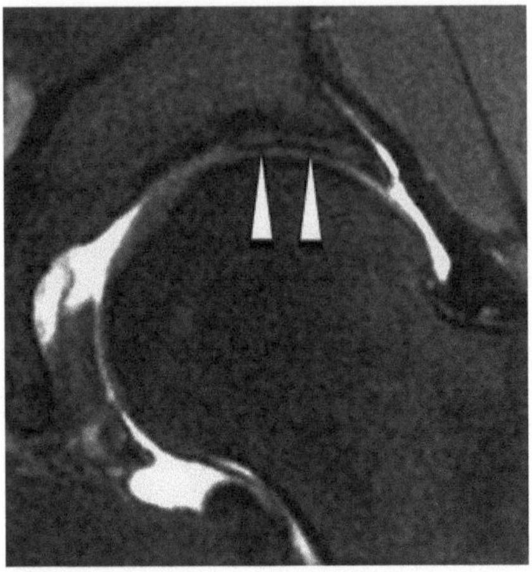

FIGURE 6.—Coronal intermediate-weighted turbo spin-echo MR image with fat saturation (2500/42) shows surgically proved delamination of acetabular cartilage. Hypointense zone (arrowheads) corresponds to delaminated cartilage. (Reprinted from Pfirrmann CWA, Duc SR, Zanetti M, et al. MR arthrography of acetabular cartilage delamination in femoroacetabular cam impingement. *Radiology.* 2008;249:236-241. Copyright by the Radiological Society of North America.)

Conclusion.—Cartilage delamination is common in patients undergoing surgery for FAI. Fluid under the cartilage delamination is a specific but rare finding. Hypointense areas in the acetabular cartilage seen on intermediate-weighted fat-saturated or T1-weighted images appear to be helpful diagnostic criteria (Figs 5 and 6).

▶ Radiologists have come to understand and readily diagnose the osseous morphologic abnormalities associated with femoral acetabular impingement. It is generally understood that it is the associated hyaline cartilage damage, which leads to early development of osteoarthritis. The degree of cartilage damage, particularly delamination of cartilage, is associated with surgical outcome, and it is valuable to be able to accurately assess cartilage damage preoperatively. Extensive damage noted preoperatively may save a patient from unnecessary osteoplasty surgery, and redirect toward total hip arthroplasty. Unfortunately, because of the tightness of the hip joint, the hyaline cartilage is the least accurately evaluated tissue in the joint, even with MR arthrography. As is described in this article, fluid demonstrated under delaminated cartilage is a specific but uncommon finding. Hip traction may improve the likelihood of observing this sign, but inconsistently. This article is valuable in describing another sign, that of hypointensity in the acetabular cartilage on T1 fat-saturated imaging following arthrography (Fig 6). Although the sensitivity is still relatively low, the accuracy improves when using this sign. The reader should look at the article carefully, though, to understand and avoid the mistake of over-calling truncation artifact as this focal region of low signal in the cartilage (Fig 5).

B. J. Manaster, MD, PhD

Knee

A Novel Magnetic Resonance Imaging Classification of Discoid Lateral Meniscus Based on Peripheral Attachment

Ahn JH, Lee YS, Ha HC, et al (Sungkyunkwan Univ School of Medicine, Seoul, Korea; Korea Univ Ansan Hosp, Korea)

Am J Sports Med 37:1564-1569, 2009

Background.—In the symptomatic discoid lateral meniscus, the effectiveness of preoperative magnetic resonance imaging (MRI) is not well documented.

Hypothesis.—Magnetic resonance imaging classification will provide more information to the surgeon in choosing the appropriate treatment methods with the help of arthroscopic findings.

Study Design.—Cohort study (diagnosis); Level of evidence, 2.

Methods.—Sixty-seven patients (82 knees) were reviewed. The preoperative MRI was checked in 76 of 82 knees. The Lysholm and Ikeuchi grading scales were evaluated. Images were analyzed from MRI, and findings were classified into 4 categories: no shift, anterocentral shift, posterocentral shift, and central shift. Tear pattern classifications were based on arthroscopic findings: horizontal tear, peripheral tear, horizontal and peripheral tear, posterolateral corner loss, and others. The correlations between MRI classification tear patterns and surgical methods were analyzed using the chi-square test or the Fisher exact test. The sensitivity, specificity, and accuracy of shift in preoperative MRI–according to the existence of peripheral tear when corroborated with arthroscopy–were also analyzed with the chi-square test. Inter-and intraobserver reliability was statistically analyzed by producing the inter- and intraclass correlation coefficient.

Results.—The mean preoperative Lysholm score was 77.3 (range, 43-97), and the last follow-up Lysholm score had increased to 96.8 (range, 84-100; $P < .001$). At last follow-up (100% follow-up), the Ikeuchi grading scale scored 48 knees as excellent, 30 as good, and 4 as fair. According to the MRI classification, 43 knees were no shift; 6, anterocentral shift; 15, posterocentral shift; and 12, central shift. Shift-type knees had a significantly larger number of peripheral tears, and repairs were performed in the shift-type knees (55%) more frequently than in the no-shift-type knees (28%). Among 82 knees, 31 were repaired simultaneously after a central partial meniscectomy.

Conclusion.—Magnetic resonance imaging classification provides more information to surgeons in choosing the appropriate treatment methods, although the final decision regarding procedure is made during arthroscopy after thorough analysis of the tear.

▶ Large complete discoid menisci are easily recognized by MR imaging, although it may be more difficult to identify incomplete discoid menisci because they maintain their "C" shape at least in part. Radiologists are also quite familiar with the concept of symptomatic locking and the expected meniscal tears in these menisci. However, it is possible that radiologists are less familiar with the criteria orthopedic surgeons use to determine whether a torn discoid meniscus is treated by subtotal meniscectomy or central partial meniscectomy, with or without repair of the tear. Arthroscopic classifications stress the importance of peripheral rim instability as an important factor in therapeutic decision making. This article proposes a classification of discoid meniscal tears based on displacement as noted on MR. The important parameters are the presence of peripheral tear and both the direction and amount of meniscal displacement. Although the system is not infallible (displaced meniscal fragments may reduce in knee extension, the position for MR examination), it may afford the radiologist a new understanding of the orthopedic surgeon's perspective and may promote use of terminology in the radiologic description that matches that of the surgeon.

B. J. Manaster, MD, PhD

Only MR can safely exclude patients from arthroscopy

Vincken PWJ, ter Braak BPM, van Erkel AR, et al (Leiden Univ Med Ctr, The Netherlands; et al)

Skeletal Radiol 38:977-982, 2009

Objective.—The aim of this study was to determine in patients with subacute knee complaints and normal standardized physical examination the fraction of magnetic resonance imaging (MRI) studies showing arthroscopically treatable intra-articular pathology.

Material and Methods.—There were 290 consecutive patients (between 16 and 45 years) with at least 4 weeks of knee complaints and low clinical suspicion of intra-articular pathology based on physical exam. Two hundred seventy-four patients were included. Sixteen patients with prior knee surgery, rheumatic arthritis, or severe osteoarthritis were excluded. MRI was used to assign patients to group 1 (treatable abnormalities) or group 2 (normal or no treatable findings), depending on whether MR demonstrated treatable pathology. Arthroscopy was performed in group 1 patients. If symptoms persisted for 3 months in group 2 patients, cross over to arthroscopy was allowed.

Results.—MR showed treatable pathology in 73 patients (26.6%). Arthroscopy was performed in 64 patients of 73 patients (group 1). In 52 patients (81.3%, 95% confidence interval (CI) 71.4–91.1%), arthroscopy was therapeutic. Of the 13 arthroscopies (6.5%) in group 2, four were therapeutic (30.8%, 95% CI 1.7–59.8). The highest fraction of MR studies showing treatable pathology was found in males, aged over 30 years, with a history of effusion (54.5%, six of 11 patients).

Conclusion.—Authors believe that the negative predictive value of clinical assessment in patients with subacute knee complaints is too low to exclude these patients from MR. MR should at least be considered in male patients aged 30 years and over with a history of effusion.

▶ In this era of avid discussion of cutting medical care costs by eliminating unnecessary care and specifically diagnostic imaging, this article presents a sound argument in favor of the use of MR of the knee to direct further medical care. The premise of the discussion is based on the literature, which evaluates whether a standard physical examination of the knee is reliable in its negative predictive value. The authors demonstrate that, in a series of 274 patients with a negative physical exam by a qualified orthopedic surgeon, MR showed unequivocal pathology, which required arthroscopic surgery in 25.9%. This represents a large proportion of patients seen initially by orthopedic surgeons who would otherwise have a significant delay in their medical care. The authors further showed that in a cohort of patients with high clinical suspicion of arthroscopically-treatable abnormalities (based on the same physical examination) only 50.3% of patients had corresponding abnormalities shown on MR. Therefore it seems clear that clinical assessment based on physical examination has only limited value in selecting patients for additional diagnostic or therapeutic procedures.

The authors extend this information to argue that general practitioners should be given direct access to order knee MR examinations (not currently available in the authors' health care system), bypassing the orthopedic surgeon's examination, which appears to be of limited value. The MR results then direct the selection of patients for referral to an orthopedic surgeon. The argument is made that it is a cost-effective policy to use the strong negative predictive value of MR to triage patients with knee pain; the patient group in which this policy would be particularly effective is male patients aged 30 years and over with a history of effusion and symptoms that do not subside for approximately 1 month.

This is a carefully constructed study, which could validly be used to direct health care policy. While it may initially seem surprising that more MR imaging is in fact cost-effective, there are many other examples where early imaging and prompt diagnosis lead to an overall cost savings. Other similar studies are sorely needed.

B. J. Manaster, MD, PhD

Shoulder

Abnormal Translation in SLAP Lesions on Magnetic Resonance Imaging Abducted Externally Rotated View

Chhadia AM, Goldberg BA, Hutchinson MR (Univ of Illinois at Chicago, IL)
Arthroscopy 26:19-25, 2010

Purpose.—The purpose of this study was to measure in vivo axial-plane translation of the glenohumeral joint by use of magnetic resonance imaging in patients with and without SLAP lesions between the conventional adducted neutral rotation (AD) view and an abducted externally rotated (ABER) view.

Methods.—Seven patients with an intraoperative SLAP lesion that was unstable and required repair were selected into the SLAP group. Although they did not have normal shoulders, 15 patients were selected into the control or comparison group, most of whom had rotator cuff pathology. The glenohumeral contact point (CP) and humeral head center (HHC) were calculated and compared with the glenoid surface as a relative anterior or posterior translation. The relative posterior translation between the ABER and AD views for each patient was calculated as ΔCP and ΔHHC. These values were compared between the SLAP and control groups.

Results.—There was a significant difference in ΔCP between the SLAP and control groups (3.62 *v* 0.79 mm of relative posterior translation, $P = .005$). There was not a similar significant difference found in ΔHHC between the SLAP and control groups (3.19 mm *v* 1.48 mm of relative posterior translation, $P = .14$). There was a significant difference between the mean translations of the SLAP-ABER group and the SLAP-AD group for both CP (−3.65 mm *v* −0.04 mm, $P = .008$) and HHC (−2.22 mm *v* +0.97 mm, $P = .03$). The difference between the control-ABER group and the control-AD group was not as pronounced.

Conclusions.—The magnetic resonance imaging ABER view in patients with unstable SLAP lesions requiring repair showed in vivo glenohumeral posterior translation relative to the adducted neutral rotation view of greater than 3 mm.

Clinical Relevance.—This finding furthers the understanding of the pathokinematics in SLAP lesions.

▶ The abducted externally rotated (ABER) positioning is generally not considered to be an essential sequence in MR examination of the shoulder. It is somewhat more frequently used with MR arthrography but not consistently at most institutions. Two reasons are generally provided for avoiding the ABER sequence. First, some clinicians prefer that it need not be used because their patients complain of discomfort in the position. Second, it is costly to perform this portion of the examination; the patient's arm position must be changed and surface coil placed, followed by relocalizing and imaging. The examination can be somewhat technically challenging to produce. Yet the position can provide improved visualization of the undersurface of the rotator cuff to evaluate partial articular side tears and internal impingement. The ABER positioning is also useful in demonstrating unstable anterior labroligamentous complex injuries, often displacing labral tears, emphasizing inferior chondral injuries, and demonstrating inferior glenohumeral ligament injury, whether intrasubstance or at the humeral insertion. This article demonstrates another advantage of the ABER position, showing translational instability in superior labrum anterior and posterior (SLAP) lesions. Posterior translation of the humeral head relative to the glenoid when held in this position is indicative of instability of a SLAP lesion, which in turn influences prognosis and treatment decisions. Clinical question of SLAP lesion might suggest that the addition of an MR sequence with the patient in the ABER position would be useful.

B. J. Manaster, MD, PhD

Assessment of Capsular Laxity in Patients With Recurrent Anterior Shoulder Dislocation Using MRI

Ng AWH, Chu CM, Lo WN, et al (North District Hosp 9, Sheung Shui, Hong Kong)

AJR Am J Roentgenol 192:1690-1695, 2009

Objective.—The purpose of our study was to investigate the usefulness of MRI in assessing capsular laxity in patients with recurrent shoulder dislocation.

Materials and Methods.—The records of 64 consecutive patients (the study group consisted of 58 patients, 45 male and 13 female; average age, 39.3 years; range, 13–82 years) who underwent MR arthrography between October 2002 and May 2008 were retrospectively reviewed. The patients were divided into three groups: group A, no shoulder dislocation; group B, first dislocation; and group C, recurrent dislocation. The maximum capsular widths at the anterior, anteroinferior, and inferior regions in the neutral and abducted and externally rotated (ABER) positions were measured on oblique sagittal images. The relationship of capsular width with the number of dislocations, type of capsular attachment, presence of anteroinferior labral tear, and patient sex were evaluated. The sensitivity and specificity of MRI and clinical tests in detecting capsular laxity were also calculated.

Results.—Among all three patient groups, there were significant differences ($p < 0.05$) in the widths with the shoulder in the ABER position and the degree of capsular tightening at the anterior region. The average maximum width was largest in group C and smallest in group A. There was a significant difference in the degree of anterior capsular tightening in groups B and C. A moderate correlation of capsular width and anterior tightening ($R = -0.45$) with number of shoulder dislocations was found. There was a significant difference in capsular width and capsular tightening in relation to the presence of an anteroinferior labral tear. The sensitivity and specificity to detect capsular laxity were 92% and 100%, respectively, for clinical tests and 85% and 96% for MRI.

Conclusion.—MRI is a useful and objective method to assess capsular laxity in patients with recurrent shoulder dislocation.

▶ It is well known that it is easy for radiologists to miss a finding that is not a positive presence (absent clavicles is a classic example). Such is the case with the diagnosis of capsular laxity. If associated soft tissue injuries are not present, an overly distended capsule may not be noted. This article nicely demonstrates a method to evaluate capsular laxity in neutral position as well as the abducted external rotation (ABER) position, in which the anterior capsule and inferior glenohumeral labral-ligamentous complex should be tightened. The authors demonstrate that capsular laxity may be demonstrated in patients with shoulder instability and that these patients can be shown to lack the anterior tightening when placed in the ABER position. The article provides a nice review of the topic.

B. J. Manaster, MD, PhD

Trauma

Value of Information Analysis Used to Determine the Necessity of Additional Research: MR Imaging in Acute Knee Trauma as an Example

Koerkamp BG, Nikken JJ, Oei EH, et al (Univ Med Ctr Rotterdam, The Netherlands)

Radiology 246:420-425, 2008

Purpose.—To help guide future outcomes research regarding the use of magnetic resonance (MR) imaging in patients with acute knee trauma in an emergency department setting, with use of prospective data from a randomized clinical trial and value of information analysis.

Materials and Methods.—A total of 189 patients (123 male, 66 female; mean age, 33.4 years) were randomly assigned to undergo radiography alone ($n=93$) or radiography and MR imaging ($n=96$). Institutional review board approval and informed consent (parental consent for minors) were obtained. During 6 months of follow-up, data on quality of life and 39 cost parameters were collected. Value-of-information analysis was used to estimate the expected benefit of future research to eliminate the decision uncertainty that remained after trial completion. In addition, the parameters that were responsible for most of the decision uncertainty were identified, the expected benefits of various study designs were evaluated, and the optimal sample size was estimated.

Results.—Only three parameters were responsible for most of the decision uncertainty: number of quality-adjusted life-years, cost of an overnight hospital stay, and friction costs. A study in which data on these three parameters are gathered would have an optimal sample size of 3500 patients per arm and would be expected to result in a societal benefit of €5.6 million or 70 quality-adjusted life-years.

Conclusion.—The optimal study design for use of MR imaging to evaluate acute knee trauma involves a trial in which there are 3500 patients per trial arm, and data on the number of quality-adjusted life-years, cost of an overnight hospital stay, and friction costs are collected.

▶ This article should be required reading for every radiologist, medical student, medical scientist, and politician/budget analyst in this country. It is clear that medical costs in the United States must be contained. It is equally clear that logical analysis must be applied to make the most useful decisions regarding limiting the use of medical technology. There have been admirable attempts at gathering such information (for example, the American College of Radiology Appropriateness Criteria). However, as a former member of the musculoskeletal section of that group, I can attest to the fact that the best analysis of the data is significantly imperfect, based on inadequate data. This article demonstrates the methodology of determining the parameters of an outcome and cost-analysis study. That the article is based on analysis of MR of the knee following trauma is incidental. It may be applied to any technology. Its importance lies in the methodology of developing a guide to proper future outcomes research, randomized clinical trials, and value-of-information analysis. This methodology should help guide NIH and other funding agencies in understanding whether proposed clinical research projects will in fact yield useful data. If used judiciously, it should help direct research priorities in health care.

B. J. Manaster, MD, PhD

Normal Thickness and Appearance of the Prevertebral Soft Tissues on Multidetector CT

Rojas CA, Vermess D, Bertozzi JC, et al (Univ of South Florida College of Medicine, Tampa; et al)
AJNR Am J Neuroradiol 30:136-141, 2009

Background and Purpose.—Analysis of the prevertebral soft tissue (PVST) is helpful in detecting osseous and ligamentous injuries of the cervical spine. Because the standard of care has shifted from radiographs to multidetector CT (MDCT), a re-examination of the PVST on MDCT images is needed to establish normal values for thickness appropriate for this imaging technique.

Materials and Methods.—Thickness of the PVST was measured in 192 patients undergoing screening cervical spine MDCT with multiplanar reconstructions as part of a trauma protocol. Patients included in the study were not intubated, had an immobilized cervical spine, had normal findings on cervical spine CT, and did not have a diagnosis of osseous or soft-tissue cervical injury. Exclusion criteria included patients with a congenital or acquired (nondegenerative) abnormality of the cervical spine or PVST.

Results.—The upper limits of normal for the thickness of the PVST were 8.5 mm at C1, 6 mm at C2, 7 mm at C3, 18 mm at C6, and 18 mm at C7. The upper limit of normal was not determined for C4 and C5 levels due to variable position of the esophagus and larynx. The smallest variability and calculated SDs were found at C2 and C3.

Conclusions.—The thickness of the PVST is important in the detection of underlying injuries to the cervical spine. We propose the obtained values as the upper limits of normal for PVST thickness on MDCT images in the adult population.

► Prevertebral soft tissue swelling has long been a radiographic hallmark of cervical spine injury. Measurements of mean and range have been established for radiographs. However, trauma evaluations now generally use multidetector CT (MDCT) of the spine, often without radiograph. The authors of this article perceived a need to establish mean and range measurements for CT. They provide these measurements, based on a large number of CT examinations.

Table 1 in the original journal article will be useful in trauma evaluation, and may validate further imaging with MR in cases where the osseous structures are normal by CT. However, it should not be forgotten that significant injury may be present without prevertebral soft tissue swelling.

B. J. Manaster, MD, PhD

Traumatic cervical spine injuries: characteristics of missed injuries

Rana AR, Drongowski R, Breckner G, et al (The Univ of Michigan Med School, Ann Arbor, MI; The C.S. Mott Children's Hosp, Ann Arbor, MI)

J Pediatr Surg 44:151-155, 2009

Purpose.—Computed tomographic (CT) scanning has mostly replaced x-rays as an imaging modality, but concerns exist because of excess radiation, missed injuries, and whether it is the definitive procedure for intubated patients. The purpose of this study was to characterize missed cervical spine injuries (CSIs).

Methods.—All pediatric (<18) trauma patients from 2004 to 2006 were analyzed. Age, sex, Injury Severity Score (ISS), mechanism, time, and missed injuries were reviewed. Flexion/extension views were used in patients with prolonged intubation. Descriptive statistics, χ^2, Student's *t* test, and bivariate correlation were used.

Results.—There were 1307 pediatric trauma patients admitted with 318 imaged for potential CSI. Computed tomography was the sole imaging study in 200, x-rays in 64, and both in 54. Time to C-spine clearance was similar for all modalities ($P > .05$). For CT, 34 (10.7%) were initially positive for CSI with 7 false-positives (FPs) and no false-negative (FN). There were 18 patients with CSI identified by x-ray, with 5 FPs and 5 FNs (missed injuries). The 5 FNs missed by x-ray were all positive by CT scan and required no intervention. None of the flexion/extension views revealed an additional injury. Sex, intubated patients, ISS, age, type, and injury location were not predictive of a missed injury ($P > .05$). The sensitivity of CT scan was 1.0, specificity was 0.976, and the positive predictive value was 79.4%. The sensitivity of plain x-ray was 61.5%, the specificity was 1.6%, and the positive predictive value was 61.5%.

Conclusions.—Our data suggest that CT scans should be the primary modality to image a CSI. Flexion/extension views did not add to the decision making for C-spine clearance after CT evaluation.

▶ Recent articles have definitively demonstrated that CT should be the primary modality used to evaluate for cervical spine injury in adults following trauma. However, there is greater concern regarding radiation exposure due to CT in the pediatric population. Good data providing guidance for optimal imaging of the cervical spine in children have been lacking, and many centers have continued using radiography, sometimes with flexion-extension filming, and often supplemented by CT. This article is a well-conceived retrospective study of a large pediatric trauma population ($N = 318$). The results are similar to those of the adult population, showing that radiographs have a significantly lower sensitivity relative to CT. There is also a nice discussion of radiation exposure to the pediatric population, including considerations of repeat radiographs, relative risk estimates, and pediatric-specific protocols.

B. J. Manaster, MD, PhD

Assessment of Distal Radioulnar Joint Instability After Distal Radius Fracture: Comparison of Computed Tomography and Clinical Examination Results

Kim JP, Park MJ (Dankook Univ School of Medicine, Cheonan, Korea; Sungkyunkwan Univ School of Medicine, Seoul, Korea)

J Hand Surg 33A:1486-1492, 2008

Purpose.—To compare computed tomography (CT) and clinical stress test findings in terms of the assessment of distal radioulnar joint (DRUJ) instability after distal radius fracture.

Methods.—Thirty-four patients with a healed unilateral distal radius fracture were evaluated after a mean follow-up period of 18 months. Stress test and CT were performed to detect DRUJ instability. Three CT methods were used: the modified radioulnar line, the epicenter, and the radioulnar ratio methods. The results of CT assessments were compared with stress test findings using kappa statistics. Relationships between DRUJ instability and the radiographic parameters of volar tilt, radial inclination, and radial shortening were analyzed.

Results.—Twelve patients were considered to have DRUJ instability based on the stress test. DRUJ instability was diagnosed in 15, 8, and 11 patients by the modified radioulnar line, the epicenter, and the radioulnar ratio methods, respectively. Reliability analysis of stress test and CT results showed moderate or fair agreement (kappa value: .33 for the modified radioulnar line, .56 for the epicenter, .41 for the radioulnar ratio). DRUJ instability diagnosed by the stress test and by CT showed no statistical correlation with radiographic parameters, but instability determined by CT appeared to be related to dorsal tilt deformity.

Conclusions.—After distal radius fracture, CT assessments of DRUJ instability were found to be influenced by residual deformities, such as dorsal tilt, but not well correlated with stress test findings. Although CT assessments are objective, CT scans primarily show alterations of DRUJ alignment and have several risks of bias in patients with a prior distal radius fracture.

▶ Distal radial fractures are well known to result in suboptimal function when they heal in a shortened or angulated position; radiologists and orthopedic surgeons feel confident in evaluating these parameters on radiographs. However, the association of distal radial fractures with distal radioulnar joint (DRUJ) instability is less well known. If there is concern for DRUJ dissociation, radiographs are unreliable for accurate diagnosis. The body of literature demonstrates that CT can be reliably diagnostic, particularly if both wrists are imaged in neutral, supinated, and pronated positions. This article is valuable in demonstrating that assessment of DRUJ instability using CT risks significant bias if the patient has had a previous distal radius fracture and suggests that clinical examination is more reliable.

B. J. Manaster, MD, PhD

Tumors

The role of the plain radiograph in the characterisation of soft tissue tumours

Gartner L, Pearce CJ, Saifuddin A (RNOH Stanmore, Middlesex, UK; Kingston Hosp, UK)
Skeletal Radiol 38:549-558, 2009

A radiograph is often the first investigation to be requested when a patient presents with limb pain or a mass. Whilst we do not advocate that this is the only investigation to be employed in the evaluation of such patients, a working knowledge of the variety of abnormal findings that can present in the soft tissues on radiographs remains useful. We reviewed the radiographic findings of soft tissue masses from a prospectively compiled database of all such lesions presenting to a specialist orthopaedic oncology service over the past 8 years. Of the cohort of 1,058 individuals with a proven soft tissue tumour, 454 had had a radiograph taken of the affected area. Of these, 281 (62%) patients had a positive radiographic finding. The most common findings were a visible soft tissue mass ($n=141$), the presence of calcification ($n=76$), fat ($n=32$) and evidence of bone involvement ($n=62$). More than one finding was sometimes present in the same patient. These findings were present in both benign and malignant tumours. This review article describes the incidence and diagnostic relevance of these plain film findings for suspected soft tissue tumours.

▶ I am an advocate for training residents in the art of radiographic interpretation. Eager trainees so often want to jump right into their MR training and regard radiographs as barriers to the more interesting modalities. However, establishing an appreciation of the value of a radiograph as well as understanding the difference between an adequate and an outstanding radiographic interpretation is essential. This article serves to re-establish the importance of the radiograph in the overall evaluation of a soft-tissue mass in the extremities. It demonstrates that, in a large case series, 62% of patients with an extremity mass have an abnormality demonstrated by radiograph. This itself is an important observation; it is not uncommon that a patient with nonspecific extremity pain, where a mass has not yet been discovered, has a radiograph as the first line of diagnostics. With careful attention, the radiologist should be able to note the presence of a soft-tissue mass by means of distorted fat planes, even without the presence of osseous destruction or fat/calcification within the mass. Moreover, knowledge of the presence of calcification or calcified matrix within a mass, seen by means of radiograph, often leads to a more accurate interpretation of the MR (small foci of calcification on MR may be overlooked or misinterpreted). While the authors and I of course agree that MR is required for appropriate interpretation and staging of a soft-tissue mass, we strongly advocate the use of radiographs as well.

B. J. Manaster, MD, PhD

---

^{18}F-FDG PET/CT as an Indicator of Progression-Free and Overall Survival in Osterosarcoma

Costelloe CM, Macapinlac HA, Madewell JE, et al (Univ of Texas M.D. Anderson Cancer Ctr, Houston)

J Nucl Med 50:340-347, 2009

The aim of our study was to retrospectively evaluate whether maximum standardized uptake value (SUV_{max}), total lesion gylcolysis (TLG), or change therein using ^{18}F-FDG PET/CT performed before and after initial chemotherapy were indicators of patient outcome.

Methods.—Thirty-one consecutive patients who underwent ^{18}F-FDG PET/CT before and after chemotherapy, followed by tumor resection, were retrospectively reviewed. Univariate Cox regression was used to analyze for relationships between covariates of interest (SUV_{max} before and after chemotherapy, change in SUV_{max}, TLG before and after chemotherapy, change in TLG, and tumor necrosis) and progression-free and overall survival. Logistic regression was used to evaluate tumor necrosis.

Results.—High SUV_{max} before and after chemotherapy ($P = 0.008$ and $P = 0.009$, respectively) was associated with worse progression-free survival. The cut point for SUV_{max} before chemotherapy was greater than 15 g/mL* ($P = 0.015$), and after chemotherapy it was greater than 5 g/mL* ($P = 0.006$), as measured at our institution and using lean body mass. Increase in TLG after chemotherapy was associated with worse progression- free survival ($P = 0.016$). High SUV_{max} after chemotherapy was associated with poor overall survival ($P = 0.035$). The cut point was above the median of 3.3 g/mL* ($P = 0.043$). High TLG before chemotherapy was associated with poor overall survival ($P = 0.021$). Good overall and progression-free survival was associated with a tumor necrosis greater than 90% ($P = 0.018$ and 0.08, respectively). A tumor necrosis greater than 90% was most strongly associated with a decrease in SUV_{max} ($P = 0.015$).

Conclusion.—^{18}F-FDG PET/CT can be used as a prognostic indicator for progression-free survival, overall survival, and tumor necrosis in osteosarcoma.

▶ Osteosarcoma is a common and often devastating tumor. Treatment is chosen depending on prognostic indicators which, historically, include initial tumor staging (histologic grade, tumor size, and presence of metastases).

Preoperative treatment with chemotherapy is followed by restaging, at which time the prognosis becomes part of the determination of whether the primary tumor will be treated with limb salvage, amputation, or even left without further treatment. These options may present difficult choices, as some limb salvage procedures may be particular arduous and not necessarily in the patient's best interest if she/he has only months to survive. Therefore, any additional and more accurate preoperative predictors of overall survival are welcome. This article is a retrospective evaluation of the usefulness of ^{18}F-fluorodeoxyglucose-positron emission tomography (FDG PET)/CT in this endeavor. It demonstrates that PET/CT shows excellent correlation of high standardized uptake value (SUV) and high total lesion glycolysis (TLG) with poor progression-free survival. Increase in either of these values following chemotherapy similarly is associated with poor overall survival. This tool will become a useful adjunct to the restaging of osteosarcoma and should be understood by all imagers using it.

B. J. Manaster, MD, PhD

Cortical scalloping and cortical penetration by small eccentric chondroid lesions in the long tubular bones: not a sign of malignancy?

Bui KL, Ilaslan H, Bauer TW, et al (Cleveland Clinic, OH)

Skeletal Radiol 38:791-796, 2009

Objectives.—The objective of this study was to evaluate by cross-sectional imaging the prevalence and degree of cortical scalloping by small eccentric chondromas correlated with histologic diagnosis and patient history.

Materials and Methods.—From 122 patients with histologically proven enchondromas and two patients without histology but with radiologic and clinical follow-up, 11 patients with small, eccentrically located chondromas in the long bones had cross-sectional imaging available. The lesions were evaluated for location, size, presence, and degree of cortical scalloping. The patient's medical charts and microscope slides were reviewed for relevant clinical history, clinical management, and histology.

Results.—The chondromas ranged in size from 1.6 to 3.8 cm (mean 2.3 cm). Two lesions were located in the proximal femoral diaphysis, two in the distal femoral diaphysis, six in the distal femoral metaphysis, and one in the proximal tibial epimetaphysis. The lesions were curetted due to diagnostic uncertainty, continued pain, marked radiologic cortical penetration, or due to patient insistence on biopsy. All 11 lesions were benign, nine histologically, and two by stability over 4 and 7 years. The prevalence of cortical scalloping among eccentric chondromas was 100%. Cortical scalloping or occupancy ranged from 50 to 100% (mean 75%).

Conclusions.—All small eccentric chondromas in this study were associated with an appearance of cortical scalloping of varying degree. All curetted lesions were histologically bland without nuclear atypia. Based on the benign histology of nine lesions and lack of growth of two lesions

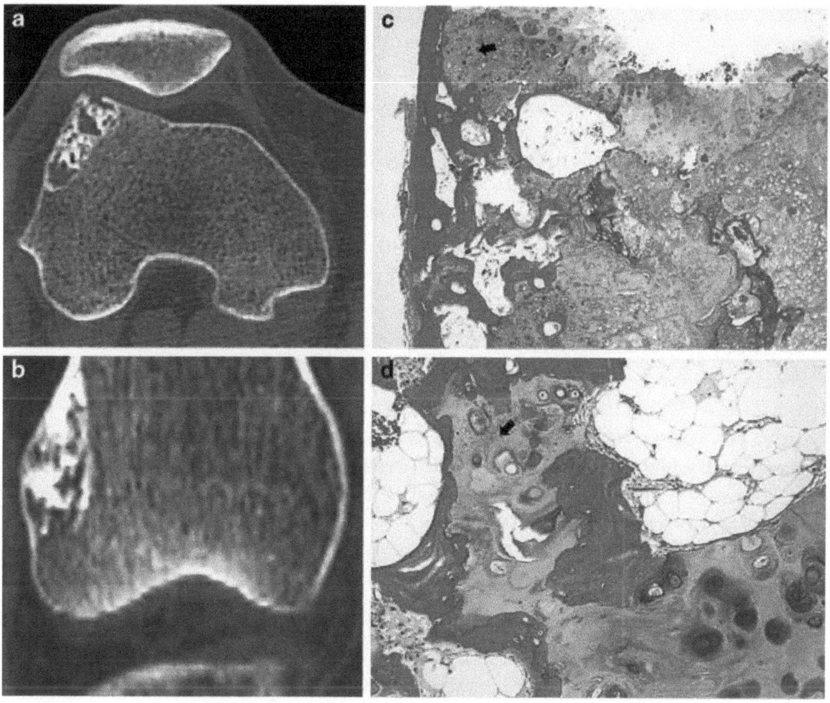

FIGURE 6.—Axial CT (**a**) and coronal reconstructed CT image (**b**) show a 2.3-cm chondroma in the anterolateral distal femoral metaphysis, which demonstrates cortical penetration. The low magnification histologic image (**c**) shows lobules of cartilage (*black arrow*) occupying the cortex. On higher magnification (**d**), there are lobules of cartilage (*black arrow*) surrounded by bone and no features of proliferating cartilage. (With kind permission from Springer Science+Business Media: Bui KL, Ilaslan H, Bauer TW, et al. Cortical scalloping and cortical penetration by small eccentric chondroid lesions in the long tubular bones: not a sign of malignancy? *Skeletal Radiol.* 2009;38:791-796.)

over several years, the degree of cortical scalloping is felt to be a result of lesion location within the endosteum rather than biological activity or malignancy.

▶ Differentiating between enchondroma and low-grade chondrosarcoma in the tubular bones is the topic of much discussion and consternation, both among musculoskeletal radiologists and pathologists. Many musculoskeletal radiologists now believe that enhancement pattern on MR may be a useful feature in differentiating these 2 lesions, but this has not been completely substantiated in the literature. One of the few signs on MR which is well-documented and often quoted is the relationship between the degree and extent of endosteal cortical scalloping in an intramedullary cartilage lesion and the likelihood of its being a low-grade chondrosarcoma. The current article demonstrates the MR characteristics of a sub-group of enchondromas, all relatively small lesions located eccentrically in tubular bones. The authors propose the term of "endosteal chondroma" for these lesions which are neither central intramedullary nor intracortical in location. These lesions nearly uniformly showed >2/3 endosteal scalloping and some even demonstrated subtle cortical penetration (Fig 6). Although these features are associated with a diagnosis of low-grade chondrosarcoma in intramedullary lesions, in this case the histology was uniformly bland and behavior nonaggressive. The authors believe that the cortical scalloping in these endosteal chondromas simply relates to their location adjacent to the cortex and associated growth behavior rather than a propensity to degenerate to chondrosarcoma. This is an important pattern to recognize because previously accepted criteria for differentiating enchondroma from low-grade chondrosarcoma to not seem to apply in these lesions.

B. J. Manaster, MD, PhD

High-Grade Soft-Tissue Sarcomas: Tumor Response Assessment—Pilot Study to Assess the Correlation between Radiologic and Pathologic Response by Using RECIST and Choi Criteria

Stacchiotti S, Collini P, Messina A, et al (Fondazione IRCCS Istituto Nazionale Tumori, Milan, Italy; et al)

Radiology 251:447-456, 2009

Purpose.—To compare radiologic response as defined according to both Response Evaluation Criteria in Solid Tumors (RECIST) and the new Choi criteria recently proposed for gastrointestinal stromal tumors with pathologic response in high-grade soft-tissue sarcomas (STSs) treated with preoperative chemotherapy and radiation therapy.

Materials and Methods.—The institutional ethical committee approved the trial in which patients were enrolled. Signed informed consent was obtained. Thirty-seven patients (21 men, 16 women; mean age, 44.2 years) enrolled in a collaborative randomized trial on preoperative chemotherapy and radiation therapy in localized high-risk STS at a single institution were selected for this retrospective analysis. Tumor response to preoperative treatment was assessed by using both RECIST and Choi criteria at computed tomography (CT) and was adapted to be used at magnetic resonance (MR) imaging. Pathologic response was assessed as either good or very good. Sensitivity, specificity, and predictive value of RECIST and Choi criteria were calculated with pathologic response as the reference standard and were reported with 95% confidence intervals.

Results.—For 28 patients without synovial sarcomas, sensitivity of RECIST versus adapted Choi criteria was 32.0% versus 88.0% for good response and 41.2% versus 82.4% for very good response, respectively; specificity for pathologic response was 100% versus 100% for not a good response and 90.9% versus 27.3% for not a very good response, respectively. In synovial sarcoma, the nontreatment-related neoplastic cystic component of the tumor was a major obstacle for both RECIST and Choi criteria.

Conclusion.—In STS treated with chemotherapy and radiation therapy, tumor size may be insufficient to render actual tumor response. Tumor attenuation at CT or tumor contrast material enhancement at MR imaging may complement tumor size, thus making Choi criteria more predictive of pathologic response.

▶ As more and more patients with sarcomas receive treatment in their local communities rather than a quaternary oncologic care center, their follow-up imaging is also obtained at their community imaging center and interpreted by radiologists who are not specifically trained in oncologic imaging. Any radiologist finding themselves in this position should read this article. The article is a fine example of a prospective study comparing imaging criteria for detecting soft-tissue tumor response with treatment, using pathology as the gold standard. While many radiologists are aware that the criterion of change in tumor size is often not applicable when evaluating treated osteosarcoma, they may not be aware that size criteria may not be as useful as other parameters in soft-tissue sarcomas. In fact, it has been recently shown that there may be dramatic histologic response to treatment of some tumors treated with molecular-targeted therapy which does not result in decreased tumor size on imaging. The article makes a preliminary argument for relying on tumor attenuation on contrast-enhanced CT or contrast enhancement on MR imaging as being more predictive of tumor response, to be used as a complement to evaluation of change in tumor size. The article also stresses the difficulty of judging tumor response in synovial sarcoma, thus reminding the radiologist to not be overly confident in his/her interpretation regarding this lesion.

B. J. Manaster, MD, PhD

Post-treatment complications of soft tissue tumours

Shapeero LG, De Visschere PJL, Verstraete KL, et al (Uniformed Services Univ of the Health Sciences, Bethesda, MD; Ghent Univ Hosp, Belgium; et al)

Eur J Radiol 69:209-221, 2009

Purpose.—To identify local and distant complications of patients with soft tissue tumours and evaluate their relationships to types of therapy.

Methods and Materials.—Fifty-one patients (29 males and 22 females, ages 14–80 years) with 34 malignant and 17 benign soft tissue tumours were evaluated for local and distant complications after resection or amputation only (26 patients) or after the addition of radiotherapy (25 patients: 17 patients had external beam therapy, 7 patients had external beam therapy and brachytherapy, and one patient had extracorporeal irradiation and reimplantation). Duration of follow-up averaged 3.75 years for malignant tumours and 2.79 years for benign tumours. Follow-up studies included radiography, T1- and T2-weighted magnetic resonance (MR) imaging, dynamic contrast-enhanced magnetic resonance imaging (DCE-MRI), computed tomography for thoracic and abdominal metastases, and 3-phase technetium-99 m-labeled-methylene-diphosphonate scintigraphy for bone metastases.

Results.—Recurrent tumours were 2.2 times more frequent in patients who had undergone their initial resection at an outside hospital as compared with those first treated at the university hospital. Nine of 11 recurrences occurred after marginal surgery. Metastases from soft tissue sarcomas, most commonly to lung (nine patients) and to bone and muscle (five patients), showed no specific relationship to type of therapy. DCE-MRI differentiated rapidly enhancing soft tissue recurrences (11 patients) and residual tumours (6 patients) from slowly enhancing muscle inflammation, and non-enhancing fibrosis and seromas that usually did not enhance. Seromas developed in 76% of patients who had postoperative radiation therapy and in 7.7% of patients who had only surgery. Subcutaneous and cutaneous oedema and muscle inflammation was at least four times more frequent after adjunct radiotherapy than after resection alone. Irrespective of the type of treatment, inflammatory changes in muscle and subcutaneous and cutaneous tissue and the majority of seromas were evident at the first follow-up study. Although seromas after resection and external beam therapy resolved with time, seromas after additional brachytherapy persisted. Inflammatory changes in muscle and cutaneous and subcutaneous tissue after resection alone disappeared by the second follow-up study, whereas these changes after radiotherapy resolved months to years after treatment. Fourteen of 51 patients showed MR findings of chronic muscular atrophy, predominantly located in the lower extremity. Heterotopic ossification was seen in three patients after resection and amputation without radiotherapy. Except for one patient with aggressive fibromatosis, bone and nerve complications occurred in patients with soft tissue malignancy. Twelve patients had osteoporosis. Six patients sustained fractures in irradiated osteoporotic bone of the lower extremity, and one patient had a vertebral fracture in radiographically normal but irradiated bone. In addition, one patient was found to have a medullary infarct in an irradiated femur. In nerve entrapment, DCE-MRI demonstrated the rapidly enhancing recurrent tumour or non-enhancing fibrosis surrounding the slowly enhancing nerve. T1- and T2-weighted MR images displayed the acute and chronic sequelae of nerve entrapment and nerve transection with denervation as T2-hyperintense acute muscle atrophy or T1-hypertense chronic fatty muscular atrophy with decrease in muscle volume.

Conclusion.—This study suggests a possible relationship between types of treatment of soft tissue tumours and subsequent complications. Postoperative radiotherapy was associated with a significant number of patients with seromas, muscle, cutaneous and subcutaneous inflammation, and fractures. Incomplete or difficult surgery resulted in residual or recurrent tumours and heterotopic ossification. Muscle atrophy and nerve entrapment were related to both treatments (resection alone or radiotherapy after resection). Diligent follow-up of patients with soft tissue tumours with recognition of these complications and their differentiation

TABLE 2.—Recurrences and Residual Tumours in Patients Treated at the University and Local Hospitals.

Location of Resection	Recurrences No.	Recurrences Percent	Residual Tumours No.	Residual Tumours Percent
University hospital	4/37[a] positive margin	10.8	3/37	8.1
	2/37 negative margin	5.4		
Local hospital	5/14[b] positive margin	35.7	3/14	21.4

[a]Thirty-seven patients had resection or amputation at the University Hospital.
[b]Fourteen patients had initial resection or amputation at a local hospital.

from recurrent or residual tumour can help guide clinical care and may negate the need for surgery when benign disease is defined (Table 2).

▶ Soft tissue sarcoma evaluation by means of imaging merits continued concern. This article is a well-conceived, prospective report of current follow-up evaluation of 51 patients treated with different types of therapy. The article has primary importance in pointing out that tumor recurrence is related to type of institution at which the patients underwent their initial surgical resection (2.2 times larger at nonuniversity hospitals) (Table 2). This underscores the importance of patients obtaining complete imaging and treatment at those institutions equipped with oncologic orthopedic surgeons, orthopedic radiologists, and bone pathologists. The article also demonstrates the time course of development of seromas, edema, and other inflammatory changes, relative to the type of treatment. This is valuable information to be used judiciously in determining whether an abnormality should be followed, ignored, or biopsied. The article would have been even more useful had it included a larger number of patients, and had it also included evaluation by means of PET CT. One hopes that more such articles are forthcoming.

B. J. Manaster, MD, PhD

Soft-tissue metastases: Their presentation and origin

Abed R, Grimer RJ, Carter SR, et al (The Royal Orthopaedic Hosp, Birmingham, England)
J Bone Joint Surg [Br] 91-B:1083-1085, 2009

In our database of 7935 patients referred for investigation of a soft-tissue mass, only 100 were found to have a soft-tissue metastasis (1.3%). Our aim was to define the clinical features of such patients and to identify the site of their primary tumour.

The most common presentation was a painful lump, deep to the fascia, ranging between 2 cm and 35 cm (mean 8.3 cm) with 78% of the lumps located deep to the fascia. The mean age of the patients at presentation was 64 years (22 to 84) and there were almost equal numbers of men and women. Of 53 patients with a history of malignancy, 52 had metastases from the same primary (lung in 12, melanoma in ten, kidney in nine, gastrointestinal track in four, breast in five, bladder in four, and others in eight). The other 47 had no history of malignancy and the metastasis was the first presentation. The primary sites in these cases were the lung in 19, gastro-intestinal track in four, kidney in two, melanoma in nine, other in three, and unknown (despite investigations) in ten. There was no correlation between the site of the metastases and the primary tumour.

Of the 7935 patients, 516 had a history of malignancy. Of these, only 10% had a soft-tissue metastasis, 29% had a benign diagnosis, 55% a soft-tissue sarcoma and 6% another malignancy. Patients with soft-tissue metastases have similar clinical features to those with soft-tissue sarcomas and should be considered for assessment at appropriate diagnostic centres for patients with suspicious soft-tissue lumps.

▶ This article presents a nice reminder that soft-tissue masses may be metastatic in origin. From a large database, these authors demonstrate that only a small, but significant, number of presenting soft-tissue masses prove to be metastatic. The metastatic masses present clinically in a similar manner to primary soft-tissue tumors. There are no distinctive imaging features. The numbers presented, and the discussion regarding possible reasons for the relative rarity of soft-tissue metastases, are interesting. However, the reader should bear in mind that these numbers represent only patients in whom the soft-tissue mass was the reason for presentation to the health care system. It is likely that numbers of soft-tissue metastases are significantly larger in patients with end-stage tumor, widely metastatic to other organs as well.

B. J. Manaster, MD, PhD

Spatial Heterogeneity in Sarcoma ^{18}F-FDG Uptake as a Predictor of Patient Outcome

Eary JF, O'Sullivan F, O'Sullivan J, et al (Univ of Washington, Seattle; Univ College Cork, Ireland)

J Nucl Med 49:1973-1979, 2008

---

^{18}F-FDG PET images of tumors often display highly heterogeneous spatial distribution of ^{18}F-FDG–positive pixels. We proposed that this heterogeneity in ^{18}F-FDG spatial distribution can be used to predict tumor biologic aggressiveness. This study presents data to support the hypothesis that a new heterogeneity-analysis algorithm applied to ^{18}F-FDG PET images of tumors in patients is predictive of patient outcome.

Methods.—^{18}F-FDG PET images from 238 patients with sarcoma were analyzed using a new algorithm for heterogeneity analysis in tumor ^{18}F-FDG spatial distribution. Patient characteristics, tumor histology, and patient outcome were compared with image analysis results using

TABLE 5.—Multivariate Analyses for Disease-Free Survival ($n = 188$, Events = 95)

Variable	Coefficient	Hazard ratio (Confidence Interval)	P
Age	0.01	1.13 (0.89, 1.44)	0.30
Sex	−0.23	0.79 (0.52, 1.21)	0.29
Grade			
High vs. low	0.59	1.80 (0.80, 4.05)	0.15
Intermediate vs. low	0.26	1.3 (0.59, 2.86)	0.52
Subtype			
Cartilage vs. bone	0.16	1.18 (0.42, 3.32)	0.75
Soft tissue vs. bone	−0.02	0.98 (0.40, 2.41)	0.97
Size (sqrt)	−0.00	0.93 (0.71, 1.21)	0.6
SUVmax (sqrt)	−0.05	0.96 (0.74, 1.23)	0.72
Heterogeneity (sqrt)	0.51	1.75 (1.28, 2.40)	**<0.001**

As in Table 4 in the original article, continuous variable, hazard ratio, and confidence intervals are interpreted as increase in survival risk associated with increase of 1 SD of covariate. Significant variables are in bold type. sqrt = square root.

univariate and multivariate analysis. Cox proportional hazards models were used to further analyze the significance of the data associations.

Results.—Statistical analyses show that heterogeneity analysis is a strong independent predictor of patient outcome.

Conclusion.—The new ^{18}F-FDG PET tumor image heterogeneity analysis method is validated for the ability to predict patient outcome in a clinical population of patients with sarcoma. This method can be extended to other PET image datasets in which heterogeneity in tissue uptake of a radiotracer may predict patient outcome (Table 5).

▶ The goal in musculoskeletal tumor work-up and staging is site evaluation for surgical resection as well as prediction of the biologic behavior for local invasiveness, metastatic potential, and to plan treatment and response-evaluation schedules. Both MR (including dynamic contrast enhanced studies) and ^{18}F-fluorodeoxyglucose positron emission tomography (FDG PET)/CT imaging have significant overlap between high-grade and low-grade malignant lesions, and occasionally even between benign and malignant lesions. With ongoing research, it has been noted that higher grade tumors have biologic heterogeneity. This is expressed in some imaging studies as regions of necrosis, others of cellular proliferation, and others of noncellular accumulations of tissue. This article offers another parameter that evaluates such heterogeneity—the spacial distribution of ^{18}F-FDG positive pixels in PET imaging (Fig 1 in the original article). This is a well-conceived study with large numbers of patients and likely will stand as the seminal article defining this new parameter of tumor heterogeneity as a strong independent predictor of survival in patients with sarcoma.

B. J. Manaster, MD, PhD

Bone Metastases in Patients with Metastatic Breast Cancer: Morphologic and Metabolic Monitoring of Response to Systemic Therapy with Integrated PET/CT

Tateishi U, Gamez C, Dawood S, et al (Univ of Texas M.D. Anderson Cancer Ctr, Houston)

Radiology 247:189-196, 2008

Purpose.—To retrospectively compare morphologic and metabolic changes in bone metastases in response to systemic therapy in patients with metastatic breast cancer (MBC) with integrated positron emission tomography (PET)/computed tomography (CT).

Materials and Methods.—The institutional review board waived the requirement for informed consent and approved this HIPAA-compliant study. A retrospective analysis was performed with 102 women (mean age, 55 years) with MBC who received systemic treatment. All patients underwent integrated PET/CT before and after treatment. Two reviewers analyzed the images in consensus. Morphologic changes, including morphologic patterns, and lesion attenuation were evaluated. Standardized uptake value (SUV) and total lesion glycolysis (TLG) were analyzed to evaluate metabolic changes. Uni- and multivariate analyses were performed to identify factors that enabled response duration (RD) to be predicted.

Results.—At baseline, the morphologic patterns of the target lesions were lytic ($n = 33$), sclerotic ($n = 22$), mixed ($n = 42$), and unclassified ($n = 5$). Progression of sclerotic change after treatment was identified in 49 patients (48%). After treatment, the mean attenuation of the lesion increased, whereas the mean SUV and TLG decreased. Increases in attenuation correlated significantly with decreases in SUV ($r = -0.510$, $P < .001$) and TLG ($r = -0.491$, $P < .001$). Univariate analysis revealed that the increase in attenuation and the decrease in SUV were potential predictors of RD. Multivariate analysis revealed that an increase in the change in SUV was a significant predictor of RD (relative risk, 2.4; $P = .003$).

Conclusion.—A decrease in SUV after treatment was an independent predictor of RD in patients with MBC who had bone metastases.

▶ Those of us following metastatic breast disease by means of radiographs have long believed that a lytic lesion, which becomes sclerotic but does not enlarge, suggests good therapeutic response. This article provides a significant advance in this hypothesis, and is much more rigorous in its approach. Positron emission tomography (PET)/CT has wonderful potential in the monitoring of therapeutic response, but there are often confounding factors which make interpretation difficult at best and unreliable at worst. One of these confounding factors is found in metastatic breast disease, where bone metastases may present as lytic, sclerotic, or mixed density. Although this is a somewhat limited retrospective analysis, it is rigorous and leads to useful information, which should provide another independent predictor of metastatic tumor response, based on a combination of metabolic and morphologic changes. This represents an excellent step forward; we would like to see more of this type of work in the field of tumor follow-up imaging.

B. J. Manaster, MD, PhD

Local changes in bone marrow at MRI after treatment of extremity soft tissue sarcoma

Hwang S, Lefkowitz R, Landa J, et al (Memorial Sloan–Kettering Cancer Ctr, New York, NY)

Skeletal Radiol 38:11-19, 2009

Objective.—To determine the prevalence and appearance of magnetic resonance imaging (MRI) signal changes that occur in local bone marrow after radiation therapy (RT) and/or chemotherapy for extremity soft tissue sarcoma (STS).

Materials and Methods.—Seventy patients with primary STS at the level of a long bone who also had undergone pretreatment MRI and at least one post-treatment MRI of the tumor bed were identified. MRIs of each patient were retrospectively reviewed for new changes in marrow signal in the region of the tumor bed and for the morphology, relative signal intensities, heterogeneity, and progression or regression of changes over time.

Results.—Focal signal changes in marrow were observed in 26/70 patients (37%) at a median of 9.5 months after RT and/or chemotherapy and diffuse changes in seven (10%) at a median of 8 months. Patients who

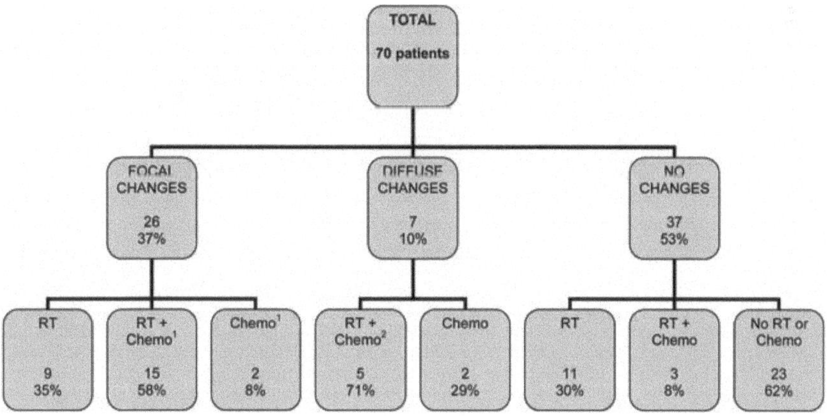

FIGURE 1.—Signal changes in marrow at MRI stratified by type of therapy. *RT* Radiation therapy, *Chemo* chemotherapy. *Superscript 1* Two patients with Ewing family tumor did not undergo surgical resection; one patient received RT and chemotherapy, and the other received only chemotherapy. *Superscript 2* One patient with rhabdomyosarcoma did not undergo surgical resection, and received RT and chemotherapy. (Reprinted from Hwang S, Lefkowitz R, Landa J, et al. Local changes in bone marrow at MRI after treatment of extremity soft tissue sarcoma. *Skeletal Radiol.* 2009;38:11-19, with kind permission from Springer Science and Business Media.)

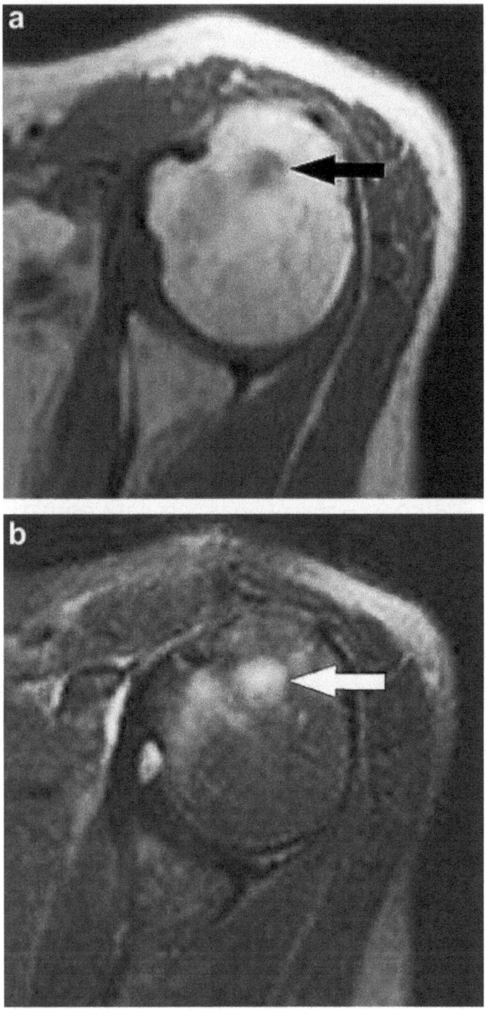

FIGURE 4.—A 76-year-old woman with leiomyosarcoma in upper arm, metastatic to lung and liver. Nodular marrow changes first became evident 9 months after RT and 4 months after first chemotherapy (without G-CSF). Patient also has had cancers of lung, breast, colon, and bladder. **a** Axial T1-weighted image demonstrates nodular foci of intermediate to high signal intensity (*arrow*) in humeral head. **b** The nodular foci (*arrow*) show heterogeneously high signal intensity in axial fat-saturated T2-weighted image. These nodular foci were considered by the readers as being indeterminate for metastasis. Subsequent follow-up MRI did not show evidence of metastasis in this region. (Reprinted from Hwang S, Lefkowitz R, Landa J, et al. Local changes in bone marrow at MRI after treatment of extremity soft tissue sarcoma. *Skeletal Radiol.* 2009;38:11-19, with kind permission from Springer Science and Business Media.)

received neither RT nor chemotherapy did not develop marrow changes. Mean RT doses in patients with changes and those without were 5,867 and 6,076 cGy, respectively. In most patients with focal changes, changes were seen in all sequences and were linear-curvilinear, patchy, or mixed at the level of the tumor bed. Predominant signal intensity of changes was between muscle and fat at T1WI and between muscle and fluid at fat-saturated T2WI or short tau inversion recovery. Most focal changes enhanced heterogeneously and increased or fluctuated in size over time.

Conclusion.—Changes in MRI appearance of long bone marrow frequently are evident after combined RT and chemotherapy for STS and most commonly increase or fluctuate in size over time. These changes have various non-mass-like configurations and often show signal intensities similar to those of red marrow and thus should not be mistaken for metastases. The marrow changes might represent an early stage of gelatinous transformation of marrow (Figs 1 and 4).

▶ Evaluation of bone marrow on MR examination is difficult; the reader must factor in patient age, gender, and various health conditions before suggesting that marrow signal suggests either focal or a generalized infiltrative disease. This difficulty is confounded in patients who have undergone chemotherapy and/or radiation therapy for adjacent soft tissue sarcoma. This study is retrospective, and does not account for many variables of treatment, and is not without reader bias. However, it nicely demonstrates the frequency of marrow abnormality in these patients (Fig 1), as well as the fact that this marrow abnormality, whether focal or diffuse, fluctuates in size over time (Fig 4). The study did not address further confounding factors such as stem cell stimulating factors or other hormonal stimulation. Nonetheless, it is an excellent review of a large number of patients following treatment and serves as a reminder to radiologists to be aware of previous treatment and make interpretations within this context.

B. J. Manaster, MD, PhD

Other

Early rheumatoid arthritis and its differentiation from other joint abnormalities

Boutry N, do Carmo CCM, Flipo R-M, et al (Hôpital Roger Salengro-CHRU de Lille, France)

Eur J Radiol 71:217-224, 2009

The introduction of disease-modifying antirheumatic drugs has created new demands on imaging to early identify patients with rheumatoid arthritis and opened new prospects in therapeutic management of patients with aggressive disease. Therefore, new imaging modalities such as magnetic resonance imaging and ultrasound have developed during the past few years in this field. In some cases, both magnetic resonance imaging and ultrasound may be also useful in making the distinction between early rheumatoid arthritis and other joints abnormalities, including early psoriatic arthritis. This article will review key aspects of

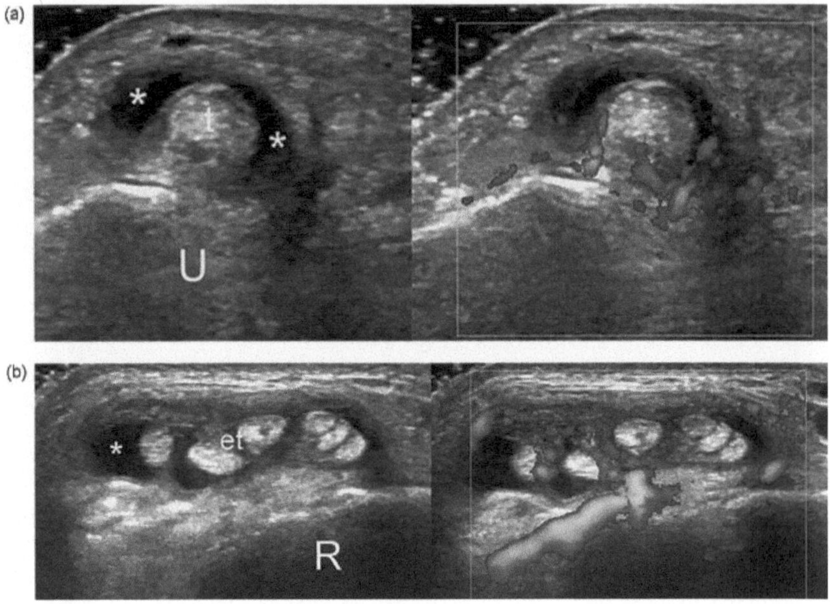

FIGURE 4.—Tenosynovitis of wrist in RA. Dorsal transverse US scans of wrist in patient with early RA (a) reveals hypo- and anechoic thickening of extensor carpi ulnaris tendon sheath (asterisks), in relation with tenosynovitis (left), and hyperemia on power Doppler imaging (right): U = distal ulna; t = extensor carpi ulnaris tendon. Note also heterogeneous appearance of tendon. Dorsal transverse US scans of wrist in another patient with early RA (b) shows extensor tenosynovitis with marked hyperemia on power Doppler imaging. Note also presence of anechoic effusion (asterisk) with no evidence of flow on Doppler imaging: R = radius; et = extensor tendon. (Reprinted from Boutry N, do Carmo CCM, Flipo R-M, et al. Early rheumatoid arthritis and its differentiation from other joint abnormalities. *Eur J Radiol.* 2009;71:217-224, with permission from Elsevier.)

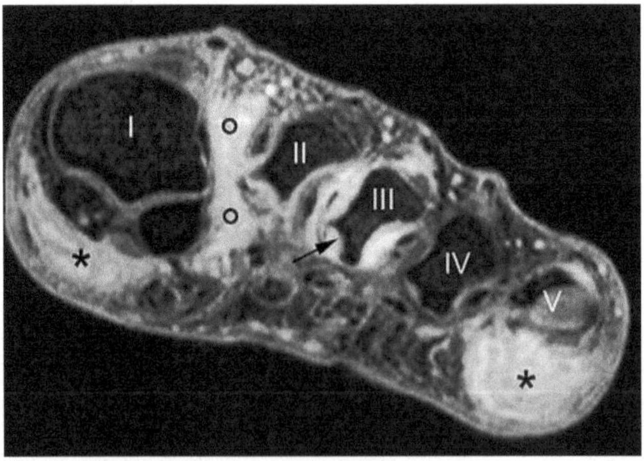

FIGURE 5.—Bursites of forefoot in RA. Transverse fat-suppressed gadolinium-enhanced 3D gradient-echo MR image of forefoot in patient with early RA shows intermetatarsal (circles) and submetatarsal (asterisks) bursites. Note also MTP joint synovitis, especially at third ray, and presence of plantar bone erosion (arrow). (Reprinted from Boutry N, do Carmo CCM, Flipo R-M, et al. Early rheumatoid arthritis and its differentiation from other joint abnormalities. *Eur J Radiol.* 2009;71:217-224, with permission from Elsevier.)

important advances in imaging in rheumatoid arthritis, particularly focusing on magnetic resonance imaging and ultrasound (Figs 4 and 5).

▶ I generally hesitate to include review articles in my selections. However, this article must prove an exception. With the effective new medical therapy now generally available for rheumatoid arthritis, early diagnosis of this age-old disease process takes on a new urgency. Both magnetic resonance imaging (MRI) and ultrasound (US) are being used much more frequently, both in the early diagnosis of disease (Figs 1 and 2 in the original article) and in evaluation of efficacy of treatment. This article is beautifully organized and illustrated and serves as a primer for radiologists starting to look at the early manifestations of rheumatoid arthritis. I highly recommend perusing it.

B. J. Manaster, MD, PhD

Diagnostic and Interventional MRI of the Sacroiliac Joints Using a 1.5-T Open-Bore Magnet: A One-Stop-Shopping Approach

Fritz J, Henes JC, Thomas C, et al (Johns Hopkins Univ School of Medicine, Baltimore, MD; Eberhard-Karls-Univ Tübingen, Germany)

AJR Am J Roentgenol 191:1717-1724, 2008

Objective.—The objective of our study was to prospectively test the hypothesis that combined diagnostic and interventional MRI of the sacroiliac joints can be performed efficiently and effectively.

Subjects and Methods.—Over a 12-month period, 60 patients (32 women and 28 men; median age, 28 years; age range, 18–49 years) with chronic lower back pain suspected to originate from the sacroiliac joints were enrolled in the study. Based on diagnostic MRI findings, MR fluoroscopy–guided sacroiliac joint injections were performed in 57 (95%) patients. Diagnostic injections (35, 58.3%) were performed if nonspecific or degenerative MRI findings were present. Therapeutic injections (22, 36.7%) were performed in patients with inflammatory arthropathy. In three (5%) patients, no injections were performed. Technical effectiveness was assessed by analyzing, first, the rate of intraarticular injection; second, the time required for the procedure; third, image quality; and, fourth, occurrence of complications and clinical outcome by analyzing pain intensity changes and volume and signal intensity of sacroiliac inflammatory changes.

Results.—The rate of intraarticular injection was 90.4% (103/114). The mean length of time for the procedure was 50 minutes (range, 34–103 minutes), with exponential shortening over time ($p \leq 0.001$). The contrast-to-noise ratios of the needle and tissues were sufficiently different for excellent delineation of the needle. No complications occurred. Diagnostic injections identified the sacroiliac joints as generating significant pain in 46.9% (15/32) of the patients. Three months after therapeutic injections, pain intensity had decreased by 62.5% ($p \leq 0.001$) and the volume and relative signal intensity of inflammatory changes had decreased by 37.5% ($p = 0.003$) and 47.6% ($p \leq 0.001$), respectively.

Conclusion.—We accept the hypothesis that combined diagnostic and interventional MRI of the sacroiliac joints can be performed efficiently and effectively for comprehensive diagnosis and therapy of lower back pain originating from the sacroiliac joints.

▶ Radiologists are more and more frequently receiving requests to perform therapeutic injections in joints, following a period of conservative treatment. These patients usually have diagnostic imaging and then must return for the injection in a different setting. With sacroiliac joint pain, the argument can be made that injections are often helpful, despite a normal diagnostic MR. The argument can also be made that radiation during a fluoroscopically or CT-guided injection is more significant for the sacroiliac joint than more peripheral joints. Therefore, it is reasonable to consider performing a combined diagnostic and interventional MR of sacroiliac joints, presuming previous failed conservative treatment. The authors of this article developed a reasonable algorithm for sacroiliac joint pain and treatment. The second requirement is to design a protocol, which is effective both in terms of successful injection and responsible use of magnet time. They have done so. It is impressive both that the average time for both diagnostic and interventional portions of this exam is 50 minutes, and that the time in magnet decreased significantly as operator experience increased. This is a useful article, which outlines a good methodology of developing such procedures.

B. J. Manaster, MD, PhD

Evaluation of the Diagnostic Utility of Spinal Magnetic Resonance Imaging in Axial Spondylarthritis

Bennett AN, Rehman A, Hensor EMA, et al (Leeds Inst of Molecular Medicine and Chapel Allerton Hosp; American Hosp Dubai, United Arab Emirates; et al)

Arthritis Rheum 60:1331-1341, 2009

Objective.—Magnetic resonance imaging (MRI) is increasingly used for the diagnosis of axial spondylarthritis (SpA), but it is unknown whether characteristic lesions are actually specific for SpA. This study was undertaken to compare MRI patterns of disease in active SpA, degenerative arthritis (DA), and malignancy.

Methods.—Fat-suppressed MRI of the axial skeleton was performed on 174 patients with back pain and 11 control subjects. Lesions detected by MRI, including Romanus lesions (RLs) and end-plate, diffuse vertebral body, posterior element, and spinous process bone marrow edema (BME) lesions, were scored in a blinded manner. An imaging diagnosis was given based on MRI findings alone, and this was compared with the gold-standard treating physician's diagnosis.

Results.—The physician diagnosis was SpA in 64 subjects, DA in 45 subjects, malignancy in 45 subjects, other diagnoses in 20 subjects, and normal in 11 subjects. There was 72% agreement between the imaging diagnosis and physician diagnosis. End-plate edema, degenerative discs, and RLs were frequently observed in patients with any of the 3 major diagnoses. Single RLs were of low diagnostic utility for SpA, but ≥3 RLs (likelihood ratio [LR] 12.4) and severe RLs (LR infinite) in younger subjects were highly diagnostic of SpA. Posterior element BME lesions of mild or moderate grade were also highly diagnostic of SpA (LR 14.5). The most common diagnostic confusion was between SpA and DA, since both had RLs present and the presence/ absence of degenerative discs did not change the diagnostic assessment.

Conclusion.—This study confirms the high diagnostic utility of MRI in axial SpA, with severe or multiple RLs evident on MRI being characteristic in younger patients and mild/moderate posterior element lesions being specific for SpA. However, MRI lesions previously considered to be characteristic of SpA could also be found frequently in patients with DA and patients with malignancy, and therefore such lesions should be interpreted with caution, particularly in older patients.

▶ It is not infrequent for radiologists to overlook the signs of early ankylosing spondylitis when evaluating spine MRI for low back pain. Too often we devote the most careful search to the disks and spinal cord, without paying attention to the vertebral bodies, posterior elements, and adjacent soft tissues. Several articles have described Romanus and Anderson lesions on MRI, demonstrating that, with care, early diagnosis of ankylosing spondylitis can be made. This article is among the best I have seen. It is a prospective article, with a large number of patients. It demonstrates nicely the value of MR in diagnosing axial ankylosing spondylitis, with excellent specificity in young patients. However, it carries further value in showing inflammatory ankylosing spondylitis lesions, which do not fit the typical location of Romanus lesions, for which other diagnoses might be considered (Fig 4 in the original article). Furthermore, it is valuable to be aware of the difficulty one might encounter in differentiating Romanus lesions from degenerative change or metastatic disease in the older patient. This article is most worthwhile, both as an advocate for the use of MR in diagnosing early ankylosing spondylitis, and as a cautionary note, showing that differentiation among the possible lesions is not always easy.

B. J. Manaster, MD, PhD

Gout in the Axial Skeleton

Konatalapalli RM, DeMarco PJ, Jelinek JS, et al (Washington Hosp Ctr; et al)
J Rheumatol 36:609-613, 2009

Objective.—Gout typically affects the peripheral joints of the appendicular skeleton and rarely involves the axial joints. The literature on axial gout is limited to case reports and case series. This preliminary study was conducted to identify the frequency and characteristics of axial gout.

Methods.—Six hundred thirty medical records with ICD codes 274.0, 274.82, and 274.9 for peripheral gout were reviewed. Ninety-two patients had clinical or crystal-proven gout, of which 64 had prior computed tomography (CT) images of the spine performed for various medical reasons. These CT images were reviewed for features of axial gout, which include vertebral erosions mainly at the discovertebral junction and the facet joints, deposits of tophi, and erosions in the vertebral body, epidural space, ligamentum flavum and pars interarticularis.

Results.—Nine of the 64 patients had radiographic changes suggestive of axial gout. Lumbar vertebrae were most commonly involved, with facet joint erosions being the most common finding. Isolated involvement of the sacroiliac joints was seen in 2 patients. Axial gout had been diagnosed clinically in only one patient.

Conclusion.—Radiologic changes of axial gout were more common than recognized clinically, with a frequency of 14%. Since not all patients had CT images, it is possible that the frequency of axial involvement was even greater. A prospective study is needed to further define this process.

▶ Gout seems ubiquitous. The incidence has increased significantly over the last 2 decades. Furthermore, the gender difference seems to be changing in the older patient population, with the prevalence in postmenopausal women approaching that of men. With greater incidence of treated end-stage kidney

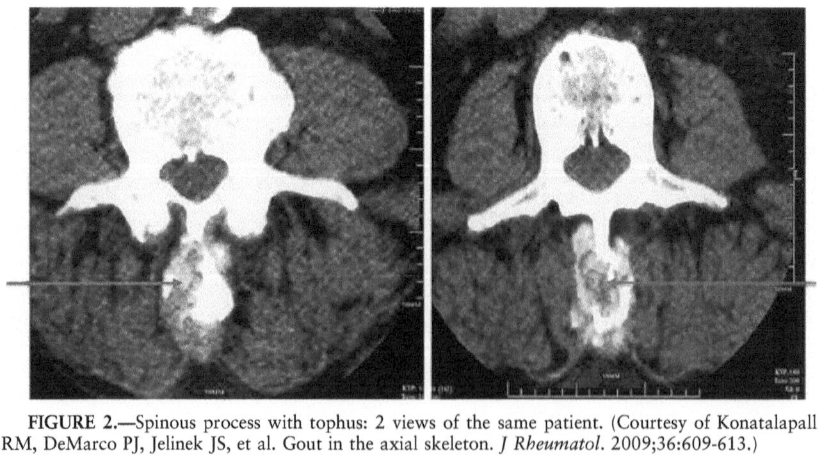

FIGURE 2.—Spinous process with tophus: 2 views of the same patient. (Courtesy of Konatalapalli RM, DeMarco PJ, Jelinek JS, et al. Gout in the axial skeleton. *J Rheumatol*. 2009;36:609-613.)

disease, we see gout more frequently in the younger patient population. Therefore, it is reasonable to reassess our understanding of the imaging of this disease. The article demonstrates that the conventional wisdom of gout occuring only infrequently in the axial skeleton is incorrect. Subtle findings of sodium urate deposits and sequelae (erosions of facets, vertebral bodies, pars interarticularis, and deposits of tophi in and around vertebral bodies and epidural space) may in fact be seen much more frequently than was previously believed when CT images are critically evaluated (Fig 2). This is a useful article in that it reminds us to seriously consider gout when evaluating the axial skeleton.

B. J. Manaster, MD, PhD

Diagnostic Utility of T1-Weighted MRI Characteristics in Evaluation of Osteomyelitis of the Foot

Johnson PW, Collins MS, Wenger DE (Mayo Clinic, Rochester, MN)

AJR Am J Roentgenol 192:96-100, 2009

Objective.—The purpose of this study was to evaluate the diagnostic utility of specific characteristics on T1-weighted MR images in the diagnosis of pedal osteomyelitis.

Materials and Methods.—The study included all adult patients who underwent MRI of the foot for evaluation of suspected osteomyelitis over an 11-month period. Images were retrospectively reviewed for specific criteria on T1-weighted images, including signal intensity of affected bone marrow (normal or decreased), distribution of abnormal signal intensity (subcortical or medullary), and pattern of involvement (hazy reticulated or confluent). The presence of osteomyelitis was presumed if the surgical pathologic result was positive or if there was lack of clinical improvement despite conservative management. Rapid clinical improvement with conservative management was considered an indication that osteomyelitis was not present.

Results.—Images from 74 examinations were evaluated. In 20 cases, osteomyelitis was considered present, and in 54 it was presumed absent. In 19 of the 20 cases (95%) in which osteomyelitis was considered present, marrow T1 signal intensity was decreased, in a medullary distribution, and in a confluent pattern in all cases. In 30 of the 54 cases (56%) in which osteomyelitis was presumed absent, T1 signal intensity was decreased, but only five cases (9%) had a medullary distribution and confluent pattern. Twenty-three cases (43%) had a hazy reticulated pattern, and two cases (4%) had only subcortical distribution. None of the cases with a subcortical distribution or hazy reticulated pattern of abnormal signal intensity had positive results for osteomyelitis. Confluent decreased T1 marrow signal intensity in a medullary distribution was 95% sensitive in the prediction of osteomyelitis with a specificity of 91%, negative predictive value of 98%, and positive predictive value of 79%.

Conclusion.—Findings on T1-weighted MR images, specifically a confluent pattern of decreased T1 marrow signal intensity in a medullary distribution, correlate highly with the presence of pedal osteomyelitis.

▶ Diabetes, and particularly complications of the diabetic foot, is a major source of morbidity in patients of developed countries. Cost of medical care and cost of lost individual productivity are high. Treatment of diabetic foot cellulitis and ulceration is significantly different than treatment for osteomyelitis; hence it is critical for the imager to differentiate between the two. Imaging results are not always clear-cut. Radiographs are insensitive for early osteomyelitis. Nuclear medicine studies are sensitive but nonspecific. MR is sensitive, but not always specific. Fluid-sensitive sequences may show extensive high signal within osseous structures adjacent to soft tissue infection, but this may be due to reactive change rather than osteomyelitis. Similarly, reactive change may cause enhancement of bone. The authors of this article have previously suggested that T1 imaging may be an important differentiating factor. This article takes this concept farther with a larger and well-conceived patient imaging study. Their conclusion that a confluent medullary abnormality seen in the bone on T1 imaging correlates with infection is one which radiologists should hail. Although for most studies, only a single T1 sequence is performed, it may be worthwhile to be certain that at least 2 planes of T1 imaging are performed when the clinical question is of osteomyelitis. This should make our MR studies more specific and worthwhile, both for identifying true positives and true negatives.

B. J. Manaster, MD, PhD

The radiological reporting of lumbar Scheuermann's disease: an unnecessary source of confusion amongst clinicians and patients

Summers BN, Singh JP, Manns RA (Princess Royal Hosp, Telford, Shropshire, UK)

Br J Radiol 81:383-385, 2008

Patients with mechanical lower back pain not uncommonly present to clinicians with anxiety following a diagnosis of "Scheuermann's disease" based on a report of a plain radiograph of the lumbar spine. In most instances, the radiographs show features of lumbar Scheuermann's, rather than the classical adolescent thoracic kyphosis described by Scheuermann in 1920. The purpose of this study was (i) to investigate how often the diagnosis of Scheuermann's disease was made at a District General Hospital in radiological reports sent to local general practitioners (GPs); (ii) to determine the type of "Scheuermann's disease" being described; and (iii) to assess what GPs understood by the term "Scheuermann's disease" in the context of a specific clinical presentation. 50 reports were identified in a computerized search for diagnoses of "Scheuermann's disease". Review of the radiographs revealed that 80% showed features of lumbar Scheuermann's disease and 20% classical Scheuermann's. A questionnaire was issued to local GPs that described a case history of an adult patient with typical mechanical lower back pain for whom a radiological report, describing degenerative changes in the lumbar spine, concluded that some of the features "were consistent with Scheuermann's disease". 86% of GPs stated that they would inform their patients that they had "Scheuermann's disease" (using that exact term), but only 51% understood the meaning of the diagnosis in the context of the case history. We recommend that the term "Scheuermann's disease" be avoided in radiological reports to GPs and other non-specialist clinicians except when reporting on classical adolescent thoracic kyphosis.

▶ There has been confusion in the radiologists' use of the term "Scheuermann's disease." The classic term, as originally described by Scheuermann, described an adolescent development of painful thoracic kyphosis with endplate irregularity and vertebral body wedging. Some specificity of the description is lacking, leading occasionally to overdiagnosis of the process. Furthermore, a type of lumbar "Scheuermann's disease" has been described, which most often is seen in the lumbar spine or thoracolumbar junction of adults, associated with Schmorls nodes and some irregularity of the endplates. This latter appearance has been variably termed "lumbar," "atypical," or "type 2" Scheuermann's disease. However, there is no vertebral wedging or kyphosis in these cases. These patients most frequently have mechanical low back pain and degenerative disk disease. In most cases, they are clearly distinct from the traditional Scheuermann's diagnosis. Unfortunately, the label of "Scheuermann's disease" made in the radiologist's report may be misunderstood by the referring clinicians, which in turn may result in inappropriate subspecialty referral and patient anxiety. This article documents the latter concerns and presents a cogent argument for restricting the term "Scheuermann's disease" to the originally described disease.

B. J. Manaster, MD, PhD

4 Pediatric Radiology

Introduction

I am amazed that a year has gone by since I first began making regular contributions to the YEAR BOOK, and what a year it has been!

On a personal level, I took the plunge and moved my family 300 miles north from Great Ormond Street Children's Hospital in London to take up an academic position at the University of Sheffield and Sheffield Children's Hospital.

Dr Oestreich was to have made a far longer journey, to Berlin, but a number of circumstances led him reluctantly to remain in Cincinnati. Quel dommage!

While we are both enjoying the new lives that have developed this past year, pediatric radiology has been enjoying a changing face, with increasing interest in two specialist areas—namely, antenatal and forensic imaging.

Dr Oestreich and I have selected articles for the 2010 YEAR BOOK related to these relatively new areas, but we have also included articles that aim to develop new or validate existing methods for diagnosis of the usual suspects such as trauma (including child abuse), cancer, infection, etc. In the field of child abuse, we have reviewed writings on the reluctance of some radiologists to testify and the overzealousness of others to testify inappropriately. Concern for unnecessary or excessive radiation exposure remains important, including better understanding of the location of the ovaries.

Our selected articles range from those that confirm scientifically what many of us instinctively believe to those that are somewhat controversial. We (the "AO/AO team") hope that all of them will provide you with food for thought, that some of them will impact positively on your day-to-day practice, and that you share our excitement at the prospect of the articles the coming year will bring.

Amaka C. Offiah, BSc, MBBS, MRCP, FRCR, PhD
Alan E. Oestreich, MD

Gastrointestinal

CT Predictors for Differentiating Benign and Clinically Worrisome Pneumatosis Intestinalis in Children beyond the Neonatal Period

Olson DE, Kim Y-W, Ying J, et al (Cincinnati Children's Hosp Med Ctr, OH; Univ of Cincinnati College of Medicine, OH)

Radiology 253:513-519, 2009

Purpose.—To determine whether computed tomographic (CT) findings can help differentiate between benign and clinically worrisome causes of pneumatosis intestinalis (PI) in children.

Materials and Methods.—This retrospective study was approved by the institutional review board, and requirement for informed consent was waived. Data were stored in a secured and HIPAA-compliant fashion. CT reports from an 8-year period (July 2000–July 2008) were reviewed to determine all cases with a diagnosis of PI. In these cases, demographic information, clinical presentation, underlying medical condition, and CT findings were reviewed. The cases were grouped into one of two final diagnostic groups: clinically worrisome versus benign PI (diagnosis of exclusion, resolution documented at serial imaging without therapeutic intervention). In each case, the CT findings reviewed included distribution (small bowel, large bowel), extent (mild, moderate, extensive), and morphologic characteristics (linear, cystic, both) of the PI and associated findings such as soft-tissue bowel wall thickening, periintestinal soft-tissue stranding, free air, free fluid, portal venous gas, small-bowel obstruction, and bowel dilatation. Associations between CT findings and benign or clinically worrisome PI were assessed with logistic regression models.

Results.—There were 44 cases identified. Final diagnostic categories for PI included benign ($n = 15$) and associated underlying bowel disease ($n = 29$; definitive in 26 and suspected but not defined in three). The following findings were significant (expressed as percentage of clinically worrisome PI vs percentage of benign): soft-tissue bowel wall thickening (51.2% vs 13.3%, $P = .0167$), free peritoneal fluid (82.8% vs 33.3%, $P = .002$), extent of PI (extensive 17.2% vs 69%, $P < .001$), and periintestinal soft-tissue stranding (55.2% vs 20.0%, $P = .0228$). Distribution, free peritoneal air, and characteristic morphology (linear vs cystic) were not associated with clinically worrisome PI (all $P > .05$).

Conclusion.—The cystic or linear pattern of pneumatosis in children is not a useful CT sign to differentiate benign from clinically worrisome PI. CT findings that include soft-tissue thickening of the bowel wall, free fluid, periintestinal soft-tissue stranding, and the extent of PI can be useful in differentiating these entities.

▶ Pneumatosis intestinalis I have always considered like the blueness of cyanosis (the blueness does not hurt the patient [except in the movie, Yellow Submarine], it's the oxygenation level of the blood that hurts). The question then is whether the pneumatosis in an individual is considered likely to be a sign that something serious is amiss. This article nicely addresses that question in children over 6 months of age (it's mostly always a sign of something bad in neonates). The most impressive finding is that thick or thin bowel wall, as shown in Fig 1 in the original article, is a major distinguishing feature: 'thin' is good! Whether the air in the bowel wall is linear or cystic in pattern was not found to be a significant discriminator between benign and clinically worrisome. The findings of this study should be applicable to the non-radiation imaging modalities ultrasound (US) and MRI, since they are both good at determining bowel wall thickening. Portal vein air, incidentally, is also a 'blueness' feature, not dangerous in itself but indicative that something may be awry. As an interesting example, both pneumatosis and portal vein air were reported in a United States case of pigbel enteritis necroticans in a 12 year old boy who consumed tainted chitterlings (pig intestines),[1] one of the worrisome causes of peumatosis seen more typically in Papua New Guinea, as well as other developing countries, but also war-torn areas.

A. E. Oestreich, MD

Reference

1. Petrillo TM, Beck-Sagué CM, Songer JG, et al. Enteritis necroticans (pigbel) in a diabetic child. *New Engl J Med.* 2000;342:1250-1253.

Evaluation of portal venous gas detected by ultrasound examination for diagnosis of necrotising enterocolitis

Dördelmann M, Rau GA, Bartels D, et al (Hannover Med School, Germany)
Arch Dis Child Fetal Neonatal Ed 94:F183-F187, 2009

Background.—Early diagnosis of necrotising enterocolitis (NEC) is difficult but essential for timely therapy. The diagnostic hallmarks and specific radiological signs for NEC are pneumatosis intestinalis (PI) and portal venous gas (PVG), but PVG in abdominal ultrasound (PVG-US) has been proposed as an effective tool in the diagnosis of NEC as well.

Objective.—To prospectively assess the value of PVG-US for the diagnosis of NEC.

Methods.—The study screened 352 neonates for PVG-US (n = 796 routine examinations) and performed 48 additional screenings in 34/352 neonates with suspected (stage I, n = 28) or definite NEC (stage ≥II, n = 20). Sensitivity and specificity of PVG-US for detection of NEC were computed by using NEC stage ≥II as the reference standard.

Results.—PVG-US was only present in cases of suspected or definite NEC. The study observed PVG-US in 4/28 NEC stage I and in 9/20 NEC stage ≥II episodes corresponding to a 86% specificity and a 45% sensitivity for diagnosis of NEC stage ≥II. However, 7/20 patients with NEC stage ≥II showed intraoperative findings other than NEC and another 3/20 infants presented with radiologically unspecific intestinal dilatation. None of these 10 infants had detectable PVG-US. Thus, with

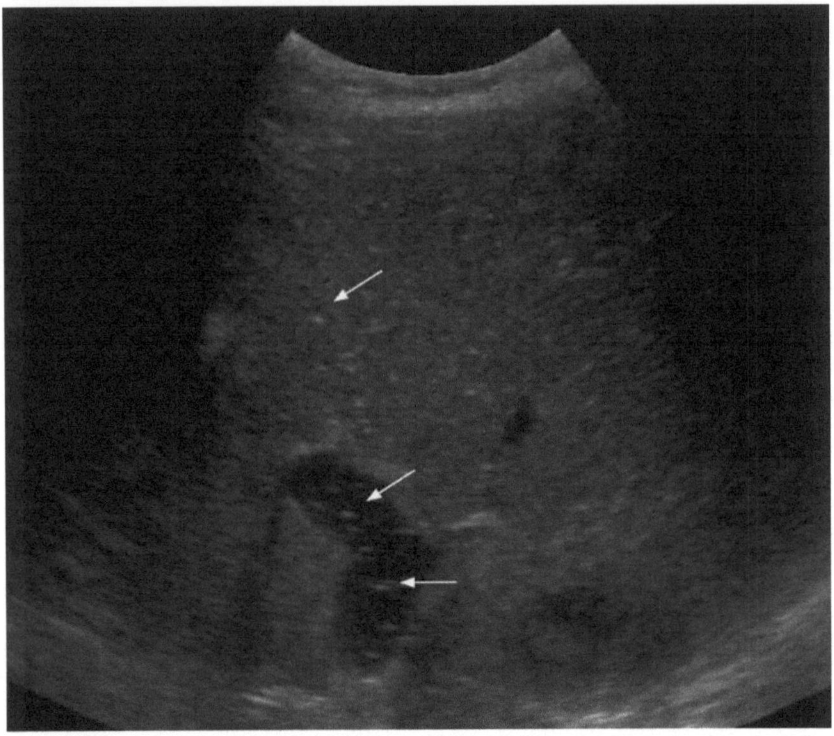

FIGURE 1.—Appearance of portal venous gas in abdominal ultrasound as continuous streaming echogenic dots in the bloodstream of the portal vein and scattered hepatic echogenic spots. (Reprinted from Dördelmann M, Rau GA, Bartels D, et al. Evaluation of portal venous gas detected by ultrasound examination for diagnosis of necrotising enterocolitis. *Arch Dis Child Fetal Neonatal Ed.* 2009;94:F183-F187, with permission from BMJ Publishing Group Ltd.)

application of specific radiological signs the sensitivity of PVG-US for diagnosis of NEC stage ≥II increased to 90%.

Conclusion.—Screening for PVG-US is a useful, easy and quick bedside test with a high specificity for NEC. Moreover, these results question the value of the Walsh criteria in the diagnosis of NEC.

▶ This is a well-conducted study showing acceptable specificity for ultrasound diagnosis of necrotising enterocolitis (NEC) based on a finding of portal venous gas (Fig 1). A strength of this article is its prospective design. However, rather disappointingly, based on standard diagnostic criteria, the sensitivity of the technique was only 43% (although this did rise to 90% when the diagnosis of NEC was based on modified criteria).

Despite the operator dependence of ultrasound, this application of ultrasound may be worth pursuing for the following reasons: the relative ease with which portal venous gas can be recognised, the acceptable scanning time (average of 4.5 minutes per scan), and the low prevalence of portal venous gas in other conditions.

To aid future studies of interobserver and intraobserver reliability, it is hoped that an acceptable (and accepted) gold standard for the diagnosis of NEC will be available.

A. C. Offiah, BSc, MBBS, MRCP, FRCR, PhD

Transient hepatic attenuation differences in neonates

Towbin AJ, Ying J, Fleck R (Cincinnati Children's Hosp Med Ctr, OH; Univ of Cincinnati, OH)

Pediatr Radiol 39:798-803, 2009

Background.—A transient hepatic attenuation difference (THAD) is a hepatic perfusion anomaly seen on contrast-enhanced CT scans caused by an alteration in the dual blood supply of the liver. Although THADs have been described in adolescents and adults, they have not previously been described in neonates.

Objective.—We describe the appearance and evaluate the frequency of THADs in neonates ≤1 month of age compared to other infants younger than 2 years.

Materials and Methods.—A retrospective study was performed looking at all CT angiograms from 2000 to 2007 in infants <2 years of age. The incidence of THADs was compared among four age groups. Significance was determined using a logistic regression model.

Results.—The study included 128 CT angiograms. A THAD was seen in 9/26 infants <1 month of age, in 3/50 infants 1 to 6 months of age, in 1/23 infants 6 months to 1 year of age, and in 1/29 infants 1 to 2 years of age. A THAD was found significantly more frequently in infants <1 month of age than in the older age groups ($P < 0.05$).

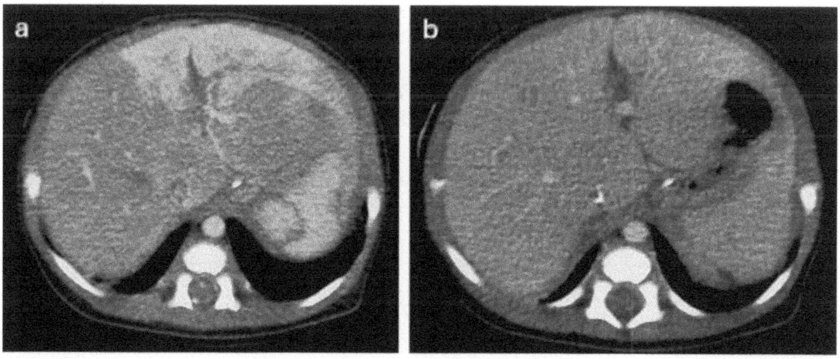

FIGURE 1.—Axial CT images of the liver in a 13-day-old boy with a history of respiratory distress. a On the arterial phase image there is a wedge-shaped area of hyperenhancement spanning segments 2 and 4a. This area is smoothly marginated and abuts the anterior surface of the liver. b On the portal venous phase image the liver has a homogeneous enhancement pattern. (With kind permission from Springer Science and Business Media: Towbin AJ, Ying J, Fleck R. Transient hepatic attenuation differences in neonates. *Pediatr Radiol.* 2009;39:798-803.)

Conclusion.—THADs are benign entities that can be seen normally in the neonatal age group. When the characteristic appearance is seen on CT, no further imaging is needed.

▶ Any imaging publication that leads to avoiding unnecessary further imaging is a contribution to lowering the monetary, and often radiation exposure (Image Gently), costs for individual patients and for society. Knowing about this unusual localized inhomogeneity, so well demonstrated in Fig 1, may obviate further imaging in the young neonate. However, beyond 1 month of age ultrasound (US) or MRI should be considered to rule out an associated lesion if not evident on the CT angiogram itself. This article is a fine example of applying information known to adult radiologists when something unexpected is seen in the very young. In the authors' plausible explanation for the occurrence of transient hepatic attenuation difference (THAD) in the very young, they mention temporarily persistent fetal style circulation through the ductus venosus as illustrated by Schlesinger and colleagues.[1] I would quibble with one terminology matter: rather than blood flowing in the fetus from the umbilical vein through the left portal vein to the ductus venosus, I would say it flows from the umbilical vein directly into the ductus while crossing the left portal vein. The analogy to highway driving may explain—at the intersection of Highway 17 and Highway 45, the person traveling on Highway 17 remains on 17 even at the intersection (while the person traveling on 45 remains on 45 even through the crossroad). In my opinion, the left portal vein similarly crosses the higher part of the umbilical vein recess (as shown[1]) but it's still umbilical vein. Nonetheless, the message is clear: THAD happens, and these authors demonstrate it elegantly.

A. E. Oestreich, MD

Reference

1. Schlesinger AE, Braverman RM, DiPietro MA. Pictorial essay. Neonates and umbilical venous catheters: normal appearance, anomalous positions, complications, and potential aid to diagnosis. *AJR Am J Roentgenol.* 2003;180:1147-1153.

Sonographic findings following appendectomy for uncomplicated appendicitis in children

Serour F, Herman A, Witzling M, et al (Tel Aviv Univ, Israel; The E. Wolfson Med Ctr, Holon, Israel)

Pediatr Radiol 39:926-932, 2009

Background.—Little is known about 'normal' local sonographic changes occurring in the postoperative period after an uneventful appendectomy.

Objective.—To analyse the local changes on US examination occurring after uneventful open (OA) or laparoscopic (LA) appendectomy in children with normal histology and with nonperforated acute appendicitis.

Materials and Methods.—US was prospectively performed in 82 children (54 boys and 28 girls) aged 1–16 years (mean 11.6 ± 3.2 years), 3 days following LA (n=51, 62%) or OA (n=31, 38%) for nonperforated appendicitis. Multivariate analysis was performed using stepwise logistic regression, with the following starting variables: surgical technique, gender, pathological finding, appendix location, and histology.

Results.—Of the 82 patients, 35 (42.7%) had postoperative pathological US findings such as peritoneal fluid, oedematous mesenteric fat and thickening of the bowel wall. While the overall incidence of pathological US findings between OA and LA groups was not significantly different, multivariate logistic regression analysis showed that OA is associated with a reduction by a factor of 0.35 in the odds ratio of postoperative pathological US findings (P=0.007).

Conclusion.—Pathological US findings are common in children after appendectomy, particularly after LA. Awareness of these pathological findings might prevent unnecessary postoperative treatment.

► So that's what happens 3 days after a child's appendectomy in which either a normal appendix or a nonperforated appendicitis is found. A slight correction to the stated objectives in the abstract, it should state "in children with normal histology OR with a nonperforated appendicitis." It is of interest that the smaller number of their study patients who had had preoperative ultrasound, only 5 out of 25 had postoperative secondary findings that were not present preoperatively. In other words, in that subset, a large majority of the findings discussed in the article were not a consequence of the laparoscopic or open surgery. Also note that all of their patients received pre- and postoperative antibiotics. So the take home point of the article for me is to not be alarmed when the described secondary findings appear on ultrasound. Not addressed is the question of postoperative later adhesions, which can occur in children even after open appendectomy in which no perforation had occurred,[1] albeit less frequently than after perforated.[1] (Oh, and it's Morison's pouch, not Morrison's, because it is named for Mr Morison.)

A. E. Oestreich, MD

Reference

1. Tsao KJ, St Peter SD, Valusek PA, et al. Adhesive small bowel obstruction after appendectomy in children: comparison between the laparoscopic and open approach. *J Pediatr Surg*. 2007;42:939-942.

MR Imaging Evaluation of the Normal Appendix in Children and Adolescents

Baldisserotto M, Valduga SG, da Cunha CFJS (Pontifícia Universidade Católica do Rio Grande do Sul, Porto Alegre, Brazil)

Radiology 249:278-284, 2008

Purpose.—To evaluate asymptomatic children and adolescents to determine the rate of detection of the normal appendix at unenhanced magnetic resonance (MR) imaging.

Materials and Methods.—This study was approved by the medical ethics committees of the authors' institutions, and informed consent was obtained from participants' parents or guardians. Forty asymptomatic volunteers underwent MR imaging to detect the appendix. Mean age (± standard deviation) for the 18 boys was 12.8 years ± 2.4 (range, 9–17 years) and for the 22 girls was 12.7 years ± 2.7 (range, 8–17 years). The MR imaging protocol consisted of axial T1-weighted fast spin-echo (SE), axial and coronal T2-weighted fast SE, and axial T2-weighted fat-saturated fast SE sequences. Two observers interpreted results independently, and κ coefficients were calculated to evaluate agreement. The rate of detection was defined as the percentage of times that the appendix was visualized with at least one sequence. Body mass index was recorded for all participants.

Results.—The normal appendix was detected by one observer in 19 cases (48%) and by the other in 20 cases (50%). The rate of appendix detection after observers reached a consensus was 48% (95% confidence interval: 32%, 64%). Interobserver agreement for appendix detection was good for most sequences analyzed. The greatest rate of normal

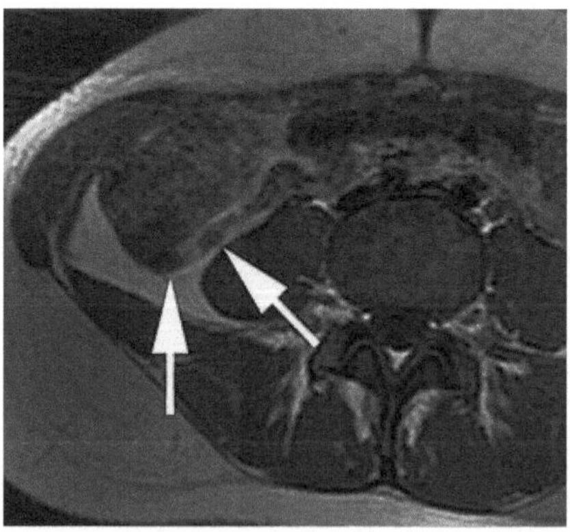

FIGURE 2.—Pelvic MR image of a 16-year-old adolescent girl. Axial T1-weighted fast SE image (700/8.7; section thickness, 4 mm; matrix, 224×192) shows pelvic appendix (arrows) that extends medially from cecum. (Courtesy of Baldisserotto M, Valduga SG, da Cunha CFJS. MR imaging evaluation of the normal appendix in children and adolescents. *Radiology*. 2008;249:278-284. Copyright by the Radiological Society of North America.)

appendix detection was found with the axial T2-weighted fast SE sequence (48%).

Conclusion.—The rate of normal appendix detection in children and adolescents at unenhanced MR imaging was 48%, which is similar to computed tomographic but lower than ultra-sonographic detection rates.

► A sigh of relief accompanies the discovery of a normal appendix whatever imaging method is used in the investigation of a child with abdominal pain (although then findings suggestive of other diagnoses still must be sought). CT does it rather fast, but with radiation exposure, especially high in careless hands that do not image gently. Ultrasound does it well when the examiner is experienced, skilled, and, often, lucky. Some sonographers have a consistent high rate of detection, and some have a low. These authors tackle MRI as an alternative, which does not use radiation, but is expensive (in lands where expense is a consideration) and may require sedation. Unlike ultrasound, no probe is used to check for rebound tenderness. But MRI images can be quite demonstrative, as shown in Fig 2, for example. These were normal volunteer children (the child who declined to participate out of fear was excluded). No intravenous contrast was used; mean examination time was 20 minutes. In actual patients, once an appendix is cleared, it may be helpful to then do additional sequences. The authors agree that the learning curve for appendix recognition may be steep—reminds one of the early ultrasound days. Then comes the question of appendicitis being unlikely if neither a normal nor an abnormal one is perceived. Authors of a CT study thought nonvisualization had a high negative predictive value against acute appendicitis.[1] MRI also permits evaluation of signal intensity of appendix content and wall, which may prove to be diagnostically helpful (including recognition of stones, foreign bodies, parasites, and gas).

A. E. Oestreich, MD

Reference

1. Garcia K, Hernanz-Schulman M, Bennett DL, Morrow SE, Yu C, Kan JH. Suspected appendicitis in children: diagnostic importance of normal abdominopelvic CT findings with nonvisualized appendix. *Radiology.* 2009;250:531-537.

Validation of US as a first-line diagnostic test for assessment of pediatric ileocolic intussusception

Hryhorczuk AL, Strouse PJ (Univ of Michigan Health System, Ann Arbor)
Pediatr Radiol 39:1075-1079, 2009

Background.—From the early 1980s onward, US has been considered a possible primary modality to assess patients for ileocolic intussusception. Since 2001, our institution has routinely used US to assess patients for ileocolic intussusception.

Objective.—We analyzed 7 years of institutional experience to assess the value of US as a primary diagnostic test for intussusception.

Materials and methods.—This study was IRB-approved. From 1 January 2001 through 16 December 2007 814 US examinations for intussusception were performed in children aged 10 years and younger. Clinical records and radiological reports were reviewed for each patient, and cases were classified as true-positive, true-negative, false-positive, or false-negative.

Results.—Of the 814 US examinations, 112 (13.8%) were interpreted as positive for intussusception, and of these, 96 were confirmed by enema, 1 was confirmed by surgery, and 15 (13.4%) were false-positive. Of the 814 examinations, 700 (85.9%) were interpreted as negative for intussusception, and of these, 698 (99.7%) were true-negative, and 2 were false-negative. Less than 1% of studies were non-diagnostic. The overall sensitivity of US for detecting intussusception was 97.9% and specificity was 97.8%. The positive predictive value of the test was 86.6% and the negative predictive value was 99.7%.

Conclusion.—US is a sensitive and specific test for detecting ileocolic intussusception and should be utilized as a first-line examination for assessment of possible pediatric intussusception.

▶ When intussusception in a child is strongly suspected clinically or based on plain images of the abdomen, I generally proceed directly to air contrast enema

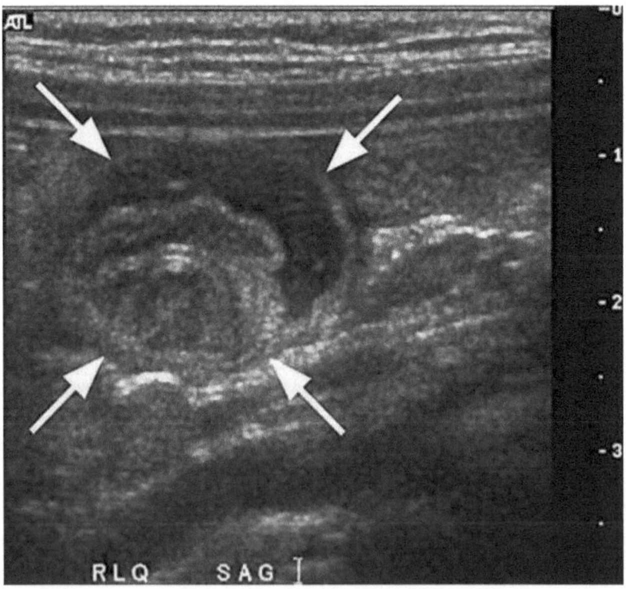

FIGURE 3.—False-negative US examination in a 15-month-old boy with bloody stools. In retrospect, a portion of the intussusception is apparent in one of the US cine clips (*arrows*). This boy's intussusception was diagnosed on CT scan and subsequently reduced with an enema. (With kind permission from Springer Science and Business Media: Hryhorczuk AL, Strouse PJ. Validation of US as a first-line diagnostic test for assessment of pediatric ileocolic intussusception. *Pediatr Radiol.* 2009;39:1075-1079.)

for diagnostic confirmation and then treatment by fluoroscopic air enema reduction. However, if the suspicion of intussusception is low or moderate, I recommend ultrasound imaging, which avoids radiation and the discomfort of rectal tube placement. This report reaffirms and documents the high reliability of ultrasound imaging to rule out or rule in ileocolic intussusception. Their uncommon false positives and false negatives are instructive. On one of the few positives not initially diagnosed, a review of the ultrasound (US) cine clip revealed the intussusceptum in retrospect as shown in Fig 3, pointing out the importance of careful viewing of those clips. Among the few false positives might be ileoileal intussusceptions, but they tend to be fleeting and smaller in diameter. Although we have not been doing US-guided hydrostatic reduction, it is a reliable and safe non-radiation way to treat as well as diagnose in the same sitting[1] — "one-stop shopping" that I would recommend you consider for these children.

A. E. Oestreich, MD

Reference

1. Bai YZ, Qu RB, Wang GD, et al. Ultrasound-guided hydrostatic reduction of intussusceptions by saline enema: a review of 5218 cases in 17 years. *Am J Surg*. 2006;192:273-275.

Nonfluoroscopic pressure colostography in the evaluation of genitourinary fistula of anorectal malformations: experience in a resource-poor environment

Abdulkadir AY, Abdur-Rahman LO, Adesiyun OM (Univ of Ilorin, Kwara State, Nigeria)

Pediatr Radiol 39:132-136, 2009

Background.—Radiological imaging is paramount for defining the genitourinary fistulae commonly associated with anorectal malformations prior to definitive surgery. The imaging options are resource-limited in many parts of the world. Nonfluoroscopic pressure colostography after colostomy is a cheap method for the evaluation of anorectal malformations.

Objective.—To describe our experience with nonfluoroscopic pressure colostography in the evaluation of anorectal malformations in boys.

Materials and Methods.—The study included 12 boys with anorectal malformation who had colostomy and nonfluoroscopic pressure-augmented colostography with water-soluble contrast medium between January 2006 and December 2007.

Results.—Patient ages ranged from 2 days to 1 year. The types of genitourinary fistula were rectovesical (7.7%) and rectourethral (92.3%). Oblique radiographs were of diagnostic value in all patients. The types of anorectal malformations were high, intermediate and low in 75%,

8.3% and 16.7%, respectively. Short-segment urethral constriction was a common feature of rectourethral fistula (75%, $n = 9$).

Conclusion.—Our experience has shown that genitourinary fistulae associated with anorectal malformations can be demonstrated reliably by nonfluoroscopic pressure colostography with two oblique radiographs, providing an option in resource-poor settings where fluoroscopic equipment is scarce.

▶ In regard to this often highly complex congenital clinical problem, it is refreshing that lower technology imaging can offer pertinent diagnostic evaluation in a resource-limited environment. Indeed, lower cost solutions that solve questions adequately are commendable, although the radiation cost to the child and the child's gonads remain a concern. Anorectal malformation (ARM) is a complex problem, but as the authors quote, not very rare (perhaps 1 in 5000 live births), and deserving of meticulous preoperative evaluation. With ingenuity, these authors report their success with a nonfluoroscopic technique—in larger centers as well, antegrade colostography is an important part of the work-up, albeit usually with the added control (and exposure) of fluoroscopy. MRI and (low dose, please) plain abdominal radiographs also serve such children.

One added advantage of pressure colostography (fluor- or no) is looking for the occasional child, sometimes with Down syndrome, who also has Hirschsprung disease,[1] in which case the procedure can help assess the distensibility of the distal bowel.

A. E. Oestreich, MD

Reference

1. Raboei EH. Patients with anorectal malformation and Hirschsprung's disease. *Eur J Pediatr Surg.* 2009;19:325-327.

Magnetic resonance imaging of the small bowel in children with idiopathic inflammatory bowel disease: evaluation of disease activity

Alexopoulou E, Roma E, Loggitsi D, et al (Natl and Kapodistrian Univ of Athens, Greece)

Pediatr Radiol 39:791-797, 2009

Background.—Examinations using ionizing radiation are frequently used in the evaluation of disease activity in children affected by idiopathic inflammatory bowel disease (IBD).

Objective.—To develop an MR imaging protocol without the need for fluoroscopic insertion of an enteral tube and to assess the disease activity in children with IBD.

Materials and Methods.—Included in the study were 37 children (22 girls and 15 boys; age range 7–15 years, mean 11.67 years) with IBD who underwent MR imaging of the small bowel. Of these 37 children,

32 had Crohn disease and 5 had indeterminate colitis. A water solution containing herbal fibres was administered orally or through a nasogastric tube. Patients were imaged on a 1.5-T MR scanner with T1-weighted and T2-weighted sequences followed by a dynamic study using 3-D T1-W images after intravenous administration of gadolinium.

Results.—The percentage enhancement of the bowel wall was significantly increased in patients with abnormal C-reactive protein (CRP) values compared to patients with CRP values in the normal range ($P<0.001$). A relatively weak but significant correlation between percentage enhancement of the bowel wall and CRP values was noted during all phases of enhancement.

Conclusion.—This MR imaging protocol is a safe and well-tolerated method for evaluating disease activity and extraintestinal manifestations of IBD in children.

▶ In support of using MRI rather than radiation methods to evaluate the activity of inflammatory bowel disease, these authors (and their abstract above) concentrate on percentage of enhancement of bowel wall. They do also look at the other MRI findings of inflammatory disease, as shown in Fig 1, for example, including thick wall, dilatation, enlarged mesenteric nodes, and creeping fat, but the statistically significant result (enhancement) that they emphasize may be especially helpful to distinguish active from quiescent

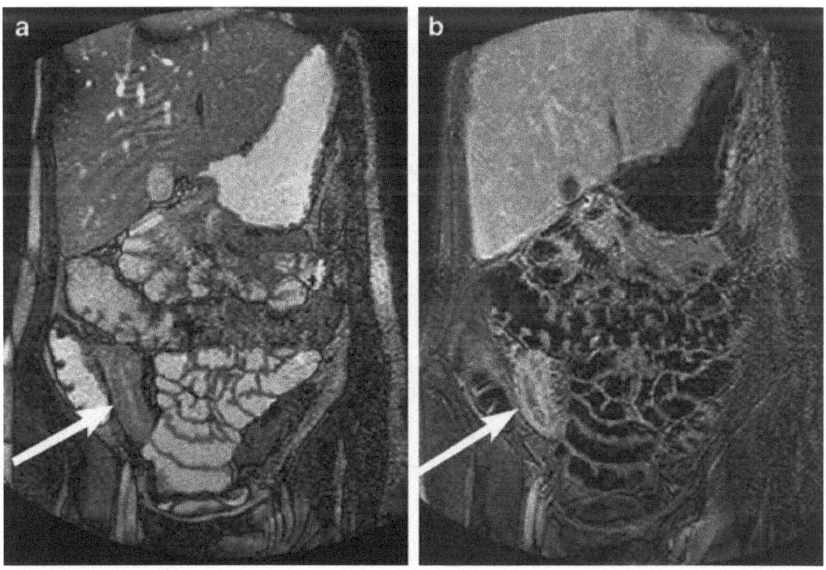

FIGURE 1.—A 12-year-old girl with known CD. **a** Coronal T2-W image shows wall thickening and oedema (*arrow*) of the terminal ileum. **b** Coronal T1-W image obtained after gadolinium administration during the arterial phase shows early and intense enhancement (*arrow*). (With kind permission from Springer Science and Business Media: Alexopoulou E, Roma E, Loggitsi D, et al. Magnetic resonance imaging of the small bowel in children with idiopathic inflammatory bowel disease: evaluation of disease activity. *Pediatr Radiol.* 2009;39:791-797.)

disease. The visualization was aided by prone position to spread small bowel loops and decrease scan volume. The herbal product used in the water solution was psyllium (also known as ispaghula; and is a component of Metamucil in the United States and Fybogel in the United Kingdom). Interestingly, psyllium may be helpful in treating irritable bowel,[1] so why not inflammatory bowel as well? Apparently some allergic reactions to psyllium have been reported, but not in the current or the quoted study. MRI enteroclysis is now an actively investigated technique for Crohn disease, so that articles elaborating and supporting the imaging findings are welcomed.

A. E. Oestreich, MD

Reference

1. Bijkerk CJ, de Wit NJ, Muris JW, et al. Soluble or insoluble fibre in irritable bowel syndrome in primary care? Randomised placebo controlled trial. *BMJ*. 2009;339: b3154.

Neonatal colour Doppler ultrasound study: normal values of abdominal blood flow velocities in the neonate during the first month of life

Papacci P, Giannantonio C, Cota F, et al ("Sacro Cuore" Catholic Univ, Rome, Italy; et al)

Pediatr Radiol 39:328-335, 2009

Background.—Doppler US to measure abdominal blood flow velocities (ABFV) is increasingly used to investigate intestinal haemodynamics in several clinical conditions in neonates. Studies that provide reference values of ABFV during the entire neonatal period are currently lacking.

Objective.—To make available normal reference values of ABFV and Doppler indices in the coeliac trunk and superior mesenteric artery during the first month of life in term and healthy preterm infants.

Materials and Methods.—ABFV were obtained with colour Doppler US in 69 neonates (12 term, 57 preterm) divided into four gestational age groups (25–28 weeks, 29–32 weeks, 33–36 weeks, and 37–41 weeks).

Results.—ABFV increased with increasing gestational and postnatal age. We also provide normal reference values of ABFV and Doppler indices to compare with measurements of abdominal blood flow changes during the neonatal period for diagnostic, therapeutic and prognostic purposes.

Conclusion.—These longitudinal reference values provide a useful tool for assessing possible alteration in ABFV secondary to neonatal pathologies.

▶ We start with 2 premises: (1) blood flow in the superior mesenteric artery and the celiac axis is important to the abdominal gastrointestinal tract in infants, including prematures, and (2) measurement of blood flow in those vessels provides an indicator of the outlook for health of the newly born. It is both self-evident and freely acknowledged that feeding affects that blood flow;

recent articles[1,2] strongly suggest that impairment to that flow will predict difficulties with handling feeds, and in particular the prediction of necrotizing enterocolitis when flow does not normally respond to feeding.[2] With all the emphasis on blood flow response or lack or response to feeding, I especially miss the details of the measurements in the current report with respect to when the babies were fed. The values provided by Papacci and colleagues are nonetheless potentially useful in relating blood flow values to gestational age and postnatal age, and they indeed help us understand the important, yet fragile, relationships affecting the blood flow. Whether or not all babies or prematures should be screened for aberrations in abdominal blood flow is also a matter for consideration.

A. E. Oestreich, MD

References

1. Murdoch EM, Sinha AK, Shanmugalingam ST, Smith GC, Kempley ST. Doppler flow velocimetry in the superior mesenteric artery on the first day of life in preterm infants and the risk of neonatal necrotizing enterocolitis. *Pediatrics.* 2006;118: 1999-2003.
2. Krimmel GA, Baker R, Yanowitz TD. Blood transfusion alters the superior mesenteric artery blood flow velocity response to feeding in premature infants. *Am J Perinatol.* 2009;26:99-105.

A semi-automated method for non-invasive internal organ weight estimation by post-mortem magnetic resonance imaging in fetuses, newborns and children

Thayyil S, Schievano S, Robertson NJ, et al (UCL Inst of Child Health & Great Ormond Street Hosp for Children, London, UK; Univ College London, UK; et al)
Eur J Radiol 72:321-326, 2009

Magnetic resonance (MR) imaging allows minimally invasive autopsy, especially when consent is declined for traditional autopsy. Estimation of individual visceral organ weights is an important component of traditional autopsy.

Objective.—To examine whether a semi-automated can be used for non-invasive internal organ weight measurement using post-mortem MR imaging in fetuses, newborns and children.

Methods.—Phase 1: *In vitro* scanning of 36 animal organs (heart, liver, kidneys) was performed to check the accuracy of volume reconstruction methodology. Real volumes were measured by water displacement method. Phase 2: Sixty-five whole body post-mortem MR scans were performed in fetuses ($n = 30$), newborns ($n = 5$) and children ($n = 30$) at 1.5 T using a 3D TSE T2-weighted sequence. These data were analysed offline using the image processing software Mimics 11.0.

Results.—Phase 1: Mean difference (S.D.) between estimated and actual volumes were −0.3 (1.5) ml for kidney, −0.7 (1.3) ml for heart, −1.7 (3.6) ml for liver in animal experiments. Phase 2: In fetuses, newborns and children mean differences between estimated and actual weights (S.D.) were −0.6 (4.9) g for liver, −5.1 (1.2) g for spleen, −0.3 (0.6) g for adrenals, 0.4 (1.6) g for thymus, 0.9 (2.5) g for heart, −0.7 (2.4) g for kidneys and 2.7 (14) g for lungs. Excellent co-correlation was noted for estimated and actual weights ($r^2 = 0.99$, $p < 0.001$). Accuracy was lower when fetuses were less than 20 weeks or less than 300 g.

Conclusion.—Rapid, accurate and reproducible estimation of solid internal organ weights is feasible using the semi-automated 3D volume reconstruction method.

▶ For many pathologists, participation in research studies of postmortem MRI may be something of a conflict of interest—after all, they do not want to see their role taken over by radiologists. However, with the continued decline in autopsy rates (secondary to withheld consent), postmortem MRI is emerging as a real alternative to the conventional autopsy.

In this article the MaRIAS group have advanced the role of postmortem (minimally invasive) MRI by showing that semi-automated techniques can reliably be used to assess organ volumes. However, a time of 54 minutes per patient to assess the volume of all organs weighed at traditional autopsies must be seen as a disadvantage. In the future, this relatively long analysis time is expected to improve with operator experience, optimisation of imaging techniques, and software development.

I believe that postmortem MRI and the role of forensic radiologist (working closely with pathologists) are here to stay. Then again, I would, and here I must declare my own conflict of interest—not only am I a radiologist, I am also a member of the MaRIAS Collaborative Study Group!

A. C. Offiah, BSc, MBBS, MRCP, FRCR, PhD

Musculoskeletal

Follow-up skeletal surveys for nonaccidental trauma: can a more limited survey be performed?

Harlan SR, Nixon GW, Campbell KA, et al (Univ of Utah School of Medicine, Salt Lake City; Primary Children's Med Ctr, Salt Lake City, UT)
Pediatr Radiol 39:962-968, 2009

Background.—Studies have demonstrated the value of the follow-up skeletal survey in identifying additional fractures, clarifying indeterminate findings, and improving dating of skeletal injuries in victims of physical abuse.

Objective.—To determine whether a more limited follow-up survey could yield the same radiologic data as a full follow-up survey.

Materials and Methods.—The study cohort comprised 101 children who had follow-up surveys that met our inclusion criteria. Consensus readings of both original and follow-up surveys were performed by two pediatric radiologists. These results were compared to determine additional findings from the follow-up surveys. Limited skeletal survey protocols were evaluated to determine whether they would detect the same fractures seen with a complete osseous survey.

Results.—In the 101 children 244 fractures were identified on the initial osseous survey. Follow-up surveys demonstrated new information in 38 children (37.6%). A 15-view limited follow-up survey identified all additional information seen on the complete follow-up survey.

Conclusion.—Our data demonstrate that a 15-view limited follow-up skeletal survey could be performed without missing clinically significant new fractures and still allow proper identification of confirmed fractures or normal findings. A limited survey would decrease radiation dose in children.

▶ Although the value of follow-up radiographs is well recognized, very few studies have actually tried to determine precisely which radiographs should be repeated. The authors of this article are to be commended for determining that a 15-film limited follow-up skeletal survey yields essentially the same information as their local full 21-film follow-up survey.

Of course, a major difficulty that we all face is getting the children back for follow-up imaging in the first place—particularly if the initial survey is completely normal. It is not certain how this situation can be improved. However, when children are brought back for follow-up surveys, it is helpful to know which images should be performed. The American College of Radiology (ACR) Practise Guidelines for Skeletal Surveys in Children[1] and the British Society of Paediatric Radiology (BSPR) Standard for Skeletal Surveys in Suspected Non-Accidental Injury (NAI) in Children[2] give very little guidance when it comes to the follow-up skeletal survey. I echo the authors' recommendation that the ACR, American Academy of Pediatrics (AAP) and BSPR guidelines should be revised, and (in the context of the follow-up survey) be more explicit.

A. C. Offiah, BSc, MBBS, MRCP, FRCR, PhD

References

1. American College of Radiology Web site. Practise Guidelines for Skeletal Surveys in Children. http://www.acr.org/SecondaryMainMenuCategories/quality_safety/guidelines/pediatric/skeletal_surveys.aspx. Accessed March 18, 2010.
2. The British Society for Paediatric Radiology Web site. The British Society for Paediatric Radiology Standards for Skeletal Surveys in Suspected Non Accidental Injury (NAI) in Children. http://www.bspr.org.uk/nai.htm. Accessed March 18, 2010.

Eliminating disparity in evaluation for abuse in infants with head injury: use of a screening guideline

Rangel EL, Cook BS, Bennett BL, et al (Cincinnati Children's Hosp Med Ctr, OH; et al)

J Pediatr Surg 44:1229-1235, 2009

Purpose.—Minority and disadvantaged children are evaluated for nonaccidental trauma (NAT) at higher rates than other children. At our institution, we implemented a guideline to perform skeletal surveys to screen for occult fractures in all infants with unwitnessed head injury (UHI). The goal was to determine if this guideline decreased disparities in the screening of African American (AA) and uninsured children.

Patients and Methods.—For 54 months, rates of skeletal surveillance and abuse determination were compared between AA and white infants admitted with UHI before and after implementation of our guideline. Logistic regression was used to control for confounders.

Results.—Before the guideline, AAs underwent skeletal surveillance more than whites (n = 208; 90.5% vs 69.3%; *P* = .01), with 20% of screened infants determined to be probable victims of NAT. Whites with private insurance were less likely to be screened compared to those without private insurance (50.0% vs 88.1%; *P* < .001). After the guideline, AA and whites were surveyed equally (n = 52; 92.3% vs 84.6%; *P* = 1.0), with 22% found to be probable cases of NAT.

TABLE 2.—Frequency of Skeletal Surveillance in Head-Injured Infants

Variable	Groups	n	White	AA	*P*
Before guideline (Jul 2003-Dec 2006)					
Overall		153	69.3% (115)	90.5% (38)	.01
Insurance	Private	82	50.0% (41)[a, b]	83.3% (5)	.21
	Uninsured/government	84	88.1% (74)[a]	91.7% (33)	.75
Sex	Male	98	73.5% (72)	91.7% (22)	.07
	Female	68	63.2% (43)	88.9% (16)	.05
ISS	≤15	82	65.9% (54)	86.4% (19)	.07
	>15	84	72.6% (61)	95.0% (19)	.06
After guideline (Jan 2007-Jan 2008)					
Overall		45	84.6% (33)	92.3% (12)	1.00
Insurance	Private	20	78.9% (15)[b]	100.0% (1)	1.00
	Uninsured/government	32	90.0% (18)	91.7% (11)	1.00
Sex	Male	30	90.0% (18)	90.0% (9)	1.00
	Female	22	78.9% (15)	100.0% (3)	1.00
ISS	≤15	28	76.2% (16)	100.0% (7)	.29
	>15	24	94.4% (17)	83.3% (5)	.45

[a]Indicates frequency of skeletal surveillance significantly different between private insurance and uninsured/government insured (*P* < .01).

[b]Indicates frequency of skeletal surveillance significantly different between white patients with private insurance before the guideline and after the guideline (*P* = –.03).

Conclusions.—This is the first report of a successful policy-based intervention to decrease disparity in care. The maintenance of a stable rate of NAT determination despite increased screening suggests more victims of abuse may be identified with guideline use, and therefore, this may be an additional benefit of the guideline.

▶ In the United States it has been known for several years that disparities exist between screening rates of African American and Caucasian children for child abuse. Disparities also exist between Caucasian children from families of lower compared with those of higher socioeconomic standing (as measured by medical insurance status). There are several potential reasons for these disparities: preconceptions and bias on the part of doctors, more convincing storytelling on the part of Caucasian families, or an increased incidence of non-accidental trauma within lower socioeconomic classes and the African American popluation. Given that after introduction of the screening guideline levels were similar in both groups (Table 2), the latter possibility seems unlikely.

Whatever the reason or reasons, the authors have provided us with what appears to be an effective solution. This is a well conducted and well written article; I look forward both to the authors' publication of their longer term experience of applying the guideline and to its general dissemination.

A. C. Offiah, BSc, MBBS, MRCP, FRCR, PhD

Are UK radiologists satisfied with the training and support received in suspected child abuse?

Leung RS, Nwachuckwu C, Pervaiz A, et al (Great Ormond Street Hosp for Children, London, UK; Whipps Cross Hosp, London, UK; et al)
Clin Radiol 64:690-698, 2009

Aim.—To determine current practice and perceptions of the adequacy of training and support received for the reporting of skeletal surveys in suspected physical child abuse.

Materials and Methods.—A list of telephone numbers of UK hospitals with a radiology department was obtained from the Royal College of Radiologists. One hundred hospitals were then randomly selected for inclusion in the survey. An 18-item questionnaire was successfully administered to consultant radiologists from 84 departments.

Results.—Sixty-one percent of departments had a named radiologist to report their skeletal surveys, 16% assigned surveys to a random radiologist, and 23% referred them elsewhere. Only 52% of departments had a dedicated paediatric radiologist, thus in a significant proportion of departments (25%) initial reports on skeletal surveys for physical abuse were provided by non-paediatric radiologists. Fifteen percent did not have ready access to a paediatric radiology opinion. Sixty-one percent thought that the service could be improved. Expert evidence was provided by 5% of respondents. Seventy-three percent would never consider providing expert evidence, even if given adequate radiology and/or legal training.

Conclusion.—The survey shows significant dissatisfaction amongst consultant radiologists with the current service, confirms a low number of paediatric radiologists taking on this work, and suggests the potential to increase numbers of radiology child abuse experts by 27% if given improved training and support. Appropriate service and education strategies should be implemented.

▶ How about you? I am a pediatric radiologist with special interests in musculoskeletal imaging and in metabolic bone disease, and I am willing (but not anxious) to provide expert testimony in legal cases involving possible child abuse. I also get upset when other apparently expert witnesses appear to provide misleading testimony in legal cases involving possible child abuse. And I believe it is my obligation and that of all radiologists to provide radiographic reports that reflect the suspicion of child abuse to the best of our ability. We make mistakes. A few years ago, for example, I strongly suggested abuse as the cause of fractures, but the reason turned out to be leukemia. So we must constantly strive for accuracy with an open mind and not merely sidestep the issue. The article here reviewed paints a disturbing picture of reluctance or inability to provide expert testimony (which could, in some instances, save the life of a child). As suggested, solutions include more training and more cooperation. It returns to the joint obligations of the radiologist: not to miss the suggestion of child abuse when present and not to diagnose child abuse when another cause of the fractures is evident. We labor mightily, including in this format, to disseminate useful information. Please do your share.

A. E. Oestreich, MD

Pseudointercondylar notch sign: manifestation of osteochondritis dissecans of the trochlea

Pruthi S, Parnell SE, Thapa MM (Univ of Washington, Seattle)

Pediatr Radiol 39:180-183, 2009

Osteochondritis dissecans (OCD) is an idiopathic condition affecting the articular epiphysis. Initially described in the knee, this entity affects several other parts of the body such as the talar dome, tarsal navicular, and femoral capital epiphysis. OCD of the elbow primarily involves the capitellum. OCD involving the trochlea has rarely been reported. We describe an unusual and interesting case of OCD affecting the trochlea, mimicking a pseudointercondylar notch.

▶ Two important articles appeared in 2009 concerning an important but little known plain image finding pertinent to a specific injury in the young athlete or other children. I have chosen for review the shorter article with highly characteristic X-ray images (Fig 1 in the original article), including a second case, and then I reference the more in-depth article with a larger number of cases and more orthopedic and scientific finesse.[1] The index case is a 13-year-old gymnast; the age is typical, but more cases have occurred in baseball pitchers and other repeated throwers.[1] The typical deeper than normal concavity at the lateral (radius-side) margin of the trochlea on frontal images should be easily recognized once one is aware of the entity. Many of us at the 2008 RSNA meeting "saw" the finding on a case of the day, but did not yet know what it implied. An even deeper notch, described as "fishtail" occurs following avascular necrosis of that region of the trochlea.[1] If needed, MRI or CT can allow a fuller investigation for, say, loose bodies related to the osteochondrosis/osteochondritis. All who image the elbow of the child with elbow pain should review both these articles to be ready for the next one appearing on your "viewbox."

A. E. Oestreich, MD

Reference

1. Marshall KW, Marshall DL, Busch MT, Williams JP. Osteochondral lesions of the humeral trochlea in the young athlete. *Skeletal Radiol.* 2009;38:479-491.

The carpal bones in Poland syndrome

Friedman T, Reed M, Elliott AM (Univ of Manitoba, Winnipeg, Canada)
Skeletal Radiol 38:585-591, 2009

Objective.—Classical Poland syndrome is represented by unilateral aplasia of the sternocostal head of the pectoralis major muscle and ipsilateral simple syndactyly and brachydactyly. Various classifications of the severity of hand involvement have been proposed. Since its initial description, numerous studies have been made of the bony, soft tissue, organ, and hematological disturbances. However, carpal bone involvement has been largely overlooked. The purpose of this study was to evaluate the carpal bones in patients with Poland syndrome from a local (Manitoba) cohort as well as those from the literature.

Materials and Methods.—Hand radiographs from local patients and cases identified from the literature with confirmed Poland syndrome were examined for evidence of carpal bone involvement. Only cases with radiographs of adequate quality were included in the analysis. Clinical information (including gender and age) was necessary for evaluation of bone maturation. In total, seven local patients and 23 patients from the literature were evaluated. Ethics approval for study of the local patients was obtained by the Research Ethics Board of the University of Manitoba.

Results.—Of the 23 literature patients, 12 patients (52%) had abnormal findings. Of the abnormal patients, four of 12 (33%) had carpal fusions, eight of 12 (67%) showed disharmonious ossification between the carpal and tubular bones and seven of 12 (58%) showed delay of carpal ossification. Of the local cohort, three patients were too young to characterize carpal involvement. Of the four remaining patients, two (50%) had abnormal carpal morphology, three out of four had disharmonious ossification and all four had delay of ossification of carpal bones. Carpal fusions, particularly of the scaphoid and trapezium, were common in both groups.

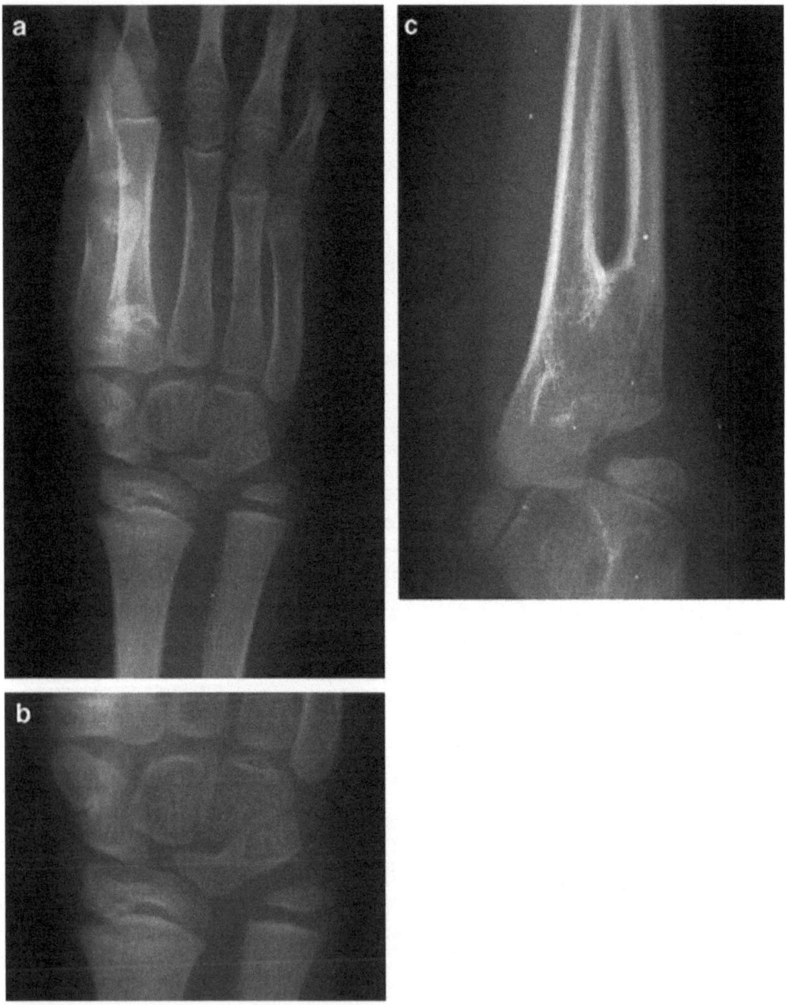

FIGURE 1.—**a,b** Patient 1. Female patient at 9 years 6 months with affected left upper extremity. There is decreased bone age of all bones. There is a pseudoepiphysis at the base of the second metacarpal. The carpal height is decreased and there is scaphoid–trapezium fusion, lunate–triquetrum fusion and a hypoplastic trapezoid. There is an ulnar plus deformity. **c** This patient also had radioulnar synostosis proximally. (Reprinted from Friedman T, Reed M, Elliott AM. The carpal bones in Poland syndrome. *Skeletal Radiol.* 2009;38:585-591, with kind permission from Springer Science+Business Media.)

Conclusion.—Carpal bone anomalies (delay, disharmony, and/or fusions) are frequent in Poland syndrome and can occur in patients with either mild or severe hand involvement. Imaging of the unaffected hand is helpful in determining the extent of carpal findings.

▶ I must admit that having seen several patients with Poland syndactyly through the years, I never noticed scaphotrapezoid fusion, a finding seen in many cases in this report (Fig 1, for example); indeed, they found it more often than can be accounted for by mere coincidence. My congratulations to the authors. Because discordant maturation of carpal bones compared with the tubular bones of the hand are so common in general as to lead most of us pediatric radiologists to ignore carpal bones in calculating bone age, I would still discount that discordance (disharmony) as a significant finding in Poland syndrome. Ah, and I will probably now have to be more skeptical of the theory proposed by Professor Josef Warkany for the syndrome: he proposed that in utero repeated pounding of the chest by a closed fist would explain the pectoral aplasia, the shortness to absence of middle phalanges, occasional ipsilateral breast hypoplasia, perhaps the soft tissue syndactyly, and especially the several cases of dextroposition of the heart associated with left-sided Poland syndrome. Fitting the scaphotriquetral congenital fusion into that model would be quite a stretch unfortunately. Now, scaphotriquetral congenital fusion can occur without Poland syndrome, witness its bilateral occurrence in a pair of twins if you want to see a lot of them[1] (but maybe a responsible Hox gene lies near a Polandness site). But remember at least this unusual carpal fusion in at least some of individuals who have the syndrome complex named for Alfred Poland (who had described it in a convict long before the discovery of the X-ray).

A. E. Oestreich, MD

Reference

1. Barnes CL, Frazier GT, Hixson ML. Bilateral congenital fusion of the scaphoid and trapezium in identical twins. *Orthopedics*. 1992;15:739-741.

Imaging Chronic Sclerosing Osteitis of the Diaphysis of Tubular Bones

Costelloe CM, Murphy WA Jr, Madewell JE (The Univ of Texas M. D. Anderson Cancer Ctr, Houston)
AJR Am J Roentgenol 192:736-742, 2009

Objective.—The purpose of this study is to report the radiographic changes seen in chronic sclerosing osteitis of the diaphysis of tubular bones during progression or healing, with CT and MRI correlation. Chronic sclerosing osteitis is an inflammatory condition of bone that can be considered a pediatric subset of the synovitis, acne, pustulosis, hyperostosis, osteitis (SAPHO) spectrum of disease. Previous descriptions of long-bone involvement focus predominately on metaphyseal lesions.

Eleven diaphyseal lesions in seven patients (age range, 13 months–14.5 years) were followed.

Conclusion.—The marked degree of cortical hyperostosis, medullary narrowing, and cortical lucencies seen in eight of 11 lesions is helpful for recognition of diaphyseal chronic sclerosing osteitis. Early identification of the disorder can prevent unnecessary repeat invasive procedures.

▶ In the past, certain patients with diaphyseal cortical sclerosis similar to that documented in this report were considered to have "Ribbing disease," a localized form of the dysplasia progressive diaphyseal dysplasia (Engelmann disease). I must say that I have had images (films!) in my teaching file that were labeled as Engelmann or Ribbing disease that look just like those shown in this article with the more modern names of chronic recurrent multifocal osteomyelitis (CRMO) and synovitis, acne, pustulosis, hyperostosis, osteitis (SAPHO). Boy, the symmetric diaphyseal changes in Engelmann disease sure look similar (with the locus of inheritance 19q13.1; mutations in TBGF1—see most recently[1] for example). Not to mention other diaphyseal dysplasia such as the Ghosal type. This article does not delve into genetics. Otherwise, it is a masterful discussion and portrayal of the cortical hyperostosis, which belongs to the CRMO/SAPHO clinicoradiologic pattern, with progression shown so nicely in Fig 1 in the original article. The CT and MRI slices in the original article are also quite instructive. Their subacute lesions showed the plasma cells on biopsy, which I have come to regard as characteristic of CRMO. The authors conclude the most likely etiology of their chronic sclerosing osteitis is a sterile inflammatory process (perhaps sharing a final common pathway with Engelmann and even Ribbing disease).

A. E. Oestreich, MD

Reference

1. Park SJ, Yoon CS, Park HW, Choi JR, Chung JS, Lee KA. The first Korean case of Camurati-Engelmann disease (progressive diaphyseal dysplasia) confirmed by TGFB1 gene mutation analysis. *J Korean Med Sci.* 2009;24:737-740.

Neurologic/Vertebral

Canadian C-spine Rule and the National Emergency X-Radiography Utilization Low-Risk Criteria for C-spine radiography in young trauma patients

Ehrlich PF, Wee C, Drongowski R, et al (The Univ of Michigan Med School and The C.S. Mott Children's Hosp, Ann Arbor)
J Pediatr Surg 44:987-991, 2009

Purpose.—The Canadian C-spine (cervical spine) Rule (CCR) and the National Emergency X-Radiography Utilization Low-Risk Criteria (NLC) are criteria designed to guide C-spine radiography in trauma patients. It is unclear how these 2 rules compare with young children.

Methods.—This study retrospectively examined case-matched trauma patients 10 years or younger. Two cohorts were identified—cohort A where C-spine imaging was performed and cohort B where no imaging was conducted. The CCR and NLC criteria were then applied retrospectively to each cohort.

Results.—Cohort A contained 125 cases and cohort B with 250 cases. Seven patients (3%) had significant C-spine injuries. In cohort A, NLC criteria could be applied in 108 (86.4%) of 125 and CCR in 109 (87.2%) of 125. National Emergency X-Radiography Utilization Low-Risk Criteria suggested that 70 (58.3%) cases required C-spine imaging compared to 93 (76.2%) by CCR. National Emergency X-Radiography Utilization Low-Risk Criteria missed 3 C-spine injuries, and CCR missed one. In cohort B, NLC criteria could be applied in 132 (88%) of 150 and CCR in 131 (87.3%) of 150. The NLC criteria identified 8 cases and CCR identified 13 cases that would need C-spine radiographs. Fisher's 2-sided Exact test demonstrated that CCR and NLC predictions were significantly different ($P = .002$) in both cohorts. The sensitivity of CCR was 86% and specificity was 94%, and the NLC had a sensitivity of 43% and a specificity of 96%.

Conclusions.—Although CCR and NLC criteria may reduce the need for C-spine imaging in children 10 years and younger; they are not sensitive or specific enough to be used as currently designed.

▶ Once again we are reminded that children are not simply small adults.

Although a retrospective case note review with medical records being analyzed by nonmedical (but trained) research assistants, this article clearly demonstrates that neither the National Emergency X-Radiography Utilization Low-Risk Criteria (NLC) nor the Canadian C-spine (cervical spine) Rule (CCR) criteria (developed for adults) can safely be applied to children. We are not surprised by this finding—a cursory glance at the criteria suggests that their applicability is in direct proportion to the age of the child.

While it behooves us all to limit the use of nonionizing radiation, the devastating effects of undiagnosed cervical spine injuries cause us to err on the side of caution—without a doubt we are performing more cervical spine radiographs than is warranted. However, until a set of sensitive and specific criteria have been designed and validated for pediatric patients, I fear that this will continue to be the case.

A. C. Offiah, BSc, MBBS, MRCP, FRCR, PhD

Magnetic resonance imaging and proton magnetic resonance spectroscopy of the brain in the diagnostic evaluation of developmental delay

Verbruggen KT, Meiners LC, Sijens PE, et al (Univ of Groningen, The Netherlands)

Eur J Paediatr Neurol 13:181-190, 2009

Aim.—To assess the contribution of MRI and proton spectroscopy (1HMRS) in establishing an etiological diagnosis in children with developmental delay (DD) and to assess whether the chance of finding specific abnormalities correlates with the presence of neurological signs and/or abnormal head circumference (HC).

Methods.—Patients were derived from a cohort of 325 consecutive patients with DD receiving structured multidisciplinary evaluation in our centre. Patients had MRI/1HMRS if a diagnosis could not be made clinically and if additional neurological signs and/or abnormal HC and/or an IQ below 50 were present. The MRI protocol consisted of axial IR, T2, FLAIR, sagittal T1 and coronal T2 sequences. Multivoxel 1HMRS was located in a plane superior to the lateral ventricles with voxels in both grey matter and white matter.

Results.—One hundred and nine children were scanned, 80 of them because of neurological signs and/or abnormal HC. Although minor abnormalities were noted in the vast majority of patients, MRI and/or 1HMRS really contributed to an etiological diagnosis in only 10 (9%) patients, all of whom were scanned because of neurological signs. In these 10 patients, 1HMRS was diagnostic in one patient and of additional value to MRI findings in 3 patients.

Conclusions.—MRI and 1HMRS may contribute to the diagnostic evaluation of DD, especially if applied specifically to patients with neurological signs, whereas its role is very limited in children without these signs.

▶ In their article , Verbruggen et al present results that suggest only a limited diagnostic role for proton magnetic resonance spectroscopy (1HMRS) (and indeed brain MRI) for children with developmental delay, in which no neurological signs exist. They advise careful assessment and a multidisciplinary approach to the management of these children.

Over the past few years, there has (at least in Europe) been a steady increase in medical imaging techniques.[1] This comes at a cost; usually financial, but occasionally individual patients may suffer detrimental effects of a particular imaging technique.

The authors remind us to consider risk benefit ratios to individual patients before embarking on what may be unnecessary investigations (in this case the risk of general anesthetic). It is refreshing to read an article that shortens rather than lengthens the list of indications for medical imaging.

A. C. Offiah, BSc, MBBS, MRCP, FRCR, PhD

Reference

1. European Science Foundation. Medical Imaging for Improved Patient Care (Science Policy Briefing, September 2007). http://www.esf.org/fileadmin/links/EMRC/ESF_POLICY28_V09_HD.pdf. Accessed October 8, 2009.

Clinician Judgment Versus a Decision Rule for Identifying Children at Risk of Traumatic Brain Injury on Computed Tomography After Blunt Head Trauma

Palchak MJ, Holmes JF, Kuppermann N (UC Davis School of Medicine, CA)
Pediatr Emerg Care 25:61-65, 2009

Objective.—To compare a decision rule with clinician judgment for identifying children at risk of traumatic brain injury (TBI) after blunt head trauma.

Methods.—We performed an observational study of children with blunt head trauma. Emergency department physicians documented suspicion for TBI before cranial computed tomography (CT), rating suspicion as very low, low, intermediate, or high. Our outcome variable was TBI on CT. We compared clinician judgment (very low vs. higher suspicion) for TBI on CT with a decision rule derived from the same database.

Results.—Of 1865 children enrolled for whom physician suspicion was recorded, 1168 (62.6%) underwent CT and comprised the study population. Eighty-nine (7.6%; 95% confidence interval [CI], 6.2% to 9.3%) of the 1168 had TBIs on CT. The decision rule had a sensitivity of 88 (98.9%) of 89 versus 84 (94.4%) of 89 for clinician judgment (difference, 4.5%; 95% CI, −0.9% to 9.9%). The specificity of the decision rule was 288 (26.7%) of 1079 versus 329 (30.5%) 1079 for clinician judgment (difference, 3.8%; 95% CI, 0.5% to 7.1%). Application of the decision rule to the study population would have resulted in 289 (24.7%) fewer CT scans, although missing 1 child with a TBI (who was discharged home from the emergency department).

Conclusions.—A decision rule trended toward greater sensitivity than clinician judgment for identifying children with TBI on CT after blunt head trauma but was less specific. Because decisions to order cranial CT did not strictly follow clinician judgment, however, use of the decision rule would have resulted in less frequent use of CT.

▶ Most radiologists can recount an occasion when while requesting an imaging investigation, their clinical colleague has admitted, "I don't think it's very likely, but ...." The radiologist then goes on to explain to their radiographer colleague, "It will probably be normal, but...". Sometimes in fact, the requested investigation (CT head, in this case) will NOT be normal and both the clinician and radiologist vindicated.

The potential for the decision rule presented in this article to reduce unnecessary CT head scans by nearly 25% is a significant result. However, it is probably the recollection of anecdotal scenarios such as the one I have described above along with a certain intangible (clinical acumen?) and the risk/benefit of an unnecessary CT head scan compared with a missed traumatic brain injury (TBI) to the individual child that causes emergency department physicians to order head scans even in children (334 of them in this article) they have clinically classified as low risk for TBI.

Unfortunately, despite the trend shown in this article for the decision rule to have higher sensitivity than clinical acumen, until it has been validated in other (less specialized) centers—and for the reasons given above, perhaps not even then—it is likely that a significant number of unnecessary pediatric head CT scans will continue to be performed.

A. C. Offiah, BSc, MBBS, MRCP, FRCR, PhD

Bedside Sonographic Measurement of Optic Nerve Sheath Diameter as a Predictor of Increased Intracranial Pressure in Children

Le A, Hoehn ME, Smith ME, et al (Univ of Tennessee Health Sciences Ctr, Memphis, TN)

Ann Emerg Med 53:785-791, 2009

Study Objective.—We investigate the test performance of emergency physician–performed sonographic measurement of optic nerve sheath diameter for diagnosis of increased intracranial pressure.

Methods.—Children between the ages of 0 and 18 years with suspected increased intracranial pressure were prospectively recruited from the emergency department and ICU of an urban, tertiary-level, freestanding pediatric facility. Pediatric emergency physicians with goal-directed training in ophthalmic sonography measured optic nerve sheath diameter. Images were recorded and subsequently reviewed by a pediatric ophthalmologist and an ophthalmic sonographer, both of whom were blind to the patient's clinical condition. Measurements obtained by the ophthalmic sonographer were considered the criterion standard. An optic nerve sheath diameter greater than 4.0 mm in subjects younger than 1 year and greater than 4.5 mm in older children was considered abnormal. The diagnosis of increased intracranial pressure was based on results of cranial imaging or direct measurement of intracranial pressure.

Results.—Sixty-four patients were recruited, of whom 24 (37%) had a confirmed diagnosis of increased intracranial pressure. The sensitivity of optic nerve sheath diameter as a screening test for increased intracranial pressure was 83% (95% confidence interval [CI] 0.60 to 0.94); specificity was 38% (95% CI 0.23 to 0.54); positive likelihood ratio was 1.32 (95% CI 0.97 to 1.79) and negative likelihood ratio was 0.46 (95% CI 0.18 to 1.23). There was fair to good interobserver agreement between the pediatric emergency physician and ophthalmic sonographer (κ 0.52) and pediatric ophthalmologist (κ 0.64).

Conclusion.—The sensitivity and specificity of bedside sonographic measurement of optic nerve sheath diameter is inadequate to aid medical decision making in children with suspected increased intracranial pressure. Pediatric emergency physicians with focused training by a pediatric ophthalmologist familiar with ophthalmic sonography can measure optic nerve sheath diameter accurately.

▶ Albert Einstein stated that "Most of the fundamental ideas of science are essentially simple and may, as a rule, be expressed in a language comprehensible to everyone."

The premise behind this and other such studies is simple—that ultrasound (quick, noninvasive, widely available, relatively cheap with no need for sedation or anesthesia and avoiding ionizing radiation), by detecting an increase in the diameter of the optic nerve, may have a bedside role in the diagnosis of raised intracranial pressure.

Indeed, results confirmed the technique to be quick (1 to 3 minutes) with satisfactory images (Fig 1) obtained from all enrolled patients. Unfortunately, however, despite 2 hours dedicated training by an ophthalmologist, in the

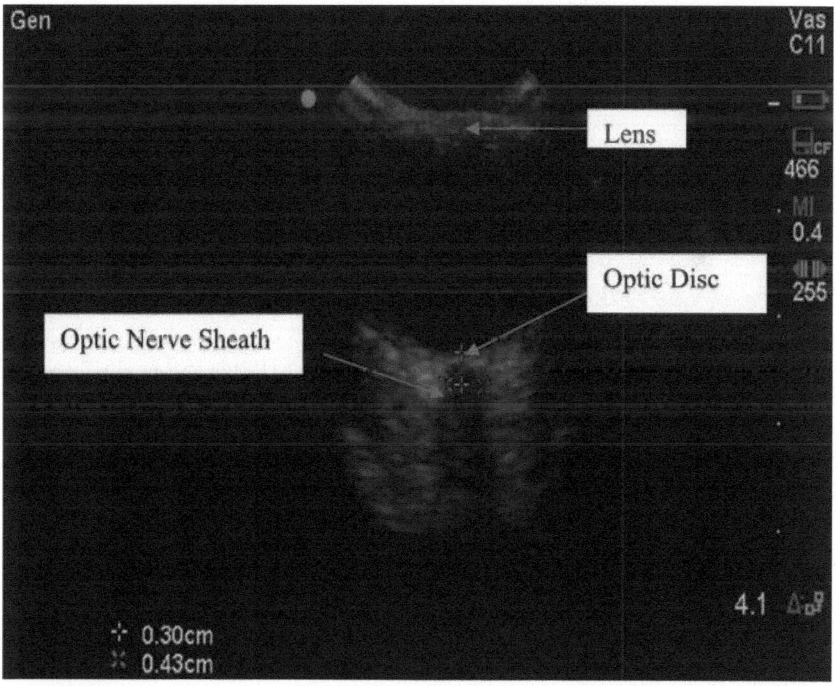

FIGURE 1.—Ultrasonographic image of optic nerve sheath. The 2 "x" marks represent the point at which the diameter of the optic nerve sheath was measured. (Reprinted from Le A, Hoehn ME, Smith ME, et al. Bedside sonographic measurement of optic nerve sheath diameter as a predictor of increased intracranial pressure in children. *Ann Emerg Med.* 2009;53:785-791, with permission from the American College of Emergency Physicians.)

hands of 2 pediatric emergency physicians who performed the scans, the sensitivity and specificity of the technique was inadequate to aid medical decision making.

Other studies in children (when ultrasound was performed by radiologists or ophthalmologists—eg, Ref 1) have, on the other hand, shown the technique to be useful. Could this be due to the known operator dependency of ultrasound—a question the authors themselves raise?

Perhaps their results would have been different had they undergone a longer training period. Can 2 hours training really have been sufficient? Before the idea is rejected altogether, I suggest that it would be valid to repeat the study—in the ER setting, but with scans also performed by radiologists and/or ophthalmologists.

In the subheading to her article "Everyone's a Radiologist Now,"[2] Jacqui Wise asks, "Are ultrasound machines truly the stethoscope of the 21st century? Should all doctors keep one in their back pocket?"

To which my simple and comprehensible answer is, "Not yet."

A. C. Offiah, BSc, MBBS, MRCP, FRCR, PhD

References

1. Malayeri AA, Bavarian S, Mehdizadeh M. Sonographic evaluation of optic nerve diameter in children with raised intracranial pressure. *J Ultrasound Med.* 2005;24:143-147.
2. Wise J. Everyone's a radiologist now. *Br Med J.* 2008;336:1041-1043.

The acoustic hood: a patient-independent device improving acoustic noise protection during neonatal magnetic resonance imaging

Nordell A, Lundh M, Horsch S, et al (Karolinska Univ Hosp, Stockholm, Sweden; Karolinska Institutet and Karolinska Univ Hosp, Stockholm, Sweden; et al)

Acta Pædiatr 98:1278-1283, 2009

Background.—Magnetic resonance imaging (MRI) is today the imaging modality of choice to investigate the neonatal brain. However, the acoustic noise during scanning is very loud, often exceeding 100 dBA.

Aim.—To reduce the acoustic noise during MRI for neonatal patients. If effective, this would create a safer environment and also result in fewer aborted examinations due to poor image quality from patient motion.

Methods.—A passive acoustic noise protector, the acoustic hood, was built out of dampening material. Sound pressure measurements with and without the acoustic hood were performed using our clinical neonatal scan protocol, consisting of eight imaging sequences. The acoustic hood is placed over the newborn inside the MR scanner tunnel during the examination to absorb acoustic noise.

Results.—The acoustic noise level was substantially reduced using the acoustic hood. Peak sound pressure was reduced 16.18–22.21 dBA depending on the pulse sequence. For the entire frequency spectra, reduction were between 4–13.59 dBA again varying with the pulse sequence.

Conclusion.—Acoustic noise can be reduced further than before by using the patient-independent acoustic hood in addition to other noise protection. We recommend the use of three passive hearing protections during neonatal MRI: (1) dental putty, (2) paediatric ear muffs, and (3) the acoustic hood.

▶ Compared with CT, one of the advantages of magnetic resonance imaging is that it avoids the use of non-ionizing radiation. However, in this study the authors highlight a potential complication of MRI that is not a problem with CT.

Perhaps only long-term population studies will tell us whether the noise associated with neonatal MR scanning is a cause of hearing loss. Nevertheless, if we consider that a noise level below 60 dB is recommended during neonatal transportation,[1] while some MR scanners produce acoustic noise in excess of 100 db, then the math is not difficult. Any technique that safely reduces noise levels should be embraced.

The acoustic hood is simple, effective, and not costly. If this (and other similar technologies) do no harm (and this appears to be the case), then if we don't use them, the possibility exists that neonates and infants undergoing MRI will have something to lose.

A. C. Offiah, BSc, MBBS, MRCP, FRCR, PhD

Reference

1. Macnab A. Vibration and noise in pediatric emergency transport vehicles: a potential cause of morbidity? *Aviat Space Environ Med.* 1995;66:212-219.

What physicians think about the need for informed consent for communicating the risk of cancer from low-dose radiation

Karsli T, Kalra MK, Self JL, et al (Children's Healthcare of Atlanta, GA; Emory Univ and affiliates, Atlanta, GA; et al)
Pediatr Radiol 39:917-925, 2009

Background.—The National Institute of Environmental Health Sciences, a subsidiary of the Food and Drug Administration, has declared that X-ray radiation at low doses is a human carcinogen.

Objective.—The purpose of our study was to determine if informed consent should be obtained for communicating the risk of radiation-induced cancer from radiation-based imaging.

Materials and Methods.—Institutional review board approval was obtained for the prospective survey of 456 physicians affiliated with three tertiary hospitals by means of a written questionnaire. Physicians were asked to state their subspecialty, number of years in practice, frequency of referral for CT scanning, level of awareness about the risk of radiation-induced cancer associated with CT, knowledge of whether such information is provided to patients undergoing CT, and opinions about the need for obtaining informed consent as well as who should provide information about the radiation-induced cancer risk to patients. Physicians were also asked to specify their preference among different formats of informed consent for communicating the potential risk of radiation-induced cancer. Statistical analyses were performed using the chi-squared test.

Results.—Most physicians stated that informed consent should be obtained from patients undergoing radiation-based imaging (71.3%, 325/456) and the radiology department should provide information about the risk of radiation-induced cancer to these patients (54.6%, 249/456). The informed consent format that most physicians agreed with included modifications to the National Institute of Environmental Health Services report on cancer risk from low-dose radiation (20.2%, 92/456) or included information on the risk of cancer from background radiation compared to that from low-dose radiation (39.5%, 180/456).

Conclusion.—Most physicians do not know if patients are informed about cancer risk from radiation-based imaging in their institutions. However, they believe that informed consent for communicating the risk of radiation-induced cancer should be obtained from patients undergoing radiation-based imaging.

▶ This is a topic about which every radiologist should have an opinion.

That CT scans account for approximately one tenth of the imaging procedures performed in Radiology Departments and yet contribute two thirds of population medical radiation dose is evidence of the relatively high radiation exposures associated with CT. On the other hand, many children will have a large number of follow-up radiographs (eg those with scoliosis, leg length, and other reconstructive procedures, cystic fibrosis etc).

While the idea of informed consent for these low-radiation investigations is attractive, in practice it would be difficult to apply. Do we (as radiologists) really know the risk/benefit to an individual patient of that nth plus one radiation-based investigation? Perhaps if we were trying to exclude pin track infection of a lengthened tibia, we might try harder to persuade parents to give consent than if we were checking for limb alignment—but we cannot be definite as to the actual risk of cancer to child X; and indeed the cosmetic effects of malalignment or leg length discrepancy might prey heavily on the minds of child X and her family.

If our views are contrary to those of the patient and family, and bearing in mind that for low-radiation procedures patients are more likely to be seen by radiographers than radiologists, then is the imaging room really the best place for this discussion? If not, and assuming (as preferred by the physicians in this study) that radiologists are to obtain consent, where and when would we seek it? Easier to arrange for hospitalized patients than outpatients.

Finally, what would we actually say? In terms of radiation-induced cancer, we give a population estimate, and we give this without knowing for certain (or why do it?) what child X's CT, bone scan, or radiograph will reveal.

For one-off acute cases (eg trauma) I suspect that neither radiologist nor responsible carer will have any difficulty. However, where do we draw the line, where does risk outweigh benefit—chest radiograph number 10, 15, 17?

So a few difficulties to overcome, but I can see one benefit; not to the patient, but to doctors and their employees—namely protection from possible litigation.

Informed consent for communicating the risk of cancer from low-dose radiation? What's your opinion—is it worth it?

A. C. Offiah, BSc, MBBS, MRCP, FRCR, PhD

Dissecting the Effects of Spinal Fusion and Deformity Magnitude on Quality of Life in Patients With Adolescent Idiopathic Scoliosis

Tsutsui S, Pawelek J, Bastrom T, et al (Rady Children's Hosp, San Diego, CA; et al)

Spine 34:E653-E658, 2009

Study Design.—A retrospective review of scores from the Scoliosis Research Society outcomes instrument (SRS-24 questionnaire).

Objective.—To quantify the isolated effects of spinal fusion and deformity magnitude on quality of life in patients with adolescent idiopathic scoliosis (AIS).

Summary of Background Data.—Significant improvements in 2-year postoperative SRS-24 questionnaire scores have been reported despite the loss of spinal motion due to instrumentation and arthrodesis. As deformity reduction may influence patient perception, it has been difficult to isolate the effect of spinal fusion on quality of life after scoliosis surgery.

Methods.—SRS-24 scores were compared between 3 cohorts of AIS patients (preoperative, postoperative, and nonoperative) using an ANOVA ($P < 0.05$) to determine the isolated effects of spinal fusion and deformity magnitude. Preoperative SRS-24 scores were collected from a group of patients with preoperative major Cobb angles greater than 40° (n = 194). Postoperative SRS-24 scores were collected from patients with preoperative major Cobb angles greater than 40° and 2-year postoperative major Cobb angles between 20° and 40° (n = 196). Finally, SRS-24 scores were collected from a nonoperative group of patients with major Cobb angles between 20° and 40° (n = 112).

Results.—Spinal fusion was found to have a negative isolated effect on the Activity domain (−0.3) and on the Total score (−0.2) ($P = 0.001$) of the SRS-24 questionnaire (score range: 1–5). A smaller deformity magnitude, on the other hand, was found to have a significantly positive isolated effect on all 4 preoperative domains ($P < 0.001$) and on the Total score ($P < 0.001$). The combined effect of surgery (spinal fusion and deformity correction) was found to be significantly positive for the Total score ($P < 0.001$) and for the domains of Pain, Self-Image, and Function ($P < 0.001$).

Conclusion.—Spinal fusion has an isolated negative effect on AIS patients' quality of life (Total score) mostly due to a decrease in scores of the Activity domain. The overall positive effect of surgery depends on the individual effects of spinal fusion (slight reduction in quality of life) and deformity reduction (modest improvement in quality of life).

▶ There are very few quality of life (QoL) articles in the field of pediatric radiology—and although the study group fell within the pediatric age group (range 9-23 years, mean 16 ± 2 years), sadly this is not one of them—all the authors being orthopedic surgeons.

I selected this article though for 2 reasons: firstly, because I found the balancing act between reduction in mobility and improvement in deformity that the surgeons have to manage an interesting one. It is almost invariable in medicine that we must weigh the pros and cons of a given intervention (I use the term intervention loosley here, including also diagnostic tests).

Secondly, I felt that here was a wasted opportunity for radiologists to be involved in a QoL study. Work is being done in adults to create tools for classifying the impact of diagnostic tests on functional health.[1] It would have taken very little tweaking of their research protocol for the authors of this well written article to have determined what impact the repeated imaging that many of these patients with adolescent idiopathic scoliosis undergo has on their QoL. Maybe this is a question for another research project—maybe it would even include (or be led by) pediatric radiologists.

A. C. Offiah, BSc, MBBS, MRCP, FRCR, PhD

Reference

1. Swan JS, Ying J, Stahl J, et al. Initial development of the Temporary Utilities Index: a multiattribute system for classifying the functional health impact of diagnostic testing [published online ahead of print]. *Qual Life Res.* 2010; 2010.

Discrepancy in reporting among specialist registrars and the role of a paediatric neuroradiologist in reporting paediatric CT head examinations

Nagaraja S, Ullah Q, Lee KJ, et al (Royal Hallamshire Hosp, UK; Univ Hosp Freiburg, Germany; et al)

Clin Radiol 64:891-896, 2009

Aim.—To evaluate the discrepancy rate among specialist registrars (SPR) to assess whether seniority had a bearing on the discrepancy rate. To investigate which were the commonly missed abnormalities and the consequences for teaching purposes. To investigate the role of a specialist consultant neuroradiologist in reporting paediatric head computed tomography examinations.

Materials and Methods.—The study was carried out over a 9-month period at the regional paediatric hospital during which time 270 CT head examinations were reported. Reporting in the department is carried out by one of the five general paediatric radiologists (GR) and also a specialist paediatric neuroradiologist (NR). The NR was considered the reference standard, who corroborated in areas of discrepancy with a second senior NR for this study. Of the 270 examinations, 260 were reported by the paediatric NR, 160 were reported by the SPR, GR, and NR, and 51 were reported by an SPR and the NR. In addition, four were reported by the GR and the NR, 45 by the NR only, seven by the GR only, and three cases were reported by the GR and an SPR. The discrepancy rates were calculated for GR versus NR, and SPR versus NR. All the discrepancies were re-evaluated by a second senior NR and confirmed in all cases. The reports of the SPR were further scrutinized. The trainees of training years 1–3 were considered junior and 4–5 were considered senior.

Results.—There was a discrepancy in 26/164 cases (15.9%) reported by the GR and NR. There was a discrepancy in 59/211 cases (28%) reported by an SPR and NR. The chi-squared test (two-sided) showed a significant difference ($p = 0.005$) between the two groups. There was a discrepancy in 36/118 cases (30.5%) reported by the junior SPR and NR. There was a discrepancy in 23/93 cases (24.7%) reported by a senior SPR and NR. The chi-squared test (two-sided) showed a non-significant difference ($p = 0.353$) between the two groups.

Conclusion.—The performance of the SPR was considered to be significantly different than the GR in this study. The year of training did not have a statistically significant bearing on the discrepancy rates. This study has been useful in guiding SPR with regards to paediatric CT head examination reporting. The NR played an important role, particularly in picking up subtle fractures and congenital abnormalities, which were missed by both the SPR and GR.

▶ Beyond a basic level of radiological training, observer characteristics are more important than years of experience. This has been shown by numerous researchers studying different imaging modalities and patient groups, for example, chest radiographs,[1] skeletal surveys in suspected child abuse,[2] and (Nagaraja et al) pediatric CT head scans (Table 2).

If that is the case, then why is it that senior radiologists with specialist interest in a given field invariably score highest? Is it that initially there is a steep learning curve that then flattens out, but never completely fails to rise? Or is it that there is something in their characters that produces specialists out of general radiologists? This something could be a desire to learn, to be "the best," a type A competitive nature. Or it could be a real love for the job. Even if not the latter, it would do no harm if we concentrated our training efforts

TABLE 2.—Discrepancy Numbers Between Junior and Senior Specialist Registrars (SPR) in Comparison with the Specialist Paediatric Neuroradiologist (NR)

	Junior (1–3 years)	Senior (4–5 years)
Agree with NR	82 (69.5%)	70 (75.3%)
Disagree with NR	36 (30.5%)	23 (24.7%)

not only on robust training programs, but also on inspiring our trainees—creating in them an enthusiasm to continuously improve and an understanding that experience is simply what we call our mistakes.

A. C. Offiah, BSc, MBBS, MRCP, FRCR, PhD

References

1. Herman PG, Hessel SJ. Accuracy and its relationship to experience in the interpretation of chest radiographs. *Invest Radiol.* 1975;10:62-67.
2. Offiah AC, Moon L, Hall CM, Todd-Pokropek A. Diagnostic accuracy of fracture detection in suspected non accidental injury: the effect of edge enhancement and digital display on observer performance. *Clin Radiol.* 2006;61:163-173.

In Utero Eye Development Documented by Fetal MR Imaging

Paquette LB, Jackson HA, Tavaré CJ, et al (Univ of Southern California, Los Angeles)

AJNR Am J Neuroradiol 30:1787-1791, 2009

Background and Purpose.—To date, very limited attention has been given to ocular abnormalities or growth parameters detected by fetal MR imaging. Our objective was to retrospectively determine the relationship between different parameters of eye development and estimated gestational age in the human fetus by use of fetal MR imaging.

Materials and Methods.—A retrospective study was performed to measure the transverse diameter, interocular distance, and lens diameter of the globes of 127 fetuses who had a morphologically normal central nervous system. Multiple single-shot T2 fast spin-echo images were obtained with a 1.5T magnet by use of contiguous 3-mm intervals in at least 2 orthogonal planes. Loess curves were fitted to explore the relationship between gestational age and each of the 3 measurements of interest. Different models were compared statistically to determine the model of best fit.

Results.—For each variable of interest, the "best" model of eye growth was a quadratic function. Specifically, lens growth seems to plateau after 36 weeks of gestation, interocular distance plateaus after 36 weeks of gestation, and globe growth plateaus after 42 weeks of gestation.

Conclusions.—The lens, orbit, and interocular distance growth of the fetus can be demonstrated on fetal MR imaging. All 3 measurements suggest a quadratic model of growth, which indicates slowing of growth toward the end of gestation.

▶ Traditionally used for the in utero diagnosis of brain abnormalities, antenatal MRI is increasingly being used to image other systems and for the development of normative data. In some instances, it may be more sensitive than antenatal ultrasound; for example in this article the authors state that because of its superior soft tissue characterisation, the placement of calipers for ocular measurements are probably more accurate with MRI (Fig 1 in the original article) than with ultrasound. They also show that the relationship between their various measurements and gestational age is quadratic that is, growth of the orbit plateaus toward the end of pregnancy.

The authors have mainly concentrated their discussion on the use of MRI to diagnose microphthalmia. There are of course other indications, such as the diagnosis of hypertelorism associated with malformation syndromes and skeletal dysplasias. Although the results presented are promising, larger studies are required to fully develop normative data spread as evenly as possible across the 42 weeks gestational period. Finally, because so far different authors have used different measurement parameters, it will be important to establish those measurements that are the most sensitive and specific.

A. C. Offiah, BSc, MBBS, MRCP, FRCR, PhD

Thoracic/Airway/Vascular

Ultrasound-guided percutaneous insertion of Hickman lines in children. Prospective study of 500 consecutive procedures

Arul GS, Lewis N, Bromley P, et al (Birmingham Children's Hosp, UK)
J Pediatr Surg 44:1371-1376, 2009

Aim.—The ultrasound-guided percutaneous technique of Hickman line insertion has not been widely adopted in pediatric surgical practice. We wished to review our own experience of using this technique for insertion into the internal jugular vein.

Methods.—Our vascular access team consists of a consultant surgeon and 2 consultant anesthetists. All procedures were prospectively recorded on a database and were either performed or directly supervised by our team.

Results.—Five hundred consecutive Hickman lines were inserted between June 2004 and October 2006. Patients' ages ranged from 14 days to 19 years (median, 44 months). Patients weighed between 600 g to more than 100 kg. Lines inserted were all tunneled silicone Hickman lines with a Dacron cuff (size 2.7F-10F, with 1-3 lumens), of which 60% were 7F double-lumen lines. Successful cannulation occurred in 99.8%. Perioperative complications (within 30 days) occurred in 12 patients (2.4%) and were all treated conservatively with no need for either blood transfusion or chest drain. Catheter-related sepsis rate was 3.16 per 1000 line days.

Discussion.—(1) The technique of ultrasound-guided percutaneous insertion of Hickman line to the internal jugular vein is safe and is applicable to all children regardless of size, age, or diagnosis. (2) Pediatric surgeons and anesthetists can learn this technique without specific training in interventional radiology. (3) A learning curve does exist, and we recommend concentrating pediatric vascular access procedures to a specialist team.

▶ According to the American social reformist, Charlotte Perkins Gilman (1860-1935), "The first duty of a human being is to assume the right functional relationship to society—more briefly, to find your real job, and do it."

The 4 authors of this article were either surgeons or anesthetists. As far as we can tell, none trained as an (interventional) radiologist. Nevertheless, their results are impressive—of 500 ultrasound guided Hickman line insertions, they achieved a success rate of 99.8%, a 30-day perioperative complication rate of 2.4% and a catheter-related sepsis rate of 3.16 per 1000 line days.

Although they did not present their results on a yearly basis so that we might judge for ourselves, they caution that a steep learning curve is associated with the procedure and advise the development of specialist teams. Depending on individual institutions, these teams may or may not involve interventional radiologists.

In this article, Arul et al have demonstrated that the make up of the team is relatively unimportant; once the job has been decided upon (and given adequate training), pediatric surgeons, anesthetists, and interventional radiologists can all do it and do it well.

A. C. Offiah, BSc, MBBS, MRCP, FRCR, PhD

Computer-aided detection for the identification of pulmonary nodules in pediatric oncology patients: initial experience

Helm EJ, Silva CT, Roberts HC, et al (Churchill Hosp, Headington, Oxford, UK; Yale Univ School of Medicine, New Haven, CT; Women's College Hosp, Toronto, Ontario, Canada; et al)

Pediatr Radiol 39:685-693, 2009

Background.—Computer-aided detection (CAD) has been shown to increase the sensitivity for detection of pulmonary nodules in adults. This study reports initial findings utilizing a CAD system for the detection of pediatric pulmonary nodules.

Objective.—To assess the performance of CAD and pediatric radiologists in the detection of pediatric pulmonary nodules.

Materials and Methods.—CT scans from a series of pediatric patients with known primary tumors and lung nodules were analyzed by four radiologists and a commercially available CAD system. IRB approval was obtained. Sensitivities were calculated for detection according to nodule size and location.

Results.—In 24 children (age 3–18 years) 173 nodules were identified. Overall the sensitivity of CAD was 34%, but the sensitivity of CAD for detection of nodules 4.0 mm or larger was 80%. Overall radiologist sensitivity ranged from 68% to 79%. There were 0.9 CAD false-positives and 0.3–2.4 radiologist false-positives per study.

Conclusion.—CAD in our pediatric oncology patients had good sensitivity for detection of lung nodules 4 mm and larger with a low number of false-positives. However, the sensitivity was considerably less for nodules smaller than 4 mm.

▶ The use of computer-aided detection (CAD) for the detection of pulmonary nodules is far less established in pediatric than it is in adult patients. The authors of this article show an acceptable sensitivity for CAD in the detection of lung nodules in children of 80% for lesions over 4 mm in diameter. However, when considering all lesions (ie, including those less than 4 mm in diameter), sensitivity was 34%—rather disappointing given that small nodule size in children is not necessarily predictive of benignity.[1] It is worth pointing out that the software employed in this study was optimized for use in adults and that despite the overall low sensitivity, CAD was nevertheless able to detect nodules smaller than 3 mm. It would seem then that the technique is worth pursuing.

Of considerable interest was the effect of observer reading style and experience on diagnostic acumen. For instance, of the 4 radiologists, the one with the highest false positive rate read most scans by scrolling down on the left lung and up on the right lung only once (the others scrolled up and down several times). Furthermore, the most experienced reader had the lowest false positive rate and one of the highest sensitivities despite having one of the fastest reading times.

Therefore, even without CAD, our diagnostic acumen can be improved; let us depend on, and learn from, our more experienced colleagues!

A. C. Offiah, BSc, MBBS, MRCP, FRCR, PhD

Reference

1. McCarville MB, Lederman HM, Santana VM, et al. Distinguishing benign from malignant pulmonary nodules with helical chest CT in children with malignant solid tumors. *Radiology.* 2006;239:514-520.

Estimation of Optimal CPR Chest Compression Depth in Children by Using Computer Tomography

Braga MS, Dominguez TE, Pollock AN, et al (Dartmouth-Hitchcock Med Ctr, Lebanon, NH; Children's Hosp of Philadelphia, PA; et al)
Pediatrics 124:e69-e74, 2009

Objective.—Pediatric consensus-driven cardiopulmonary resuscitation guidelines target chest compression (CC) depths of one third to one half anterior-posterior (AP) chest depth. Estimates for this target as assessed by computed tomography (CT) measurements of internal and external AP chest dimensions could direct future pediatric cardiopulmonary resuscitation guidelines.

Methods.—A total of 280 consecutive chest CT scans in permuted blocks of 20 for each of 14 age divisions between 0 and 8 years were reconstructed and analyzed. External and internal AP depths were measured at midsternum, and residual chest depth was calculated at simulated one-third and one-half AP compressions.

Results.—After a simulated compression calculation, one-half external AP depth CC would result in residual internal depth of <10 mm for 94% (263 of 280) of children 3 months to 8 years. For a one-third external AP CC, only 0.4% (1 of 280) of children 3 months to 8 years had a calculated residual internal chest depth <10 mm.

Conclusions.—By using CT reconstruction estimates of chest dimensions across the developmental spectrum from 0 to 8 years of age, we demonstrated that a simulated CC targeting approximately one-third external AP chest depth seems radiographically appropriate for children aged 3 months to 8 years, whereas simulated CC targeting approximately one-half external AP chest depth seems radiographically to be too deep, resulting in residual internal chest depth of <10 mm for most patients of this age.

▶ There are 2 studies from 2 different groups both using CT to answer the same question and published in the same issue of the same journal. I could do nothing less than include them both for review.

Kao et al[1] present their results in terms of absolute numbers, while Braga et al present results in terms of relative depth. Irrespective of how the results have been presented, it must be gratifying for both groups that they essentially reach the same conclusions. The conclusion being that current guidelines for basic life support in children recommend depths of cardiopulmonary resuscitation (CPR) that have been shown by CT to be excessive. In light of their results, both groups advise review of the guidelines.

In the past, BLS guidelines for children have been based on results from adult studies, from animal models and on expert opinion. It is good to see that with the aid of CT, BLS guidelines for children can now be reviewed based on evidence obtained from children.

A. C. Offiah, BSc, MBBS, MRCP, FRCR, PhD

Reference

1. Kao PC, Chiang WC, Yang CW, et al. What is the correct depth of chest compression for infants and children: a radiological study. *Pediatr.* 2009;124:49-55.

What Is the Correct Depth of Chest Compression for Infants and Children? A Radiological Study

Kao P-C, Chiang W-C, Yang C-W, et al (Natl Taiwan Univ Hosp, Taipei, Taiwan)

Pediatrics 124:49-55, 2009

Objective.—For infant and child resuscitation, current basic life support guidelines recommend a compression depth of one third to one half of the anteroposterior chest diameter. This study was conducted to assess the actual compression depths in infants and children when current guidelines are strictly followed.

Patients and Methods.—Chest computed tomography scans of 36 infants (<1 year old) and 38 children (1–8 years old) were reviewed. Patient demographic data were collected from medical records. Measurements of the anteroposterior diameter from chest computed tomography scans were taken from the anterior skin at either the internipple line or the middle of the lower half of the sternum, perpendicular to the skin on the posterior thorax.

Results.—In the infant group (25 boys, 11 girls), the mean age was 3.6 months. In the child-age group (21 boys, 17 girls), the mean age was 4.0 years. Compression depths were 3.4 to 5.1 cm in the infant group and 4.1 to 6.6 cm in the child group when current guidelines were followed. There was no difference in compression depths measured at internipple line versus in the lower half of the sternum. The intrathoracic structures observed beneath these 2 suggested that compression landmarks were similar.

Conclusions.—Radiological assessment of infants' and children's chests indicates similar or higher compression depths for infants and children versus the recommended compression depths for adults (3.8–5.1 cm) according to current guidelines. More evidence is needed to guide the proper depth of chest compression in pediatric populations.

▶ There are 2 studies from 2 different groups both using CT to answer the same question and published in the same issue of the same journal. I could do nothing less than include them both for review.

Kao et al present their results in terms of absolute numbers, while Braga et al[1] present results in terms of relative depth. Irrespective of how the results have been presented, it must be gratifying for both groups that they essentially reach the same conclusion. The conclusion being that current guidelines for basic life support in children recommend depths of cardiopulmonary resuscitation (CPR) that have been shown by CT to be excessive. In light of their results, both groups advise review of the guidelines.

In the past BLS guidelines for children have been based on results from adult studies, from animal models and on expert opinion. It is good to see that with the aid of CT, BLS guidelines for children can now be reviewed based on evidence obtained from children.

A. C. Offiah, BSc, MBBS, MRCP, FRCR, PhD

Reference

1. Braga MS, Dominguez TE, Pollock AN, et al. Estimation of optimal CPR chest compression depth in children by using computer tomography. *Pediatr.* 2009; 124:e69-e74.

Screening for Aortic Root Dilation in Marfan Syndrome Using the Ratio of the Aortic Root to Descending Aortic Diameters in Children

Kemna MS, Murphy DJ, Silverman NH (Seattle Children's Hosp and Univ of Washington; Lucile Packard Children's Hosp and Stanford Univ, CA)
J Am Soc Echocardiogr 22:1109-1113, 2009

Background.—The aim of this study was to develop a simpler echocardiographic screening tool for aortic root dilation in pediatric patients with Marfan syndrome. Aortic root dilation represents the most common cause of morbidity and mortality in patients with Marfan syndrome. It is traditionally defined by nomograms, relating aortic root dimensions to height or body surface area. In this study, the descending aorta (DAo) was used as an internal reference, thereby eliminating the need for nomograms when screening for aortic root dilation.

Methods.—The diameters of the aortic root at the sinus of Valsalva and the DAo were measured in 35 patients with Marfan syndrome and 52 normal controls in the parasternal long-axis echocardiographic view. The root/DAo ratio was calculated.

Results.—A root/DAo ratio ≥2 provided optimal sensitivity for root dilation (100%), whereas a root/DAo ratio ≥2.3 provided optimal specificity (100%).

Conclusion.—The root/DAo ratio provides a simple, rapid, and valuable screening test for aortic root dilation, independent of nomograms and body size.

▶ The stated rationale for seeking a method of confirming aortic root dilatation in this group of patients is that the traditional method (which requires taking measurements of patients' body weight and height and comparing with a nomogram) is tedious, prone to error, and only possible if a nomogram is available. The authors appear to have eliminated these difficulties by introducing the root/DAo ratio. As they say in their conclusion, it is simple and rapid and does not require a knowledge of patient height and weight, nor does it depend on the availability of a nomogram. Furthermore, use of the ratio allows documentation of progressive aortic root dilatation in the same patient—not possible when using nomograms.

In short this is a really good idea. Like all really good ideas one wonders, "Why hasn't it been thought of before?"

A. C. Offiah, BSc, MBBS, MRCP, FRCR, PhD

Craniofacial Venous Malformations: Magnetic Resonance Imaging Features That Predict Treatment Outcome

Jin Y, Lin X, Chen H, et al (Shanghai Jiao Tong Univ, China)

J Oral Maxillofac Surg 67:2388-2396, 2009

Purpose.—Craniofacial venous malformations (VMs) are challenging lesions. Treatment is not routinely indicated unless the benefit and loss have been weighed properly. Therefore it is crucial to predict clinical outcome before treatment. This study was performed to retrospectively determine whether pretreatment magnetic resonance imaging (MRI) findings are predictive of outcome in patients who undergo sclerotherapy for craniofacial VMs.

Patients and Methods.—MRI findings and clinical results of percutaneous sclerotherapy with ethanol and bleomycin A_5 in 69 patients with craniofacial VMs were retrospectively reviewed. Lesions were categorized with MRI based on margins; 51 were limited VMs, and 18 were infiltrating VMs. We performed *t* test analysis to assess the difference in the number of sclerotherapy sessions between the 2 groups. The Kruskal-Wallis test was used to evaluate differences in symptom improvement and satisfaction level between the groups. The difference in complication rates was assessed by χ^2 analysis.

Results.—The better response to treatment was among the patients with limited VMs. The number of sclerotherapy sessions was 1.9 ± 1.2 for limited VMs and 3.6 ± 1.5 for infiltrating VMs, with a significant difference between the groups ($P = .0001$). The patients with limited VMs were significantly more likely to have better symptom improvement ($P = .0001$) and a higher satisfaction level ($P = .0001$). The complication rates of the limited and infiltrating VMs were 18.2% (18/99) and 32.8% (21/64), respectively. A significant difference was found between the groups ($P = .033$).

Conclusions.—The feature of lesion margin on MRI before sclerotherapy is an important predictor of treatment outcome for craniofacial VMs.

▶ Fig 2 illustrates the disfiguring nature of craniofacial malformations (VMs)—also highlighted by the fact that 97.1% of patients in this article reported cosmetic problems.

Because direct percutaneous sclerotherapy is one of the few treatment options that have been shown to work, this study was not about the use of MRI to guide patient selection, but rather the use of MRI to inform the patient consenting process.

There are a number of statistically significant results that tend to support the fact that infiltrating VMs require more sclerotherapy sessions and have a poorer outcome than limited VMs. However, the authors also report that 68 of the 69 recruited patients showed improvement on physical examination and on returned questionnaires. With such a result (albeit a relatively small study population) it seems to me that sclerotherapy is indicated in their institution,

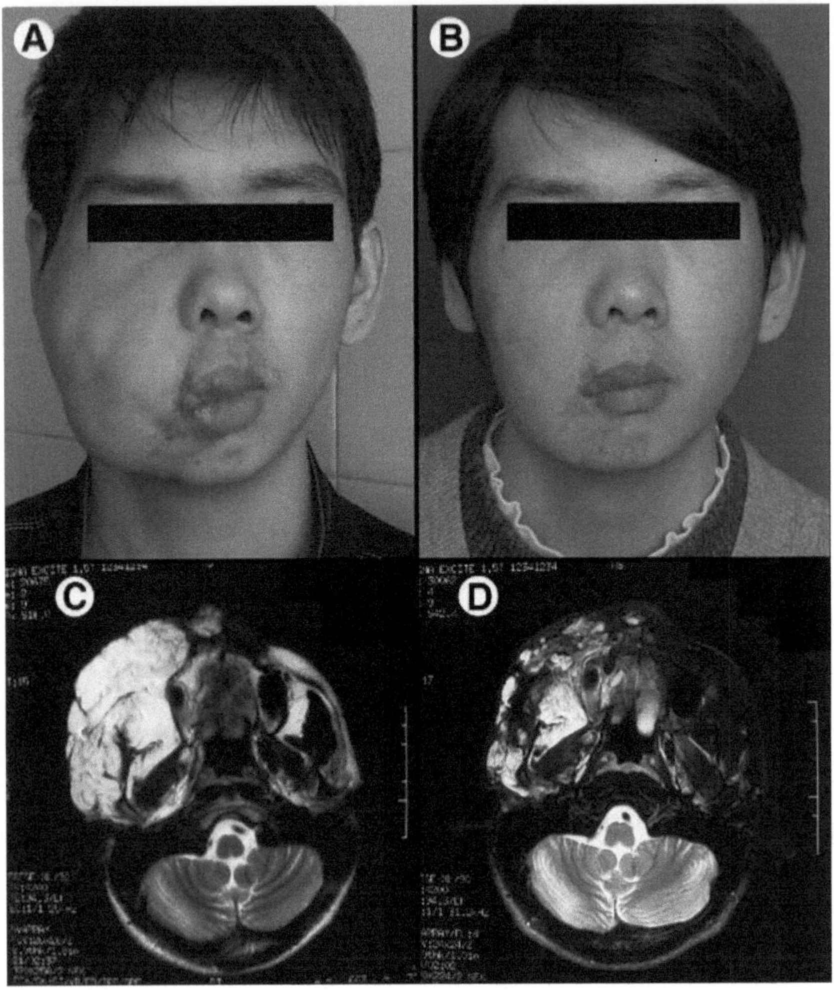

FIGURE 2.—The cosmetic problem was evident in a 22-year-old man with a VM in the cheek area. After 7 sessions of the treatment, the clinical outcome was graded as near normal. The appearance before treatment (*A*) and 18 months after the end of treatment (*B*) is shown. *C*, T2-weighted MRI scan showing high signal intensity of the large lesion, with a sharp abrupt transition from the surrounding tissue (limited VM). *D*, The lesion improved significantly on MRI after treatment. (Reprinted from Jin Y, Lin X, Chen H, et al. Craniofacial venous malformations: magnetic resonance imaging features that predict treatment outcome. *J Oral Maxillofac Surg*. 2009;67:2388-2396, with permission from American Association of Oral and Maxillofacial Surgeons.)

whether the lesion is infiltrating or not. Little or no problem then with obtaining informed consent!

The one dissatisfied patient complained of pain that was not improved by sclerotherapy. In fact, he developed local necrosis (which did subsequently improve), but felt that his condition was made worse by sclerotherapy not better. Statistical analyses aside, for me this is a significant finding, and begs the question, "Is sclerotherapy of VMs indicated when the presenting complaint is pain?"

A. C. Offiah, BSc, MBBS, MRCP, FRCR, PhD

Urogenital

Location of the ovaries in girls from newborn to 18 years of age: Reconsidering ovarian shielding

Bardo DME, Black M, Schenk K, et al (Oregon Health and Science Univ, Portland; New York Univ Med School; et al)

Pediatr Radiol 39:253-259, 2009

Background.—Guidelines for ovarian shielding are to place a lead shield in the midline of the abdomen at the level of the umbilicus. However, the ovaries are routinely seen at other locations on all types of imaging examinations.

Objective.—To determine the position of the ovaries in girls, newborn to 18 years of age, in order to assess efficacy of ovarian shield placement.

Materials and Methods.—We identified 336 girls who underwent lumbar spine and pelvic MRI. Images were reviewed noting the position of the ovaries relative to anatomic landmarks: symphysis pubis, iliac crest and anterior superior iliac spine (ASIS). In 174 girls a total of 307 ovaries were visualized. The girls were divided into three age groups and analyzed together. Right and left ovaries were analyzed together. The mean, 95% confidence interval (CI), standard deviations and range were calculated.

Results.—The ovaries lay at or below the iliac crest (the level of the umbilicus), most often just medial to the ASIS and above the pubic symphysis in girls of all ages.

Conclusion.—Current methods of shielding only the midline of the pelvis for the purpose of reducing radiation dose to the ovaries during radiographic imaging are ineffective given that the ovaries are almost always positioned laterally in the pelvis. Therefore current shielding techniques should be changed; lead ovarian shields should be placed in a lateral position or even abandoned if relevant anatomy will be obscured.

▶ In the current radiation, as low as reasonably achievable (ALARA) environment, it should be an embarrassment how poorly we as a discipline, whether pediatric radiology or general radiology, shield the ovaries of girls and young women. All too often I see a shield too small or too far afield to have the required protective effect. What is clearly needed is knowledge of where the ovaries actually lie, so that proper shielding by technologists/radiographers can be done with the least likelihood of missing them. And what better way, without ionizing radiation, to see where ovaries lie than MRI as so elegantly done in this study? (Ultrasound nicely shows ovaries, but it does not put them as well into the context of body landmarks.) A bonus of this study is finding that the degree of bladder distention has little effect on ovary location.

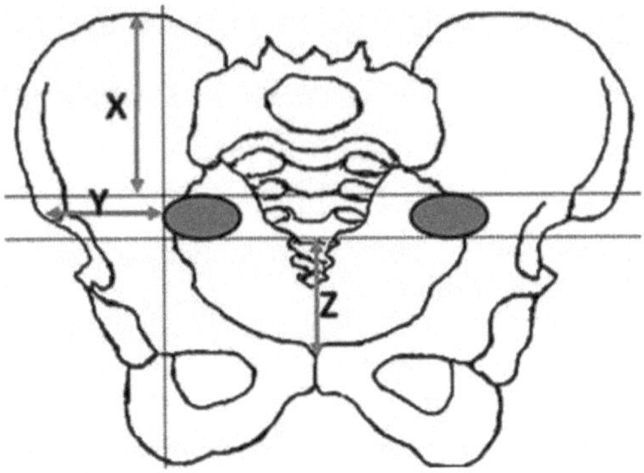

FIGURE 1.—Graphic representation of ovarian position in relation to pelvic osseous landmarks. Distances measured were as follows: superior margin of the ovaries (*gray ovals*) to the iliac crest (*X*), ASIS to the lateral margin of the ovary (*Y*), and the symphysis pubis to the inferior margin of the ovaries (*Z*). (With kind permission from Springer Science and Business Media: Bardo DME, Black M, Schenk K, et al. Location of the ovaries in girls from newborn to 18 years of age: reconsidering ovarian shielding. *Pediatr Radiol.* 2009;39:253-259.)

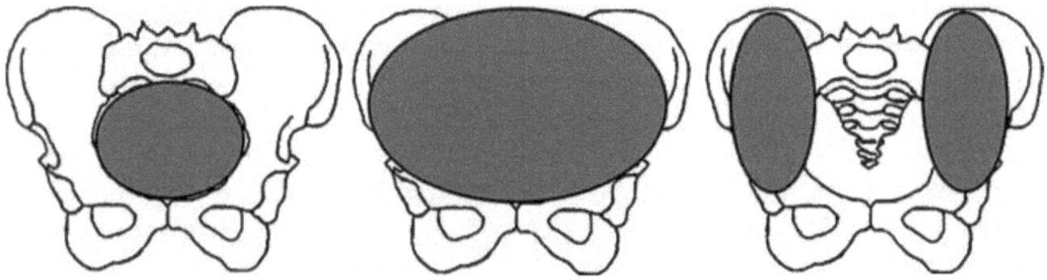

FIGURE 5.—Graphic representation of common, theoretical and proposed ovarian shield placement (*gray ovals*). *Left* Typical placement of a shield with limited pelvic coverage. *Center* Theoretical shield coverage of the entire pelvis, extending cephalocaudal from the symphysis pubis to the iliac crests and mediolaterally from ASIS to ASIS, would protect all ovaries. *Right* Proposed ovarian shielding technique: the lateral pelvis is shielded, protecting the majority of ovaries but allowing midline pelvic and spinal structures to be seen radiographically. (With kind permission from Springer Science and Business Media: Bardo DME, Black M, Schenk K, et al. Location of the ovaries in girls from newborn to 18 years of age: reconsidering ovarian shielding. *Pediatr Radiol.* 2009;39:253-259.)

So, their Fig 1 effectively summarizes where they are; and Fig 5 proposed how to shield them, with the proposed shielding on the (reader's) right of Fig 5 being the putative ideal shield. Naturally, the hip and the acetabular roof need to be seen, so that twin shielding should keep that part of the location in mind. Regardless of whether you follow their recommendation, a careful reading of this major article by all who do radiographs of the pelvis region ought to improve the current unacceptable placements of shields.

A. E. Oestreich, MD

Uterine position in adnexal torsion: specificity and sensitivity of ipsilateral deviation of the uterus

Harmon JC, Binkovitz LA, Stephens J (Nationwide Children's Hosp, Columbus, OH; The Ohio State Univ, Columbus)

Pediatr Radiol 39:354-358, 2009

Background.—Ipsilateral deviation of the uterus (IDU) has been suggested to be specific for torsion.

Objective.—To investigate the sensitivity and specificity of IDU in diagnosing torsion.

Materials and Methods.—We retrospectively reviewed the pelvic US examinations in girls performed at Nationwide Children's Hospital in 2007 and compared the uterine position in girls with a normal US scan (group 1) or with adnexal masses (group 2) with the uterine position in girls with torsion (group 3).

Results.—Uterine deviation was found in 7 of 297 girls (2%) in group 1 and in 6 of 47 (13%) in group 2, and was attributed to stool in the colon splaying the adnexa and uterus. In group 3, 18 of 44 girls (41%) had moderate or marked uterine deviation and none was associated with stool displacing the uterus. Uterine deviation was more common and severe in girls in group 3 than in girls in group 1 ($P<0.001$) and group 2 ($P=0.010$). The sensitivity and specificity of IDU for torsion were 41% and 98%, respectively.

Conclusion.—IDU is more common in adnexal torsion than in other causes of adnexal masses. With supporting clinical and imaging findings, IDU appears to be a specific marker for torsion.

▶ Anything that helps identify torsion of the ovary and/or fallopian tube in girls is worth knowing. This study puts into statistical terms a strong tendency for uterine deviation from the midline to be helpful in such a diagnosis. Intellectually, it may be disturbing that contralateral deviation of the uterus was found in 11% of the 44 girls and young ladies (included in the review were female pediatric patients younger than 22 years old); however, in the search for torsion diagnosis, anything that works, works. Of the 44, 30 had ovarian torsion, 6 had isolated fallopian tube torsion, and 8 had both ovarian and tubal, with the uterine deviation sign useful across the spectrum. I also note that associated with the torsion were 7 teratomas and 2 cystadenoma, as well as 17 cysts (masses and cysts are frequently associated with torsion—but surgery becomes more urgent when torsion is present). Mixing common sense with statistics, the authors point out that bulky stools can also push the uterus, and one should not rely on the uterine deviation sign in that circumstance. Unfortunately, symptoms associated with the surgically important diagnosis of adnexal torsion can be quite misleading as in a recent case initially mistaken for intussusception[1]; fortunately, ultrasound imaging is frequently indicated in girls for a whole host of symptoms, and the lessons of the current article are pertinent.

A. E. Oestreich, MD

Reference

1. Smith CJ, Bey T, Emil S, Wichelhaus C, Lotfipour S. Ovarian teratoma with torsion masquerading as intussusception in 4-year-old child. *West J Emerg Med.* 2008;9:228-231.

Ultrasonography and clinical findings in children with epididymitis, with and without associated lower urinary tract abnormalities

Karmazyn B, Kaefer M, Kauffman S, et al (Indiana Univ School of Medicine, Indianapolis; et al)

Pediatr Radiol 39:1054-1058, 2009

Background.—Epididymitis is most commonly idiopathic but can also be associated with urinary tract abnormalities (UTAs). The distinctive clinical and imaging findings of children with epididymitis and underlying UTAs are not known.

Objective.—To describe clinical and imaging findings in children with epididymitis and the association with UTAs.

Material and Methods.—The study group included all children evaluated for epididymitis confirmed by scrotal US in a 6-year period. The clinical and imaging findings and disease recurrence were compared between children with and without UTAs.

Results.—A total of 47 boys (mean/SD 9.61/4.40 years, range 0.1 to 17.1 years) met the entry criteria, of whom 17 had UTAs. The most common UTAs were hypospadias, neurogenic bladder, and functional bladder abnormality (six each). Age at presentation and likelihood of testicular swelling or hydrocele was not different between children with and without UTAs. Marked epididymal swelling was more common in children with UTAs (9/17, 53%) than in those without UTAs (5/30, 17%; $P=0.02$), as was recurrent epididymitis (with UTAs, 9/17, 53%; without UTAs, 5/30, 17%; $P=0.02$). Chronic epididymitis (five children), presentation with scrotal mass (four), and epididymal abscess (two) occurred only in children with UTAs.

Conclusion.—Children with epididymitis who have UTAs are more likely to present with marked epididymal swelling, develop recurrent disease, and have a more protracted course.

▶ For me, the most relevant question about epididymitis has been whether to image the kidneys in boys presenting with that diagnosis—since the likelihood of finding something renal seems quite low. The urinary tract abnormalities (UTAs) described in this study are all of the lower urinary tracts and not the kidneys, so my question is not here directly answered. But surely, it should be of no difficulty to screen the kidneys briefly with ultrasonography (US) whenever US does reveal epididymitis. Now as to their article—marked epididymal swelling, recurrence, and protracted course would all then suggest that if a lower UTA is not yet known, then look for it. May I go back and cite my own example where upper tract imaging was pertinent in an acute epididymitis? This mentally challenged adolescent had an acute swollen scrotum, and I was asked to do an intravenous urogram (this was way back in the days before US imaging had caught on) to check the kidneys. The kidneys were fine, but dense, round, small foreign bodies of various sizes were seen in the distribution of the bowel—he had swallowed part of a mercury-containing thermometer. I presume US would today also reveal mercury droplets to the astute sonologist.

A. E. Oestreich, MD

Cellular mesoblastic nephroma (infantile renal fibrosarcoma): institutional review of the clinical, diagnostic imaging, and pathologic features of a distinctive neoplasm of infancy

Bayindir P, Guillerman RP, Hicks MJ, et al (Ege Univ Med Faculty, Izmir, Turkey; Texas Children's Hosp, Houston)

Pediatr Radiol 39:1066-1074, 2009

Background.—Cellular mesoblastic nephroma has been associated with a more aggressive course than classic mesoblastic nephroma, including local recurrences and metastases.

Objective.—To define the clinicopathologic and imaging features distinguishing cellular from classic mesoblastic nephroma.

Materials and Methods.—Retrospective review of clinical charts and imaging studies of ten children with mesoblastic nephroma from 1996 to 2007 at a large children's hospital.

Results.—In six children the mesoblastic nephroma was pure cellular, in two mixed, and in two classic. The mean ages at diagnosis were 107 days for those with the cellular form, and 32 days for those with the classic form. Hypoechoic or low-attenuation regions representing necrosis or hemorrhage were found in all children with the cellular form and in none of those with the classic form. Hypertension was present in 70% and hypercalcemia in 20% of the children and resolved following nephrectomy. Two cellular tumors encased major abdominal vessels. Local recurrence and metastases occurred within 6 months of tumor resection in two children with the cellular form. Intraspinal extension and intratumoral pseudoaneurysm were seen in one child with the cellular form. The cellular tumors shared histopathologic features with infantile fibrosarcoma (IFS), and RT-PCR testing in two children with the cellular form revealed the t(12;15) ETV6-NTRK3 gene fusion common to IFS.

Conclusion.—Distinct from the classic form, cellular mesoblastic nephroma is more heterogeneous in appearance on imaging, tends to be larger and present later in infancy, and can exhibit aggressive behavior including vascular encasement and metastasis. Intraspinal extension and intratumoral pseudoaneurysm are previously unreported findings encountered in our cellular mesoblastic nephroma series. The shared histopathology

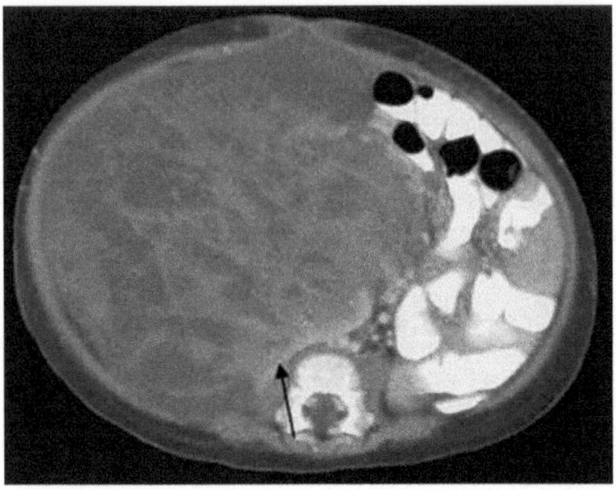

FIGURE 1.—Abdominal mass in a 5-month-old girl. a Axial contrast enhanced abdominal CT image demonstrates a large mass of heterogeneous attenuation centered in the right hemiabdomen. The mass crosses the midline and displaces the midline vascular structures. A rim of normal renal parenchyma is visible at the posteroinferior portion of the mass (*arrow*). (With kind permission from Springer Science and Business Media: Bayindir P, Guillerman RP, Hicks MJ, et al. Cellular mesoblastic nephroma (infantile renal fibrosarcoma): institutional review of the clinical, diagnostic imaging, and pathologic features of a distinctive neoplasm of infancy. *Pediatr Radiol.* 2009;39:1066-1074.)

and translocation gene fusion support the concept of cellular mesoblastic nephroma as the renal form of IFS.

▶ This article and a companion article in the same issue of the journal[1] discuss and illuminate an important bit of information that I am afraid I was unaware of until now. I was consciously present about the time that the most common tumor of the newborn kidney was revealed to be mesoblastic nephroma (at that time it was so "new" that a prominent pediatric radiologist with renal interest missed it as an unknown at a national conference). Let the current relatively new information become widespread as well: 2 separate forms of mesoblastic nephroma must be distinguished. The prognosis, the treatment, and the nature of the tumors differ considerably. The inhomogeneity of the cellular form is quite striking on cross sectional imaging (including ultrasound [US]) as demonstrated in Fig 1a. Since the usual treatment of classic mesoblastic nephroma is typically watch and wait (for them to shrivel up eventually), articles with pathologic correlation will be more heavy weighted in numbers to the cellular mesoblastic nephroma. Because of the important implications for the care of the involved infant, knowing the details of these 2 articles is important for all who care for newborns or interpret their imaging—the ultrasound and MRI/CT patterns seem likely to make the distinction between them.

A. E. Oestreich, MD

Reference

1. Chaudry G, Perez-Atayde AR, Ngan BY, Gundogan M, Daneman A. Imaging of congenital mesoblastic nephroma with pathological correlation. *Pediatr Radiol.* 2009;39:1080-1086.

Clinical and Imaging Features in 16 Infants Exposed to Food Contaminated with Melamine and Cyanuric Acid

Ren F-Y, Wang Y, Hou X-B, et al (General Hosp of Lanzhou Command PLA, Ganshu, P. R. China; Hosp of Ganshu Armed Police Forces, Lanzhou, P. R. China)

AJR Am J Roentgenol 192:707-710, 2009

Objective.—The objective of our study was to describe the clinical and imaging features of infants exposed to milk formula contaminated with melamine and cyanuric acid.

Conclusion.—The imaging feature of patients exposed to food contaminated with melamine and cyanuric acid was multiple calculi of the urinary tract that varied greatly in size and shape. Most patients presented with symptoms of urinary obstruction.

▶ The wave of many unanticipated cases of renal and urinary calculi in infants in (mainland) China in the latter half of 2008 has been firmly established as being associated with (and almost certainly due to) contamination of infant milk formula with melamine. Unlike this article, most medical and imaging articles discuss melamine alone as the contaminant[1] and not cyanuric acid. Deaths have also been recorded. Of the current report's 16 cases, all underwent ultrasound imaging and 14 had unenhanced, unfortunately high dose, CT. The other quoted study[1] restricted imaging to radiation-free ultrasound. In the current report 12 of 16 had bilateral renal stones; exposure to the contaminant was 3 to 16 months; 13 had urinary obstruction symptoms and only 1 was asymptomatic; renal calculi were staghorn, oval, or punctiform; ureteral calculi were oval; those calculi that were passed spontaneously fragmented with slight pressure, indicating they were loose and fragile. Treatment by withdrawing the offensive milk leads to gradual resolution (this study and the study by Zhu et al).[1] Once discovered, the association was widely found in China and widely reported, leading to resolution of stones in most affected infants and to elimination of the public health hazard. Here in the United States, we have been asked to ultrasound children adopted from China who had been so exposed; fortunately, our recent patient showed no current stones.

A. E. Oestreich, MD

Reference

1. Zhu S-L, Li J-H, Bao Z-X, et al. Conservative management of pediatric nephrolithiasis caused by melamine-contaminated milk powder. *Pediatrics*. 2009;123: e1099-e1102.

Clinical Courses of Children With Acute Lobar Nephronia Correlated With Computed Tomographic Patterns

Cheng C-H, Tsau Y-K, Chen S-Y, et al (Chang Gung Memorial Hosp, Taoyuan, Taiwan; Natl Taiwan Univ Hosp, Taipei; et al)
Pediatr Infect Dis J 28:300-303, 2009

Background.—Acute lobar nephronia (ALN) is a complicated acute renal infection that is part of the progression of the inflammatory process of acute pyelonephritis that can progress to renal abscess if left untreated. We examined the place of ALN in the spectrum of upper urinary tract infections.

Methods.—The medical records of all patients diagnosed with ALN by computed tomography (CT) were reviewed retrospectively in terms of their demographic characteristics, clinical presentations, and laboratory findings. The patterns of the ALN lesions were checked on all contrast-enhanced CT films and the volume of the CT lesions (the extent of renal involvement) was estimated in every patient. Then, the correlation between the clinical presentation and CT lesions was examined.

Results.—Two distinct CT patterns of ALN lesions were observed in 127 patients: 94 simple ALN and 33 complicated ALN. Simple ALN appeared as striated or wedge-shaped, poorly defined regions of homogeneously decreased nephrographic density, whereas the lesions of complicated ALN showed heterogeneously decreased nephrographic density after contrast enhancement. The volume fraction of the CT lesions correlated well with the duration of fever before ($P = 0.006$) and after ($P < 0.001$) treatment in patients with simple ALN, while only the correlation between the fever duration after treatment and the volume fraction of the CT lesions ($P < 0.001$) was significant in patients with complicated ALN. Patients with complicated ALN were generally older ($P = 0.004$) and febrile longer after starting antibiotic treatment ($P < 0.001$). All treatment failures were in the complicated ALN group.

Conclusions.—Children with ALN comprise 2 groups with different clinical presentations and treatment responses: simple ALN and complicated ALN. We suggest that simple ALN be regarded as a continuation of acute pyelonephritis and require 2 weeks of antibiotic treatment, while complicated ALN is a distinct more severe form requiring a 3-week or longer therapy regimen as the treatment of choice.

► I had thought that the term "acute lobar nephronia" (ALN), while a clever analogy to acute lobar pneumonia, had gone out of fashion in recent literature (my new fellows had not recognized the term). I had also hoped that serial CT in children, especially both with and without contrast, had become more limited in use, in order to "Image Gently" and reduce the radiation burden. So, hopefully the interesting results of this study can pass over to MR urography. With these caveats out of the way, the results presented are of interest in the guidance of therapy in childhood urinary tract infections. The complicated version of ALN as shown in Fig 2 is recognizably different from simple ALN, also

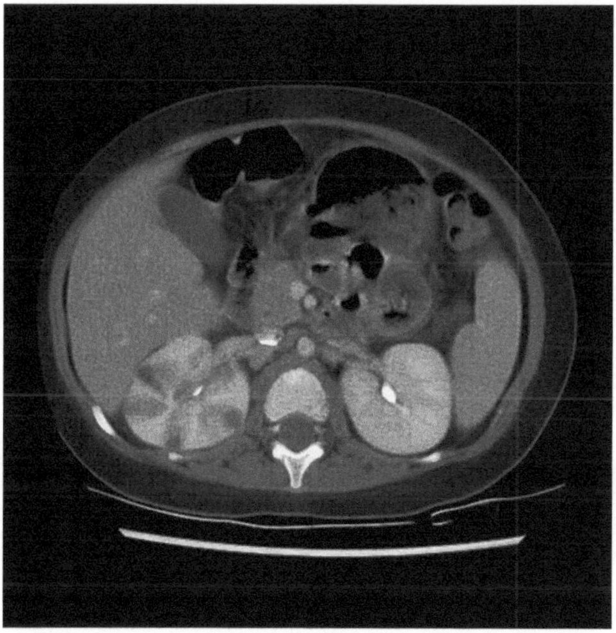

FIGURE 2.—Complicated ALN. The contrast-enhanced CT shows multiple ill-defined hypodense zones of heterogeneously decreased nephrographic density. The lesion contains both dense regions of decreased enhancement and small areas of extremely low attenuation. (Courtesy of Cheng C-H, Tsau Y-K, Chen S-Y, et al. Clinical courses of children with acute lobar nephronia correlated with computed tomographic patterns. *Pediatr Infect Dis J*. 2009;28:300-303.)

illustrated in the original article. This difference had clear implications for the duration and intensity of medical therapy. So, learn from the authors' experience, but at the least, if MR urography is not available, omit the precontrast CTs in this situation. We note from another recent article[1] that the entity sometimes known as ALN can affect a transplanted kidney as well.

A. E. Oestreich, MD

Reference

1. Tai HC, Yang PJ, Lee PH, Chung SD, Chueh SC, Yu HJ. Acute lobar nephronia in a renal allograft: a case report and literature review. *Transplant Proc*. 2008;40: 1737-1740.

Diagnosis of Wilms' tumor using near-infrared Raman spectroscopy

Wills H, Kast R, Stewart C, et al (Wayne State Univ/Detroit Med Ctr, Detroit, MI; et al)

J Pediatr Surg 44:1152-1158, 2009

Purpose.—Raman spectroscopy has distinguished malignant from normal tissues in several types of cancer. This is the first report of applying Raman spectroscopy to the diagnosis of Wilms' tumor.

Methods.—Specimens of normal kidney, Wilms' tumor, xanthogranuloma, nephrogenic rests, and rhabdoid tumor were collected fresh from the operating room. Specimens of Wilms' tumor, normal kidney, and congenital mesoblastic nephroma were retrieved from the cryobank and thawed to room temperature. At least 12 Raman spectra were collected from each tissue sample. Histologic slides of each specimen were reviewed by pediatric pathologists. A computer algorithm based on discriminant function analysis (DFA) classified the Raman spectra of Wilms' tumor and the normal sample.

Results.—Four hundred sixty-seven spectra were collected from 41 specimens. Using DFA, Raman spectroscopy differentiated Wilms' tumor from normal with 100% sensitivity and specificity and treated from untreated Wilms' tumor with 100% sensitivity and specificity. Using

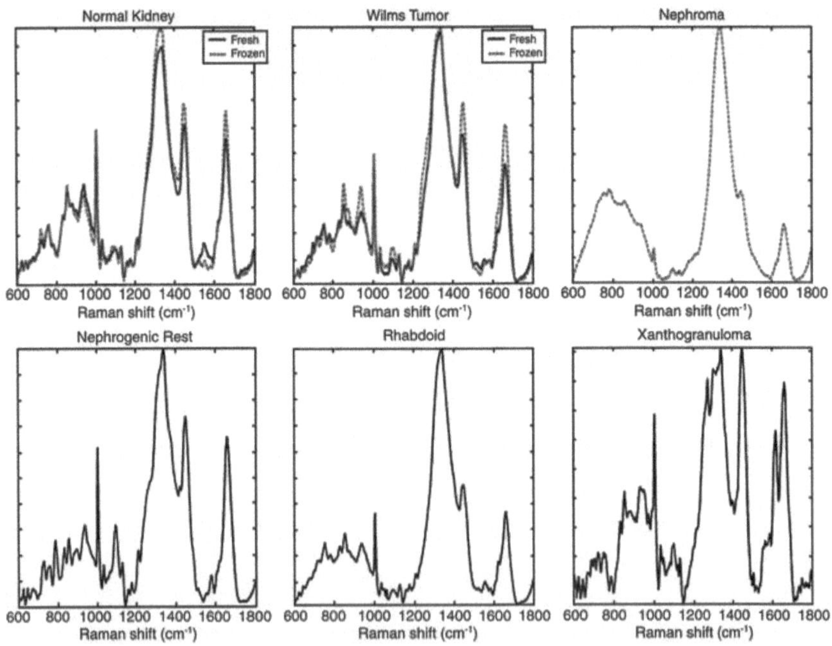

FIGURE 1.—Mean spectra of fresh (solid line) and frozen (dashed line) tissues. (Reprinted from Wills H, Kast R, Stewart C, et al. Diagnosis of Wilms' tumor using near-infrared Raman spectroscopy. *J Pediatr Surg*. 2009;44:1152-1158, with permission from Elsevier.)

a DFA model built from cryopreserved specimens but applied to fresh Wilms' and normal samples, the sensitivity and specificity were 93.3% and 90.9%, respectively.

Conclusion.—Raman spectroscopy is an accurate technique for differentiating Wilms' tumor from normal kidney and treated from untreated Wilms' tumor. It has potential to diagnose in minutes what currently takes several hours to days.

▶ Although not a radiology article, I have selected this article because the challenges faced by pathologists in the diagnosis of Wilms' tumor and in the assessment of its response to treatment are the same challenges faced by radiologists. Both radiologists and pathologists are in the process of optimising (relatively) new techniques to solve the dilemmas; namely diffusion weighted MRI and Raman spectroscopy respectively.

Specific to pediatric tumors, at present the results of DTI are less impressive[1] than the 100% sensitivity and specificity achieved by Raman spectroscopy (Fig 1).

Nevertheless, MRI is noninvasive—a clear advantage in pediatrics and when multiple assessments of the same patient may be necessary. For this reason, DTI of extracranial pediatric solid tumors is an area of research worth pursuing. Studies should be conducted along the lines advised by Padhani et al[2]—and perhaps with the inclusion of Raman spectroscopy in the research protocols?

A. C. Offiah, BSc, MBBS, MRCP, FRCR, PhD

References

1. Humphries PD, Sebire NJ, Siegal MJ, Olsen OE. Tumors in pediatric patients at diffusion-weighted MR imaging: apparent diffusion coefficient and tumor cellularity. *Radiology.* 2007;245:848-854.
2. Padhani AR, Liu G, Mu-Koh D, et al. Diffusion-weighted magnetic resonance imaging as a cancer biomarker: consensus and recommendations. *Neoplasia.* 2009;11:102-125.

Clinical and Radiological Characteristics of Patients Operated in the First Year of Life Due to Ureteropelvic Junction Obstruction: Significance of Renal Pelvis Diameter

Uluocak N, Ander H, Acar Ö, et al (Gaziosmanpasa Univ, Tokat, Turkey; Istanbul Univ)
Urology 74:898-902, 2009

Objectives.—To analyze the data of our series of patients either operated in the first year of life or managed conservatively because of prenatally detected ureteropelvic junction obstruction, with particular consideration for the significance of renal pelvis diameter.

Methods.—A total of 56 patients with prenatal ultrasonographic findings suggestive of ureteropelvic junction obstruction were enrolled in this study. Operated group and control group consisted of 35 and 21 patients, respectively. We compared postnatal imaging findings of patients managed either conservatively or operatively. The diagnostic accuracy of imaging studies in confirming the diagnosis of obstruction and determining the need for surgery was analyzed.

Results.—If relative function of the involved side being <40% was accepted as the gold standard, renal pelvis diameter was found to be a statistically significant predictor of surgical intervention based on the results of receiver operating characteristic curve analysis. The critical diameter was found to be 22 mm, at which sensitivity and specificity of identifying surgical candidates were calculated to be 78.9% and 50%, respectively. In addition, if we consider that having undergone a pyeloplasty procedure is the most accurate definition of ureteropelvic junction obstruction, the critical pelvis diameter was found to be 19 mm. Renal pelvic dilation >19 mm defines obstruction that may necessitate surgical treatment with a sensitivity of 93.9% and specificity of 71.4%.

Conclusions.—Renal pelvis diameter may indicate the need for surgical intervention or further assessment to define clinical significance more accurately.

▶ As normal, as well as potentially abnormal, pregnancies are increasingly imaged prenatally with ultrasonography (US) (and eventually MRI), wide renal pelves are increasingly discovered, raising questions of ureteropelvic obstruction. Not all prenatally wide renal pelves and not all postnatally wide renal pelves are due to obstruction (some, for example, may reflect just high flow). Nonetheless, the article attempts a guideline of size on postnatal (first year of life) imaging, as to which child with pelviectasis should be considered (not scheduled yet, however) for surgical treatment. An editorial comment following the article emphasizes that dilatation does not necessarily mean obstruction and slow drainage also does not necessarily mean obstruction (he uses the analogy of bath and sink have the same drainage pipe).[1] This article does suggest that it is clinically pertinent for the medical imager to report the width of wide pelves.

A. E. Oestreich, MD

Reference

1. Mouriquand PDE. Editorial comment. *Urology.* 2009;74:902.

Selective Use of Voiding Cystourethrography in Children Undergoing Renal Transplant Evaluation

Singer JS, Zaid U, Gritsch HA, et al (David Geffen School of Medicine at UCLA, Los Angeles, CA)

J Urol 182:1158-1162, 2009

Purpose.—Voiding cystourethrography is a routine component in evaluating children awaiting renal transplantation. We examined whether this assessment is necessary in children with renal failure due to dysplasia/aplasia/hypoplasia syndrome and unknown etiology, which account for up to 25% of those with renal failure requiring renal replacement therapies.

Materials and Methods.—We performed an institutional review board approved, retrospective review of 191 children undergoing transplantation between 2002 and 2007. We reviewed clinical factors associated with positive findings on voiding cystourethrogram. We also reviewed cystography results in children with chronic kidney disease due to renal dysplasia and unknown etiology.

Results.—We identified 113 boys and 78 girls who underwent renal transplantation during the study period. Pre-transplant voiding cystourethrography was documented in 108 children (57%). Predictors of positive pre-transplant results included history of hydronephrosis, urinary tract infections and renal failure due to urological causes. No pre transplant cystogram was positive in children with renal failure due to dysplasia or unknown etiology.

Conclusions.—We recommend selective use of voiding cystourethrography to evaluate children awaiting renal transplantation. We continue to support performing this test in children with renal failure due to urological causes and those with a history of urinary tract infection, hydronephrosis or voiding dysfunction. In the absence of these findings children with renal failure due to renal dysplasia/aplasia/hypoplasia syndrome or unknown etiology need not undergo pre-transplant voiding cystourethrography.

▶ Another step in the right direction. This article is by urologists and transplantologists, and still comes up with a category of children pretransplant in whom they recommend against pretransplant voiding cystourethrogram. They nicely cite the monetary expense, the psychological effect of stress on the patient and the patient's family, and the ionizing radiation necessary. (In Europe, the study would be done by ultrasound voiding cystourethrogram, eliminating the radiation factor only.) They also suggest the procedure may itself cause urinary tract infection in some subjects. Previous guidelines for children did not exempt any children under 8 years of age, but the current authors extended the reprieve to younger children as well. As urologists and radiologists and other health care workers consider all the costs of such cystourethrograms, questionable benefits will hopefully lead to more reasons not to do this uncomfortable imaging procedure. However, there are still voices on the other side, for example, advocating not eliminating the voiding cystourethrogram under 6 months of patient age.[1]

A. E. Oestreich, MD

Reference

1. Tse NK, Yuen SL, Chiu MC, Lai WM, Tong PC. Imaging studies for first urinary tract infection in infants less than 6 months old: can they be more selective? *Pediatr Nephrol.* 2009;24:1699-1703.

5 Economics, Research, Education, and Quality

Introduction

Social and economic dimensions of radiology practice for 2009-2010 must be viewed within a larger process that dominated the political landscape (of at least the United States) during that period—national health care reform. On numerous occasions, the expense and overuse of high-tech imaging modalities were cited by politicians as a major factor driving up health care costs.

It is not surprising, therefore, that several articles dealing with cost effectiveness of imaging strategies took on special importance and were selected for inclusion in this chapter. Renewed interest was directed at strategies such as application of the American College of Radiology's "appropriateness criteria." A renewed focus on final patient "outcomes" rather than imaging test "results" was also seen.

The second major theme of the year was radiation dose and safety. Like health care reform, this item received considerable national attention in October 2009 when a major California hospital announced that a number of its patients had received radiation burns when a misprogrammed CT scanner subjected them to 8-10 times the already high dose for brain perfusion imaging studies. In response to this recurring interest in radiation effects, some articles concerning dose and cancer risk from ionizing radiation have been included. Fortunately, organized radiology has been ahead of the curve on this issue, as witnessed by the popularity and success of their "Image Gently" campaign.

Allen D. Elster, MD

Appropriate Imaging and Cost Effectiveness

Approaches for Economic Evaluations of Health Care Technologies

Tarride J-E, Blackhouse G, Bischof M, et al (Programs for Assessment of Technology in Health Res Ins, Ontario, Canada; et al)
J Am Coll Radiol 6:307-316, 2009

Economic evaluations involve the comparison of alternative courses of action in terms of both their costs and their consequences. In response to increasing health care costs, economic evaluations of competing technologies, including radiologic interventions, are increasingly used to inform resource allocation decisions. It is therefore crucial that radiologists have a thorough understanding of the methods. The objective of this paper is to present a detailed overview of the principles and methods of economic evaluations of health technologies, including recent methodologic developments. For the purpose of this paper, the key elements of an economic evaluation are divided into 5 broad sections: 1) types of economic evaluations (eg, cost-effectiveness analysis, cost-utility analysis), 2) study perspectives (ie, temporal and cost perspectives), 3) analysis of costs and effects (eg, incremental cost-effectiveness ratios, cost-effectiveness planes), 4) conducting economic evaluations alongside trials or using decision-analytic models, and 5) dealing with the different forms of uncertainty in economic evaluations (eg, sampling uncertainty in trials, parameter uncertainty in models). Examples from the recent radiology literature are used to explain the key concepts. This review improves upon the previous educational papers published in radiologic journals, as it covers recent methodologic advances regarding the treatment of uncertainty.

▶ There is no doubt that in an era of national health care reform initiatives, radiological procedures will be subjected to increasing scrutiny from economic perspectives. As the unbridled use of expensive technologies can no longer be sustained, radiologists will need to be knowledgeable about how costs and benefits are analyzed. This article provides a useful framework for defining these dimensions of health care technologies.

The first type of economic evaluations are based on the characteristics of the consequences (Table 1). These are cost-minimization analysis (CMA), cost-effectiveness analysis (CEA), cost-utility analysis (CUA), and cost-benefit analysis (CBA).

A second useful perspective is the graphical representation of incremental costs versus outcomes which may be presented on what is called a cost-effectiveness plane (Fig 2). Here incremental costs are presented on the vertical axis (more vs less costly) and incremental outcomes on the horizontal axis (more vs less effective).

The origin is the treatment of reference (eg, existing technology). Such a plane has 4 quadrants. In the lower right quadrant, the new technology dominates the existing one (less costly and more effective), whereas the existing

TABLE 1.—Types of Economic Evaluations

Type of Study	Measurements/ Valuation of Costs	Identification of Consequences	Measurements/Valuation of Consequences
Cost minimization analysis	Dollars	Identical between alternatives	None
Cost-effectiveness analysis	Dollars	Single effect of interest, common to the alternatives, but achieved to different degrees	Natural units such as life years gained, percentage of correct diagnosis or cases detected
Cost-utility analysis	Dollars	Single or multiple effects, not necessarily common to the alternatives	Quality-adjusted life-years
Cost-benefit analysis	Dollars	Single or multiple effects, not necessarily common to both alternatives	Dollars

Source: Adapted from Drummond et al [1]. Editor's Note: Please refer to original journal article for full reference.

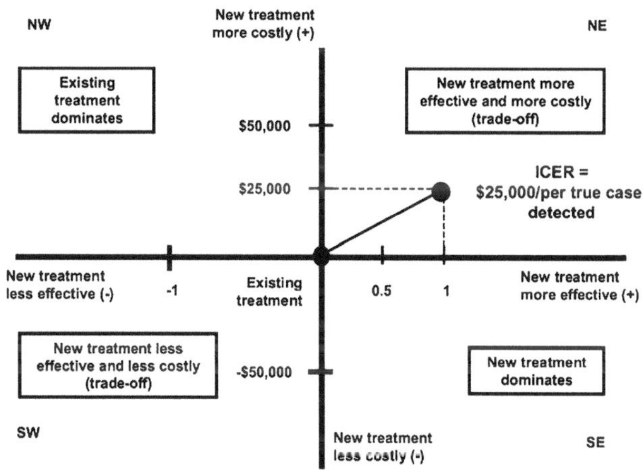

FIGURE 2.—Cost-effectiveness plane. ICER = incremental cost-effectiveness ratio. (Reprinted from Tarride J-E, Blackhouse G, Bischof M, et al. Approaches for economic evaluations of health care technologies. *J Am Coll Radiol.* 2009;6:307-316, with permission from the American College of Radiology.)

technology dominates the new technology in the upper left (more costly and less effective). The 2 remaining quadrants represent situations in which the new treatment is either less costly and less effective (lower left quadrant) or more costly and more effective than the reference treatment (upper right quadrant).

The article provides a good review of several other topics worth considering, including limitations of such analyses and cost-effectiveness acceptability curves.

A. D. Elster, MD

Cumulative Radiation Exposure and Cancer Risk Estimates in Emergency Department Patients Undergoing Repeat or Multiple CT

Griffey RT, Sodickson A (Washington Univ School of Medicine, St. Louis, MO; Brigham and Women's Hosp and Harvard Med School, Boston, MA)

Am J Roentgenol 192:887-892, 2009

Objective.—The purpose of our study was to define a conservative estimate of the number of patients undergoing repeat or multiple emergency department CT studies and to quantify their cumulative CT radiation doses and lifetime attributable risk of developing cancer.

Materials and Methods.—We identified all patients at a tertiary care adult academic medical center with at least three emergency department visits within a 1-year period that included CT of the neck, chest, abdomen, or pelvis. For this cohort, we identified all diagnostic CT studies over the previous 7.7 years. We calculated cumulative radiation doses by summing typical effective doses of the anatomic regions scanned, and we calculated lifetime attributable risk using the population-averaged dose-to-risk conversion factor of one cancer per 1,000 patients receiving a 10-mSv dose, in accordance with the seventh Biologic Effects of Ionizing Radiation (BEIR VII) report.

Results.—One hundred thirty emergency department patients met the inclusion criteria. Over the 7.7-year period, median, mean, and maximum values for the study count were 10, 13, and 70 with cumulative CT doses of 91, 122, and 579 mSv and lifetime attributable risk of one in 110, one in 82, and one in 17, respectively. Emergency department studies comprised 55% of those captured. Repeat imaging of the same study type represented at least half of the imaging for 72% of the cohort and all of the imaging for 12%.

Conclusion.—A small proportion (1.9%) of emergency department patients undergoing CT of the neck, chest, abdomen, or pelvis have high cumulative rates of multiple or repeat imaging. Collectively, this patient subgroup may have a heightened risk of developing cancer from cumulative CT radiation exposure.

▶ This is a nice companion to the *New England Journal of Medicine* article by Fazel et al.[1] The numbers are significantly smaller (130 patients vs 655 000) and the cases are confined to an emergency department at a single center; still, the conclusions and implications are the same. A small group of patients receive surprisingly high cumulative doses of radiation from routine diagnostic procedures (Fig 1 in the original article).

In the emergency room setting, body CTs account for the vast majority of radiation exposure. Typical examples include repeated chest CTs for chest pain or pulmonary embolism, and repeated abdominal and pelvic CTs for abdominal pain and urinary tract calculi evaluation.

The discussion section of this report provides a nice review of low-dose radiation effects reported in the scientific literature and putative associations with solid cancers and leukemia.

A. D. Elster, MD

Reference

1. Fazel R, Krumholz HM, Wang Y, et al. Exposure to low-dose ionizing radiation from medical imaging procedures. *N Engl J Med.* 2009;361:849-857.

Do Clinicians Use the American College of Radiology Appropriateness Criteria in the Management of Their Patients?

Bautista AB, Burgos A, Nickel BJ, et al (Robert Wood Johnson Univ Hosp and Robert Wood Johnson Med School/UMDNJ, New Brunswick, NJ)

AJR Am J Roentgenol 192:1581-1585, 2009

Objective.—The purpose of this study was to investigate the use of the American College of Radiology (ACR) appropriateness criteria by referring physicians during decision making about imaging in the management of their patients.

Conclusion.—There is a low utilization of the ACR appropriateness criteria by clinicians when ordering imaging studies for their patients. The ACR has invested a great deal of resources in these criteria and should therefore be aware of information regarding utilization. Our findings may have implications about how the ACR appropriateness criteria are reviewed, revised, and disseminated.

▶ In 1993, the American College of Radiology (ACR) developed a set of criteria, scientific-based guidelines intended to instruct referring physicians about the favorable use of diagnostic and interventional radiology for given clinical situations.[1] Since that time, the ACR appropriateness criteria have been continuously updated. This article reports the result of a mail survey to 438 nonradiology physicians to determine their knowledge and use of the ACR criteria.

Respondents were asked to rank in order the methods they use when deciding whether to order one of several tests in confusing situations. Gratifyingly, consultation with a radiologist ranked #1, cited by nearly 2/3 of respondents. The second most popular response was use of the physician's own specialty journal (48%), followed by UpToDate (41%). The ACR appropriateness criteria ranked dead last, 11/11, cited by only 3% of physicians. Well ahead of the ACR criteria were Google and other.

Not surprisingly, the authors discovered that most clinicians are either unaware of the ACR appropriateness criteria or do not depend on them in their daily decision-making. Perhaps placing these in the medical curriculum as an available resource would be useful in the future.

A. D. Elster, MD

Reference

1. Bettmann M. The ACR appropriateness criteria: view from the committee chair. *J Am Coll Radiol.* 2006;3:510-512.

Exposure to Low-Dose Ionizing Radiation from Medical Imaging Procedures

Fazel R, Krumholz HM, Wang Y, et al (Emory Univ School of Medicine, Atlanta, GA; Yale Univ School of Medicine, New Haven, CT)

N Engl J Med 361:849-857, 2009

Background.—The growing use of imaging procedures in the United States has raised concerns about exposure to low-dose ionizing radiation in the general population.

Methods.—We identified 952,420 nonelderly adults (between 18 and 64 years of age) in five health care markets across the United States between January 1, 2005, and December 31, 2007. Utilization data were used to estimate cumulative effective doses of radiation from imaging procedures and to calculate population-based rates of exposure, with annual effective doses defined as low (≤3 mSv), moderate (>3 to 20 mSv), high (>20 to 50 mSv), or very high (>50 mSv).

Results.—During the study period, 655,613 enrollees (68.8%) underwent at least one imaging procedure associated with radiation exposure. The mean (±SD) cumulative effective dose from imaging procedures was 2.4 ± 6.0 mSv per enrollee per year; however, a wide distribution was noted, with a median effective dose of 0.1 mSv per enrollee per year (interquartile range, 0.0 to 1.7). Overall, moderate effective doses of radiation were incurred in 193.8 enrollees per 1000 per year, whereas high and

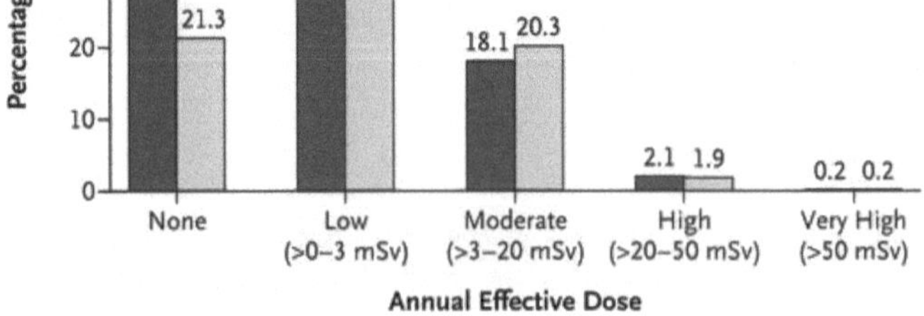

FIGURE 1.—Overall distribution of annual effective doses of radiation in the study population, stratified according to sex. Percentages may not total 100 because of rounding. (Reprinted from Fazel R, Krumholz HM, Wang Y, et al. Exposure to low-dose ionizing radiation from medical imaging procedures. *N Engl J Med.* 2009;361:849-857.)

TABLE 3.—Medical Imaging Procedures with Largest Contribution to Cumulative Effective Dose*

Procedure	Average Effective Dose (millisieverts)	Annual Effective Dose per Person (millisieverts)	Proportion of the Total Effective Dose from All Study Procedures %
Myocardial perfusion imaging	15.6†	0.540	22.1
CT of the abdomen	8	0.446	18.3
CT of the pelvis	6	0.297	12.2
CT of the chest	7	0.184	7.5
Diagnostic cardiac catheterization	7	0.113	4.6
Radiography of the lumbar spine	1.5	0.080	3.3
Mammography	0.4	0.076	3.1
CT angiography of the chest (noncoronary)	15	0.075	3.1
Upper gastrointestinal series	6	0.058	2.4
CT of the head or brain	2	0.049	2.0
Percutaneous coronary intervention	15	0.043	1.8
Nuclear bone imaging	6.3	0.035	1.4
Radiograph of the abdomen	0.7	0.028	1.1
CT of the cervical spine	6	0.020	0.8
CT of the lumbar spine	6	0.018	0.7
Chest radiograph	0.02‡	0.016	0.7
Thyroid uptake	1.9	0.016	0.7
Intravenous urography	3	0.014	0.6
CT of the neck	3	0.014	0.6
Cardiac resting ventriculography	7.8	0.014	0.6

Editor's Note: Please refer to original journal article for full references.

*Average effective doses for these imaging procedures are based on data from Mettler et al.

†Calculation of the average radiation dose for myocardial perfusion imaging with the use of single-photon-emission CT relied on dose coefficients from a detailed review of radiation dosimetry of specific cardiac radiopharmaceuticals, median injected radiopharmaceutical doses (millicuries) from the guidelines of the American Society of NuclearCardiology, and distributions of the use of various protocols in the United States.

‡This dose is the effective dose for a posteroanterior study of the chest.

very high doses were incurred in 18.6 and 1.9 enrollees per 1000 per year, respectively. In general, cumulative effective doses of radiation from imaging procedures increased with advancing age and were higher in women than in men. Computed tomographic and nuclear imaging accounted for 75.4% of the cumulative effective dose, with 81.8% of the total administered in outpatient settings.

Conclusions.—Imaging procedures are an important source of exposure to ionizing radiation in the United States and can result in high cumulative effective doses of radiation.

▶ When an article on radiology or radiation exposure hits the *New England Journal of Medicine* level, people sit up and take attention, even when the same or substantially similar information has been published elsewhere previously. Such is the case in this article, which made national television news the week it appeared. Unlike other articles discussing radiation exposure and cancer, this article did not claim to show diagnostic radiology studies were causing cancer or mutations. The purpose was to show the levels of exposure a large cohort (nearly 1 million subjects) received over a 2-year period.

Abundant epidemiologic evidence exists linking relatively low-dose ionizing radiation with the development of cancer.[1] Accordingly, workers occupationally exposed to radiation are monitored and typically restricted to effective doses of 20 mSv per year.[2,3]

The *New England Journal of Medicine* study is useful to estimate the relative numbers of patients receiving moderate or high levels of radiation for diagnostic purposes (Fig 1). Myocardial perfusion imaging and CT of the abdomen and pelvis were the biggest offenders (Table 3). In summary, this article serves as a reminder to physicians to use medical radiation procedures judiciously.

A. D. Elster, MD

References

1. National Research Council. *Health Risks From Exposure to Low Levels of Ionizing Radiation: BEIR VII Phase 2*. Washington, DC: National Academies Press; 2006.
2. The 2007 recommendations of the International Commission on Radiological Protection: ICRP publication 103. *Ann ICRP.* 2007;37:1-332.
3. Wrixon AD. New ICRP recommendations. *J Radiol Prot.* 2008;28:161-168.

National audit of provision of MRI services 2006/07

Barter S, Drinkwater K, Remedios D (Royal College of Radiologists, London, UK)

Clin Radiol 64:284-290, 2009

In 2003 the Royal College of Radiologists Clinical Radiology Audit Sub-Committee began an audit process evaluating the standards of provision of magnetic resonance imaging (MRI) services. This was prompted by the publication of the 2002 Audit Commission Report, which had identified that lack of MRI provision was responsible for more than half of the total waiting times for diagnostic imaging investigations. The audit found that the time from request to report did not meet the standard for cancer staging examinations, but nationally, was within the target set for routine orthopaedic examinations. However, national mean waiting times were longer than recommended for both cancer and orthopaedic MRI. Since then, there has been massive investment in MRI capacity, both from installation of MRI systems in NHS Trusts, and in England, from outsourcing of routine MRI cases through the Department of Health contract with an independent provider. A re-audit in 2006/7 shows that there has been a significant improvement in waiting times for routine orthopaedic examinations, but the position with cancer staging examinations has deteriorated. Control chart methodology shows that underperformance is due to common cause variation, i.e., improvements need to be made to the overall process from receiving the request for MRI to the issue of the report. Follow-up with participating departments demonstrated there

TABLE 1.—Mean (SD) Percentage of Cancer Staging Cases Reported Within 14 Days and Mean (SD) Percentage of Routine Orthopaedic Cases Reported Within 13 Weeks Comparing 2003 and 2006

	2003	2006/07
Cancer staging cases reported within 14 days of request	62%	62% (31.3)
Routine orthopaedic cases reported within 13 weeks of request	66%	81% (28.6)

were some common themes for underperformance, and suggestions for improvement are made from departments with best performance.

▶ The United States is often accused of having an excess of high-technology imaging devices, which has led to overuse of services and increasing cost to the health care system. The situation is often claimed to be the reverse in countries such as the United Kingdom where there is much greater national regulation of the number and distribution of such devices. The important question raised by this article is whether patient care has been adversely affected by a relative shortage of MR scanners in the United Kingdom.

The situation came to a head in 2002, when an Audit Commission documented a number of significant problems, including unequal access and huge variations in waiting times for both urgent and nonurgent studies. The median MRI waiting time was 20 weeks, but in 1 in 4 departments it exceeded 34 weeks.

The Department of Health responded by doubling the capacity of MRI scanners from 42 to 92, and this study was undertaken to determine whether the effect had been beneficial. The goals for success were surprisingly modest by United States standards: (1) cancer staging cases should be reported within 14 days of request in 95% of cases, and (2) routine orthopedic scans should be reported within 13 weeks of request in 50% of cases.

Even with such low standards, compliance was not obtained nationwide (Table 1). Only 57% of cancer staging cases were reported within 14 days of request and 82% of routine orthopedic cases were reported within 13 weeks of request. Thus, compliance fell short of the target for cancer staging cases (95%) by 38%, but exceeded the target for routine orthopedic cases (50%) by 32%.

Would physicians in the United States be happy or disgusted by such response rates? I will leave that conclusion to the reader to judge the success of the British system in managing imaging needs.

A. D. Elster, MD

Prospective Evaluation of Whole-Body Cancer Screening With Multiple Modalities Including [^{18}F]Fluorodeoxyglucose Positron Emission Tomography in a Healthy Population: A Preliminary Report

Nishizawa S, Kojima S, Teramukai S, et al (Hamamatsu Med Imaging Ctr; Development Bureau and Central Res Laboratory, Shizuoka; Translational Res Informatics Ctr, Hyogo; Kyoto Univ Hosp, Japan)

J Clin Oncol 27:1767-1773, 2009

Purpose.—To prospectively evaluate the utility of whole-body cancer screening with multiple modalities including [^{18}F]fluorodeoxyglucose positron emission tomography (FDG-PET) in a healthy population. This report summarizes the results of the first three annual screenings.

Participants and Methods.—A total of 1,197 healthy volunteers ≥ 35 years old were enrolled between August 2003 and July 2004 and offered annual cancer screening for 5 years with subsequent long-term follow-up. Screening modalities included were whole-body FDG-PET, chest and abdominal computed tomography (CT), brain and pelvic magnetic resonance imaging, several tumor markers, and fecal occult blood testing.

Results.—As of the end of 2006, 22 primary cancers were pathologically confirmed. Nineteen of 22 were detected by the screening; 18 in the initial, one in the second, and none in the third. Three were diagnosed after development of symptoms. Of the 18 detected in the initial screening (six thyroid, four lung, three prostate, three breast, one endometrial, and one thymic), 12 were at stage I and 11 were PET positive. PET-negative cancers were detected by CT or the prostate-specific antigen (PSA) test. Sensitivity and specificity were 50.0% (11 of 22) and 93.2% (1,095 of 1,175), respectively, for FDG-PET alone and 81.8% (18 of 22) and 82.0% (963 of 1,175), respectively, for the combination of imaging modalities and PSA.

Conclusion.—While FDG-PET alone is insufficient, whole-body cancer screening with selected modalities including FDG-PET has initial performance supporting possible utility by detecting a wide variety of early-stage cancers with reasonable sensitivity. However, the detection of many indolent cancers and false positives necessitate continuing study for appropriate evaluation.

▶ Whole-body screening for cancer using imaging techniques remains a controversial topic. Previous studies have evaluated the roles of CT and MRI to determine their cost-effectiveness for this purpose. Although no uniformly accepted conclusions have resulted, there is general agreement that some new cancers, often in their early stages, are detected by screening techniques. However, it remains unclear whether patient outcome is improved by earlier diagnosis. Moreover, there is morbidity and expense of working up the numerous incidental noncancerous lesions that are commonly encountered during screening.

This article, based on a cohort of patients in Japan, is interesting because it is the first large-scale prospective screening study to include PET/CT. A few previous Japanese studies have also screened patients for cancer with PET only, claiming detection rates in the 1% to 3% range.[1,2] The most commonly detected occult cancers identified by [^{18}F]fluorodeoxyglucose (FDG)-PET have been thyroid, colorectal, lung, and breast.

The types of cancers detected in this study somewhat differed from the previous studies in that no colorectal, breast, or FDG-PET positive cancers were detected. This likely related from selection bias, in that most participants had received annual health check-ups including chest x-ray, upper gastrointestinal series, and mammography for many years—all effective for detecting cancers.

Due to its incremental high expense and the relatively poor yield, PET does not appear to have a role in population screening for cancer, except perhaps in some very specific subpopulations.

A. D. Elster, MD

References

1. Yasuda S, Shoutsu A. Cancer screening with whole-body 18F-fluorodeoxyglucose positron-emission tomography. *Lancet.* 1997;350:1819.
2. Yasuda S, Ide M, Fujii H, et al. Application of positron emission tomography imaging to cancer screening. *Br J Cancer.* 2000;83:1607-1611.

Recurrent CT, Cumulative Radiation Exposure, and Associated Radiation-induced Cancer Risks from CT of Adults

Sodickson A, Baeyens PF, Andriole KP, et al (Brigham and Women's Hosp, Boston, MA; et al)

Radiology 251:175-184, 2009

Purpose.—To estimate cumulative radiation exposure and lifetime attributable risk (LAR) of radiation-induced cancer from computed tomographic (CT) scanning of adult patients at a tertiary care academic medical center.

Materials and Methods.—This HIPAA-compliant study was approved by the institutional review board with waiver of informed consent. The cohort comprised 31 462 patients who underwent diagnostic CT in 2007 and had undergone 190 712 CT examinations over the prior 22 years. Each patient's cumulative CT radiation exposure was estimated by summing typical CT effective doses, and the Biological Effects of Ionizing Radiation (BEIR) VII methodology was used to estimate LAR on the basis of sex and age at each exposure. Billing ICD9 codes and electronic order entry information were used to stratify patients with LAR greater than 1%.

Results.—Thirty-three percent of patients underwent five or more lifetime CT examinations, and 5% underwent between 22 and 132 examinations. Fifteen percent received estimated cumulative effective doses of more than 100 mSv, and 4% received between 250 and 1375 mSv. Associated LAR had mean and maximum values of 0.3% and 12% for cancer

TABLE 4.—Sample Patient Detail in Non-Cancer Patients

Patient No./ Sex/Age (y)	LAR Incidence (%)	LAR Mortality (%)	No. of Examinations Total	Abdomen-Pelvis	Chest	Head	Spine	Abstracted Medical History
1/F/45	9.6	5.3	70	69	0	1	0	Recurrent pyelonephritis, lithotripsy, stone extraction, ulcerative colitis
2/F/49	5.2	3.1	81	27	24	21	1	End-stage renal disease, hemodialysis, lupus
3/F/34	4.4	2.2	35	21	10	1	0	Chronic pancreatitis and Whipple operation, ventral hernias, mesh infections, empyema
4/F/58	3.9	2.5	88	20	20	35	7	Lupus, osteogenesis imperfecta, seizures, stroke
5/F/46	3.8	2.0	28	26	1	1	0	Crohn, hemicolectomy, chronic abdominal pain, heart transplant
6/M/33	3.6	2.0	36	32	2	1	0	Necrotizing pancreatitis, enterocutaneous fistula
7/M/27	2.7	1.4	37	7	22	4	3	Cystic fibrosis, lung transplants
8/F/42	2.5	1.4	38	11	10	12	0	Lupus, asthma, seizures, pulmonary embolus, pulmonary hypertension
9/F/34	2.4	1.3	16	15	1	0	0	Gastric bypass, small bowel obstructions
10/F/24	2.2	1.1	45	2	5	36	0	Acute disseminated encephalomyelitis, ventriculoperitoneal shunt
11/M/54	2.0	1.2	37	19	2	12	0	Renal transplant, hemodialysis, abdominal abscess
12/F/44	1.9	1.1	48	3	21	5	0	Recurrent chest pain, sickle-β thalassemia, emphysema, pulmonary embolus
13/F/55	1.9	1.2	37	11	5	18	1	End-stage renal disease, human immunodeficiency virus, ruptured middle cerebral artery aneurysm
14/F/81	1.7	1.4	40	30	8	0	0	Perforated diverticulitis, congestive heart failure, pulmonary hypertension
15/F/62	1.6	1.1	29	11	10	5	0	Bronchioloitis obliterans organizing pneumonia, perinephric hematoma
16/F/49	1.5	0.9	17	9	1	3	2	Diverticulitis, colectomy, ventral hernia repair, recurrent abdominal pain
17/F/53	1.4	0.9	13	4	0	0	8	Lumbar fusion, back pain
18/F/58	1.4	0.9	18	12	2	1	3	Diverticulitis, trauma
19/F/57	1.3	0.9	19	10	4	4	0	Sarcoidosis, pulmonary hypertension, intraductal papillary mucinous neoplasm
20/F/56	1.2	0.8	24	6	11	2	0	Endocarditis, aortic valve replacement, hepatitis C
21/F/58	1.1	0.7	25	5	9	5	2	Chronic obstructive pulmonary disease, chronic pancreatitis
22/F/31	1.1	0.5	9	5	1	2	0	Flank pain, asthma, migraines
23/F/47	1.1	0.6	21	6	1	12	1	Perforated appendicitis and complications, stroke, ventriculoperioneal shunt
24/M/55	1.1	0.6	19	8	7	2	1	Type A aortic dissection repair
25/M/53	1.0	0.6	36	6	4	19	5	Alcoholic, frequent trauma

incidence and 0.2% and 6.8% for cancer mortality, respectively. CT exposures were estimated to produce 0.7% of total expected baseline cancer incidence and 1% of total cancer mortality. Seven percent of the cohort had estimated LAR greater than 1%, of which 40% had either no malignancy history or a cancer history without evidence of residual disease.

Conclusion.—Cumulative CT radiation exposure added incrementally to baseline cancer risk in the cohort. While most patients accrue low radiation-induced cancer risks, a subgroup is potentially at higher risk due to recurrent CT imaging.

▶ This is our second companion piece to the *New England Journal of Medicine* report by Fazel et al, "Exposure to Low-Dose Ionizing Radiation from Medical Imaging Procedures,"[1] also reviewed herein. The number of patients in the radiology study was 31 462, and the conclusions are similar—too many patients are receiving substantial doses of radiation from diagnostic tests, especially CT scanning of the chest, abdomen, and pelvis.

Some of the individual stories in this reported cohort are more significant and memorable than the aggregated data (Table 4). For example, a 45-year-old woman with ulcerative colitis and urinary tract infections and calculi who received 69 abdominal and pelvic CTs or a 58-year-old woman with lupus who received 88 scans.

Overall, it was concluded that about 7% of the population had received sufficient radiation exposure from diagnostic tests to place them at risk for development of cancers. The authors hope that patient-specific cumulative radiation risk estimates can be used to better inform risk-benefit decisions in balance with an individual's clinical presentation and the anticipated benefits of recurrent imaging.

A. D. Elster, MD

Reference

1. Fazel R, Krumholz HM, Wang Y, et al. Exposure to low-dose ionizing radiation from medical imaging procedures. *N Engl J Med.* 2009;361:849-857.

The Use of CT for Screening: A National Survey of Radiologists' Activities and Attitudes

Burger IM, Kass NE, Sunshine JH, et al (Johns Hopkins Bloomberg School of Public Health, Baltimore, MD; American College of Radiology, Fairfax, VA; et al)

Radiology 248:160-168, 2008

Purpose.—To investigate the activities, motivations, and attitudes of radiologists regarding specific computed tomographic (CT) screening examinations by using a survey.

Materials and Methods.—All study activities were approved by the institutional review board. A self-administered, mailed survey was used to collect data on the practices and attitudes of U.S. radiologists regarding three CT screening tests—coronary artery calcium scoring (CACS), lung cancer screening CT, and whole-body screening CT. The survey was sent to 1000 diagnostic radiologists who were randomly sampled from the American Medical Association Physician Masterfile.

Results.—A total of 398 (41.4%) of 961 eligible radiologists completed the survey. Among respondents, 33.6% reported reading CT screening studies, the most common being CACS (26.7%), followed by lung screening (19.2%) and whole-body screening (9.5%). Among respondents, 34.1% supported CACS and 29.9% supported lung CT screening for particular patients, while 1.9% supported whole-body CT screening. The most common reasons reported for reading CT screening studies were responses to requests from physicians (83.3%) or patients (75.0%), while fewer (40.8%) cited patient benefit from screening as a reason.

Conclusion.—A substantial proportion of a nationally representative sample of radiologists in the United States reads CT screening studies of the heart, lungs, and whole body and holds favorable attitudes toward CACS and lung CT screening. These attitudes may allow for the premature diffusion of new screening tests into practice before higher-level evidence demonstrates their benefits for population mortality.

▶ CT screening remains a controversial topic. Fairly good evidence demonstrates that nearly one third of patients undergoing whole-body CT screening required follow-up of important findings,[1] and results of 1 cost-effectiveness analysis[2] suggest that screening CT has minimal impact on population-based mortality at high cost. Nearly all professional societies have taken official stances against whole-body CT screening.

This study was undertaken to ascertain the opinions practicing radiologists have about limited and whole-body CT screening, yielding several interesting conclusions.

The first surprise is how frequent screening CT is. About a third of United States radiologists who read diagnostic CT images of the chest, abdomen, or pelvis report also reading CT screening studies of the same areas. Why do they do this? They report that response to referring physician and patient request more important motivations for reading CT screening studies than the belief that patients actually benefit from screening. Relatively few surveyed radiologists are in favor of whole-body CT screening, but most support coronary artery and lung cancer screening.

As experts in imaging, radiologists have an obligation to determine whether and how CT screening examinations enter practice. The diffusion of CT screening ahead of strong empirical evidence support should warrant concern and introspection on behalf of all radiologists.

A. D. Elster, MD

References

1. Furtado CD, Aguirre DA, Sirlin CB, et al. Whole-body CT screening: spectrum of findings and recommendations in 1192 patients. *Radiology*. 2005;237:385-394.
2. Beinfeld MT, Wittenberg E, Gazelle GS. Cost-effectiveness of whole-body CT screening. *Radiology*. 2005;234:415-422.

Clinical Practice

A Survey of Bedside Ultrasound Use by Emergency Physicians in California

Stein JC, River G, Kalika I, et al (Univ of California, San Francisco; et al)
J Ultrasound Med 28:757-763, 2009

Objective.—The purpose of this study was to investigate the current practice of emergency physician–performed bedside ultrasound examinations in California and to assess differences between academic and community practice.

Methods.—We queried all emergency departments (EDs) in California to determine whether bedside ultrasound was used by emergency physicians. Among EDs that were using bedside ultrasound, we administered a survey to assess use patterns, credentialing criteria, and quality assurance (QA) programs.

Results.—We contacted all eligible EDs ($n = 293$) by telephone and had a 100% response rate for our primary question: 101 EDs (34%) reported use of bedside ultrasound. Of these 101 EDs, 97 (96%) responded to the secondary survey, showing the following: (1) 48% of physicians at each site were credentialed to use ultrasound in at least 1 modality; (2) 70% of EDs used American College of Emergency Physicians (ACEP) criteria for credentialing guidelines; and (3) 33% had an ultrasound QA program. Comparing practice settings, 68% of academic departments used bedside ultrasound compared with 29% of community departments (difference, 39%; 95% confidence interval [CI], 23% to 54%; $P < .0001$). In academic departments, a mean of 60% of physicians were credentialed, compared with 41% in community EDs (difference, 19%; 95% CI, 2.5% to 35%; $P - .036$).

Conclusions.—Most California EDs do not use bedside ultrasound. Although most EDs using ultrasound report that they follow ACEP emergency ultrasound guidelines, most do not have a QA program as recommended by these guidelines. Compared with community EDs, academic EDs are more likely to use bedside ultrasound, have physicians credentialed in ultrasound use, and have QA programs.

▶ The use of bedside ultrasound by emergency physicians has increased exponentially over the last 20 years.[1] This trend has been driven primarily by academic emergency medicine environment. Ultrasound training for emergency physicians is now a mandate of the Residency Review Committee of the Accreditation Council for Graduate Medical Education. As a result of these changes, by 2003 over 95% of academic emergency medicine programs were teaching ultrasound, and 92% reported 24-hour availability of bedside sonographic services by academic emergency physicians.[2]

A number of societies have developed policy statements for both use of bedside ultrasound and practitioner training. These include the American College of Emergency Physicians (ACEP), the Society of Academic Emergency Physicians, the American Institute of Ultrasound in Medicine (AIUM), and the American Medical Association (AMA).

As academic settings account for only a small percentage of total emergency rooms in the nation, this study was designed to survey to what extent the use of bedside ultrasound had diffused from academic centers into the community environment. The results are not surprising, in that academic centers used ultrasound about twice as frequently as community emergency departments (EDs). Both types of sites were found lacking credentialing and training standards, as well as quality assurance mechanisms. With increasing numbers of emergency medicine residents graduating each year, it is certain that the use of bedside ultrasound in all EDs will continue to increase in the future.

A. D. Elster, MD

References

1. Cardenas E. Emergency medicine ultrasound policies and reimbursement guidelines. *Emerg Med Clin North Am.* 2004;22:829-838.
2. Counselman FL, Sanders A, Slovis CM, Danzl D, Binder LS, Perina DG. The status of bedside ultrasonography training in emergency medicine residency programs. *Acad Emerg Med.* 2003;10:37-42.

Adverse Events in a Freestanding Radiology Office

Shah-Patel LR, Piraner M, Silberzweig JE (St Luke's-Roosevelt Hosp Ctr, NY)
J Am Coll Radiol 6:263-267, 2009

Objective.—The aim of this study was to evaluate the frequency, types, and severity of adverse events that occur in a freestanding outpatient radiology office setting and discuss the actions taken for patient treatment and subsequent triage.

Materials and Methods.—Adverse event records from a freestanding outpatient radiology office were reviewed from January 2003 through December 2006. Data were grouped into different modalities, including computed tomography, magnetic resonance imaging, positron emission tomography/computed tomography, and interventional radiology. Adverse reactions and events were reviewed, including what, if any, immediate treatment was provided in the office, how the patient responded, whether the patient was discharged home, and whether emergency medical services were called.

Results.—A total of 106,800 patients were evaluated over 4 years, among whom 59 experienced (0.06%) adverse events. Seventy-six percent of the patients ($n = 45$) were treated in the office and were discharged home or returned to work. Thirty-one percent of the patients (n = 18) required assistance from emergency medical services.

Conclusion.—The outpatient radiology office setting is a safe arena for patients to undergo imaging studies and selected interventional procedures. Medical personnel in office settings are able to assess and treat those patients experiencing adverse events and can appropriately judge whether further aid is needed for patients in distress via emergency medical services or hospital care.

▶ This article describes the authors' personal experience over a 4-year period with adverse medical events occurring at a free-standing outpatient center in Manhattan. The facility provided a full menu of diagnostic procedures (CT, MR, PET/CT, ultrasound, mammography, DEXA). Of even greater interest is that the facility also performs a number of low-risk interventional procedures, including breast and thyroid biopsies, selective hysterosalpingography and guide-wire fallopian tube recanalization, central venous access procedures (ports, PICC, and hemodialysis catheter placement), arthrography, various hemodialysis shunt interventions (angioplasty, stent, thrombectomy), and venous insufficiency therapies (endovenous laser and sclerotherapy).

The adverse event rate for the approximately 100 000 patients treated was very low (0.6%). The most serious reactions were shortness of breath, chest pain, and intravenous (IV) infiltrations. The staff had a low threshold for calling in emergency medical personnel for transport to the hospital.

In conclusion, with good patient selection and training of personnel, a wide range of diagnostic and low risk interventional procedures can be safely performed in the outpatient arena.

A. D. Elster, MD

Financial Impact of Emergency Department Ultrasound

Soremekun OA, Noble VE, Liteplo AS, et al (Brigham & Women's Hosp and Massachusetts General Hosp, Boston; Massachusetts General Hosp and Harvard Med School, Boston)

Acad Emerg Med 16:674-680, 2009

Objectives.—There is limited information on the financial implications of an emergency department ultrasound (ED US) program. The authors sought to perform a fiscal analysis of an integrated ED US program.

Methods.—A retrospective review of billing data was performed for fiscal year (FY) 2007 for an urban academic ED with an ED US program. The ED had an annual census of 80,000 visits and 1,101 ED trauma activations. The ED is a core teaching site for a 4-year emergency medicine (EM) residency, has 35 faculty members, and has 24-hour availability of all radiology services including formal US. ED US is utilized as part of evaluation of all trauma activations and for ED procedures. As actual billing charges and reimbursement rates are institution-specific and proprietary information, relative value units (RVUs) and reimbursement based on the Centers for Medicare & Medicaid Services (CMS) 2007 fee schedule (adjusted for fixed diagnosis-related group [DRG] payments and bad debt) was used to determine revenue generated from ED US. To estimate potential volume, assumptions were made on improvement in documentation rate for diagnostic scans (current documentation rates based on billed volume versus diagnostic studies in diagnostic image database), with no improvements assumed for procedural ED US. Expenses consist of three components—capital costs, training costs, and ongoing operational costs—and were determined by institutional experience. Training costs were considered sunken expenses by this institution and were thus not included in the original return on investment (ROI) calculation, although for this article a second ROI calculation was done with training cost estimates included. For the purposes of analysis, certain key assumptions were made. We utilized a collection rate of 45% and hospitalization rates (used to adjust for fixed DRG payments) of 33% for all diagnostic scans, 100% for vascular access, and 10% for needle placement. An optimal documentation rate of 95% was used to estimate potential revenue.

Results.—In FY 2007, 486 limited echo exams of abdomen (current procedural terminology [CPT] 76705) and 480 limited echo cardiac exams were performed (CPT 93308) while there were 78 exams for US-guided vascular access (CPT 76937) and 36 US-guided needle placements when performing paracentesis, thoracentesis, or location of abscess for drainage (CPT 76492). Applying the 2007 CMS fee schedule and above assumptions, the revenue generated was 578 RVUs and $35,541 ($12,934 in professional physician fees and $22,607 in facility fees). Assuming optimal documentation rates for diagnostic ED US scans, ED US could have generated 1,487 RVUs and $94,593 ($33,953 in professional physician fees and $60,640 in facility fees). Program expenses include an initial capital expense (estimated at $120,000 for two US machines) and ongoing operational costs ($68,640 per year to cover image quality assurance review, continuing education, and program maintenance). Based on current revenue, there would be an annual operating loss, and thus an ROI cannot be calculated. However, if potential revenue is achieved, the annual operating income will be $22,846 per year with an ROI of 4.9 years to break even with initial investment.

Conclusions.—Determining an ROI is a required procedure for any business plan for establishing an ED US program. Our analysis demonstrates that an ED US program that captures charges for trauma and procedural US and achieves the potential billing volume breaks even in less than 5 years, at which point it would generate a positive margin.

▶ Radiologists should read this article just to know what their emergency department physician colleagues are thinking. This is an economic analysis that looks at the financial feasibility of ED physicians at a Harvard-based academic center of owning, performing, and billing for ultrasound services in the emergency department.

The study assumes an initial purchase for two $60 000 machines. There is also a budget for ED physician training, which involves them each taking a 2-day course to master the field. It was also assumed that the ED physicians themselves would be performing the examinations without the use of a trained technologist. The analysis is based on billing primarily for the following codes: CPT-76205 (Limited ultrasound [US] study of the abdomen), CPT-93308 (Limited echocardiography), CPT-76675 (Limited US of retroperitoneum), CPT-76937 (US for vascular access), and CPT-76942 (US for needle placement for thoracentesis, abscess aspiration, etc).

Using current volumes at their center (about 1000 diagnostic and 1000 interventional procedures per year), there was a negative return on investment (meaning a net financial loss). However, when diagnostic volumes were significantly increased (to 2500 per year) the finances reversed and a positive return on investment with 5-year payback was calculated.

To me, this study reveals that at reasonable current volumes, it is not cost efficient for ED physicians to perform and bill for their own studies. However, if volumes were to double or triple via a self-referral mechanism, suddenly the service would generate a profit.

Surprise....Surprise.

A. D. Elster, MD

Predictors of Radiologists' Perceived Risk of Malpractice Lawsuits in Breast Imaging

Dick JF III, Gallagher TH, Brenner RJ, et al (Dartmouth Med School, Hanover, NH; Univ of Washington School of Medicine, Seattle; Bay Imaging Consultants, Oakland, CA; et al)

Am J Roentgenol 192:327-333, 2009

Objective.—The shortage of radiologists in breast imaging may be related to heightened malpractice lawsuit concerns. Our objective was to examine radiologists' reported experiences and perceptions of future lawsuit risk and explore personal and professional factors that may be associated with elevated perceptions of risk.

Materials and Methods.—Radiologists who routinely interpret mammography examinations in diverse regions of the United States (Washington, Colorado, and New Hampshire) completed a mailed survey in 2002 and 2006, including questions on demographics, practice characteristics, and medicolegal experience and perceptions as well as a validated scale measuring reactions to uncertainty in clinical situations. A survey assessing the radiologists' work facilities was also completed in 2002.

Results.—Participation by eligible radiologists was 77% (139 of 181) in 2002 and 71% (84 of 118) in 2006. The percentage of radiologists reporting malpractice claims related to mammography in the previous 5 years was 8% on the 2002 survey and 10% on the 2006 survey. Radiologists' mean estimate of the probability of being sued for malpractice in the next 5 years (41% in 2002 and 35% in 2006) was markedly higher than the actual reported risk. Radiologists' age, sex, clinical experience, and workload were not associated with a higher perceived risk of being sued. Radiologists who reported higher perceived risk of lawsuits were more likely to have experienced a prior malpractice claim, to report knowing colleagues with prior lawsuits, and to score higher on a scale measuring anxiety caused by uncertainty in clinical situations. Radiologists working at facilities that did not use double reading reported higher perceived risk, but the difference was not statistically significant.

Conclusion.—Radiologists working in breast imaging substantially overestimate their risk of a future malpractice lawsuit. Radiologists with higher risk perceptions show more negative reactions to uncertainty in a clinical setting. Understanding that their actual risk of malpractice lawsuits may be substantially lower than anticipated may help reduce radiologists' fears and alleviate the manpower shortage in mammography. Programs to address the shortage of breast imagers could be targeted toward radiologists with heightened malpractice lawsuit concerns.

▶ Three reasons make this a particularly relevant review: (1) According to recent surveys, about two thirds of radiologists practicing in the United States perform some type of breast imaging; (2) According to the Physicians Insurers Association of America (PIAA), a large consortium of physician-owned liability carriers, delay in diagnosis of breast cancer is the most common reason that physicians are subject to malpractice lawsuits; and (3) In these lawsuits, radiologists are generally the principal defendants.

Radiologists perceive the risk for a lawsuit related to mammography to be about 4 times higher than the reported rate of malpractice claims—41% perceived versus 10% actual. Perception is powerful, perhaps everything, and this abnormally perceived risk of litigation may be driving radiologists out of mammography or increasing their dissatisfaction with such work.

A. D. Elster, MD

Survey of hospital clinicians' preferences regarding the format of radiology reports

Plumb AAO, Grieve FM, Khan SH (East Lancashire Hosps NHS Trust, Blackburn, UK)

Clin Radiol 64:386-394, 2009

Aim.—To determine hospital consultants' preferences for the format and content of radiology reports.

Materials and Methods.—Ninety-nine questionnaires were sent to consultant staff with responsibility for requesting ultrasound examinations. The participants were invited to rank a variety of hypothetical reports in order of preference. They were also asked whether they felt other commonly included features of a radiology report were of value. Rank data were analysed by the Friedman statistic, Fisher's multiple comparisons least significant difference test, and the Kemeny–Young method.

Results.—Forty-nine responses were received. There was a preference for more detailed reports that included a clinical comment by the radiologist, for both normal and abnormal results ($p < 0.05$). Reports presented in tables were preferred. The combination of a detailed tabular report with a radiologist's comment was the most popular single structure, preferred by 43% of respondents for normal reports and 51% for abnormal reports.

Conclusion.—Detailed reports with a radiologists' comment are preferred to briefer reports, even for normal examinations. Tabular reports are preferred to prose, with the combination of a detailed report presented in a tabular format accompanied by a radiologist's comment being the most preferred style.

▶ All radiology residents learn from faculty mentors early in their careers what constitutes the "ideal" report. Unfortunately, there is little science or psychological analysis to justify what constitutes best practice in reporting. Residents are frequently presented with several divergent viewpoints and sometimes must change their reporting style to match the desires of the mentor they happen to be working with on a particular day.

Viewed in economic terms, a radiology report is a "product" and the referring clinicians are the "consumer." Would it not make sense to ask the consumer what type of product he or she prefers? Such is the impetus behind the current investigation.

The results are relatively straightforward: tables preferred to prose; detail opposed to superficiality, even for normal studies; and comments by the radiologist. One more piece of evidence is that a structured reporting system is superior to the older prose-based methods.

A. D. Elster, MD

Contrast Agents and Safety

Frequency and Severity of Adverse Effects of Iodinated and Gadolinium Contrast Materials: Retrospective Review of 456,930 Doses

Hunt CH, Hartman RP, Hesley GK (Mayo Clinic, Rochester, MN)

AJR Am J Roentgenol 193:1124-1127, 2009

Objective.—The purpose of this study was to determine the frequency and characteristics of adverse effects of low-osmolar iodinated and gadolinium contrast agents in a single-center experience with a large number of cases.

Materials and Methods.—A retrospective review of all intravascular doses of low-osmolar iodinated and gadolinium contrast materials administered from 2002 through 2006 was conducted. Adverse effects were identified through the use of radiologist and nurse event recording.

Adverse effects were examined for type and severity of reaction, treatment required, and outcome.

Results.—A total of 456,930 contrast doses (298,491 low-osmolar iodinated, 158,439 gadolinium) were administered over the study period. A total of 522 cases of adverse effects (0.11% of total) were identified (458 low-osmolar iodinated, 64 gadolinium). The most common adverse effects were hives (274, 52.5%) and nausea (92, 17.6%). Of all adverse effects, 79 of low-osmolar iodinated and 15 of gadolinium contrast material necessitated treatment. Most treatments were performed in the radiology department alone. Only 16 cases of adverse effects necessitated transfer for further observation or treatment. Epinephrine was used to manage an adverse effect in nine instances. Thirty-two of the adverse effects of low-osmolar iodinated contrast material (6.9%) occurred in patients with a history of allergy who received premedication. Only two of these premedication reactions necessitated transfer to the emergency department. The one death in the study period occurred after administration of low-osmolar iodinated contrast material. The patient had no symptoms during the contrast administration or imaging but died suddenly within 30 minutes of receiving the dose.

Conclusion.—Both iodinated and gadolinium contrast agents are associated with a very low rate of adverse effects. Most adverse effects are mild and can be managed in the radiology department. Transfer for additional treatment or observation is rarely needed (Tables 3 and 4).

▶ This report from the Mayo Clinic describes their contrast-reaction experience over a 4-year period, giving approximately 298 000 doses of low-osmolar

TABLE 3.—Severity of Adverse Effects of Contrast Agents

Severity	Low-Osmolar Iodinated ($n = 458$)	Gadolinium ($n = 64$)	Total ($n = 522$)
Mild	374 (81.6)	49 (76.6)	423 (81.0)
Moderate	69 (15.1)	11 (17.2)	80 (15.3)
Severe	15 (3.3)[a]	4 (6.3)	19 (3.6)

Note—Values are numbers of patients with percentage in parentheses.
[a]One death.

TABLE 4.—Management of Adverse Effects of Contrast Agents

Treatment	Low-Osmolar Iodinated ($n = 458$)	Gadolinium ($n = 64$)	Total ($n = 522$)
Observation only	365 (79.7)	49 (76.6)	414 (79.3)
Diphenhydramine only	56 (12.2)	7 (10.9)	63 (12.1)
Additional medication	23 (5.0)	8 (12.5)	31 (5.9)
Received epinephrine	7 (1.5)	2 (3.1)	9 (1.7)
Transfer from radiology suite for care	10 (2.2)	6 (9.4)	16 (3.1)

Note—Values are numbers of patients with percentage in parentheses.

iodinated contrast and 158 000 doses of gadolinium contrast. Although criticism might be laid against this study in that it arises from a single medical center, the record-keeping and quality-reporting mechanisms of the Mayo Clinic are of such high standards that the conclusions can be considered reliable, complete, and generalizable.

Although lumped together, it is worth considering the reactions to iodinated and gadolinium agents separately. The summary data are displayed in Tables 3 and 4.

The rate of adverse reactions to low-osmolar iodinated contrast was approximately 0.15% (1.5 per 1000 risks). The overwhelming majority of these reactions were mild, being treated only with observation and possibly diphenhydramine. Serious reactions (cardiovascular collapse, moderate-to-severe bronchospasm, laryngeal edema, loss of consciousness, or seizure) occurred in 0.01% (1 in 10 000). One death from cardiovascular collapse was reported (0.0003% or approximately 1 per 300 000).

The rate of adverse reactions to gadolinium was much lower, overall tabulated at about 0.04% (1 in 2500). Again, the vast majority of these were mild or moderate, but a small number (0.003% or about 3 per 100 000) had severe reaction. No gadolinium-related deaths were reported.

Although the number of severe reactions to both low-osmolar iodinated contrast and gadolinium are low, continued diligence is warranted.

A. D. Elster, MD

In Emergency Situations, Should Serum Creatinine Be Checked in All Patients Before Performing Second Contrast CT Examinations Within 24 Hours?

Abujudeh HH, Gee MS, Kaewlai R (Massachusetts General Hosp, Boston)
J Am Coll Radiol 6:268-273, 2009

Objective.—The aim of this study was to determine the frequency of contrast-induced nephropathy (CIN) in patients undergoing repeat contrast-enhanced computed tomographic (CT) examinations within 24 hours, as well as associated risk factors.

Materials and Methods.—Through a search of medical and radiologic records, patients who underwent 2 contrast-enhanced CT examinations within 24 hours during a 4-year period (2003-2006), with available serum creatinine measurements before and after CT imaging were included. Medical records were reviewed for demographic data, risk factors, renal status, and CT contrast volume. The frequency of CIN was calculated, and odds ratios of risk factors were determined.

Results.—There were 164 patients (90 men, 74 women; mean age, 56.3 years), with an average baseline serum creatinine level of 1.02 ± 0.73 mg/dL (range, 0.3-6.6 mg/dL). Three hundred twenty-eight CT examinations were performed: 2 in each patient, at an average interval of 11.4 hours. The mean doses of contrast medium used for the first and second CT examinations were 126.2 and 123.4 mL, respectively.

Twenty-one patients (12.8%) developed CIN. Comparing patients with and without CIN, the only statistically significant risk factor for CIN was an increase in serum creatinine between the first and second CT examinations, with an odds ratio of 18 ($P < .005$).

Conclusion.—The incidence of CIN in patients who underwent repeat contrast-enhanced CT examinations was 12.8%. An increase in serum creatinine between the first and second CT examinations was highly associated with CIN and may serve as a risk factor for developing CIN.

▶ The answer is yes. This is a very practical problem that arises in busy emergency departments. Large academic centers often receive patients who have undergone contrast-enhanced CT studies at outside facilities which for particular reasons may need repeat examinations. Typical examples of the need for repeated studies include a poorly performed exam or new indication (such as need to exclude pulmonary embolus in a patient who recently received a contrast-enhanced CT of the abdomen).

This article, although retrospective, found an appreciable (12.8%) risk of contrast-induced nephropathy (CIN) in these patients. The use of interval creatinine clearance measurements could better identify those at risk for renal injury from a second examination (eg, the compact disc was not sent with the patient).

Other situations may arise when patients have undergone contrast-enhanced CT examinations (eg, for abdominal pain), and then different indications were found for further contrast-enhanced studies (eg, suspicion of pulmonary embolus).

This issue raises concern for the development of CIN and subsequent acute renal failure. It was suggested in older studies with relatively small samples that multiple contrast studies performed within a short-time period also confer an increased risk for developing CIN.

These studies were based largely on patients who underwent angiographic procedures with high-osmolar contrast medium administration, so their applicability to modern diagnostic CT examinations is questionable.

This is a practical problem in emergency departments, where there are pressures to expedite patients through the process. Not infrequently, second contrast examinations are requested because patients' conditions require them.

A. D. Elster, MD

Skin testing in patients with hypersensitivity reactions to iodinated contrast media – a European multicenter study

Brockow K, Romano A, Aberer W, et al (Technical Univ Munich, Germany; Rome and IRCCS Oasi Maria S.S., Troina, Italy; Med Univ of Graz, Austria; et al)

Allergy 64:234-241, 2009

Background.—Iodinated contrast media cause both immediate and nonimmediate hypersensitivity reactions. The aim of this prospective study was to determine the specificity and sensitivity of skin tests in patients who have experienced such reactions.

Methods.—Skin prick, intradermal and patch tests with a series of contrast media were conducted in 220 patients with either immediate or nonimmediate reaction. Positive skin tests were defined according to internationally accepted guidelines. Seventy-one never-exposed subjects and 11 subjects who had tolerated contrast medium exposure, served as negative controls.

Results.—Skin test specificity was 96–100%. For tests conducted within the time period from 2 to 6 months after the reaction, up to 50% of immediate reactors and up to 47% of nonimmediate reactors were skin test positive. For immediate reactors, the intradermal tests were the most sensitive, whereas delayed intradermal tests in combination with patch tests were needed for optimal sensitivity in nonimmediate reactors. Contrast medium cross-reactivity was more common in the nonimmediate than in the immediate group. Interestingly, 49% of immediate and 52% of nonimmediate symptoms occurred in previously unexposed patients. Many of these patients were skin test positive, indicating that they were already sensitized at the time of first contrast medium exposure.

Conclusions.—These data suggest that at least 50% of hypersensitivity reactions to contrast media are caused by an immunological mechanism. Skin testing appears to be a useful tool for diagnosis of contrast medium allergy and may play an important role in selection of a safe product in previous reactors.

► Controversy has long existed about the immunological nature of reactions to iodine-based contrast agents. A generation ago it was routinely taught that most of such reactions were not immunologically mediated, because the bulk of patients experiencing adverse reactions had never been exposed to these agents and a reaction does not always recur with a second exposure. However, several lines of evidence now suggest that immunological mechanisms are indeed involved, at least in immediate reactors. Such evidence includes detection of contrast media (CM)-specific IgE antibodies in sera from immediate reactors[1,2] and positive basophil activation tests.[3]

The fact that in this study approximately half of patients had positive skin tests indicates that a significant fraction of the contrast-induced hypersensitivity reactions are immunologic, involving contrast-reactive IgE antibodies and T cells, respectively. Over 30% of the skin test positive patients had been

TABLE 5.—Results From Cross-Reactivity Testing of Patients With Previous Nonimmediate Hypersensitivity Reactions After Contrast Medium Exposure*

Test Substances	Patient number 1	2	3	4	5	6	7	8	9	10	11	12	13	14	15	16	17	18	19	20	21	22	23	24	25
Nonionic																									
Iodixanol	x	x	x	x	x	x	x	x	x	x	x	x		x	x	x	x	x	x	x	x	x	x	x	x
Iohexol	x	x	x	x	x	x	x	x	x	x	x	x	x	x	x	x	x	x	x	x	x	x	x	x	x
Iopentol	x	x	x	x	x				x	x	x	x	x	x	x	x	x	x	x	x	x	x	x	x	x
Ioversol	x	x		x	x	x	x	x	x	x	x	x	x	x	x	x		x	x	x	x	x	x	x	x
Iomeprol	x	x		x	x	x	x	x	x	x	x	x	x	x	x	x	x	x	x	x	x	x	x	x	x
Iopamidol	x	x	x	x	x	x	x	x	x	x	x	x	x	x	x	x	x	x	x	x	x	x	x	x	x
Iopromide	x	x		x	x	x	x	x	x	x	x	x		x	x	x	x	x	x	x	x	x	x	x	x
Iobitridol	x	x	x	x	x	x	x	x	x	x	x	x	x	x	x	x	x	x	x	x	x	x	x	x	x
Iotrolan	x	x	x	x	x				x	x	x			x	x	x	x	x	x	x					
Ionic																									
Ioxaglate	x	x	x	x	x	x	x	x	x	x	x	x	x	x	x	x	x	x	x	x	x	x	x	x	x
Ioxithalamate	x	x	x	x	x				x	x	x		x	x	x	x	x	x	x	x					
Amidotrizoate		x	x			x	x	x	x	x	x	x	x	x	x	x	x		x	x	x	x	x	x	x

*The contrast media used for skin tested are marked ×, and the contrast media that caused the hypersensitivity reactions are highlighted with borders. (The implicated contrast medium was unknown for patients 6 and 17 or not used for skin testing in patients 11 and 16.) The shaded boxes indicate those contrast media that were skin test positive.

administered a contrast for the first time. This lack of a clear sensitization phase indicates that nonexposed patients may have already been sensitized by cross-reaction to an unknown agent.

This multicenter study is the largest to date that has investigated the usefulness of skin tests in patients with immediate and nonimmediate hypersensitivity reactions to contrast agents. With further study, such skin testing may become a practical method for determining the risk of patients for contrast-induced reactions and assessing cross-reactivity with various formulations of contrast agents (Table 5).

A. D. Elster, MD

References

1. Laroche D, Aimone-Gastin I, Dubois F, et al. Mechanisms of severe, immediate reactions to iodinated contrast material. *Radiology.* 1998;209:183-190.
2. Mita H, Tadokoro K, Akiyama K. Detection of IgE antibody to a radiocontrast medium. *Allergy.* 1998;53:1133-1140.
3. Trcka J, Schmidt C, Seitz CS, Bröcker E-B, Gross GE, Trautmann A. Anaphylaxis to iodinated contrast material: nonallergic hypersensitivity or IgE mediated allergy? *Am J Radiol.* 2008;190:666-670.

Survey of Radiologists' Knowledge Regarding the Management of Severe Contrast Material–induced Allergic Reactions

Lightfoot CB, Abraham RJ, Mammen T, et al (Dalhousie Univ, Nova Scotia, Canada; et al)

Radiology 251:691-696, 2009

Purpose.—To evaluate radiologists' knowledge of the appropriate management of severe contrast material–induced allergic reactions by means of a telephone survey.

Materials and Methods.—Institutional research ethics board approval was obtained. Following verbal consent, a telephone survey of radiologists working in Canada's 13 English-speaking and 13 U.S. university–affiliated radiology departments was performed. Participants were selected by using a multistage sampling scheme and simple random sampling within departments. Given a severe contrast material–induced allergic reaction case scenario, radiologists were first asked their initial medication of choice, then questioned specifically on the use of epinephrine. The Canadian and U.S. cohorts were compared by using the χ^2 and Fisher exact tests, as appropriate, and proportions and 95% confidence intervals (CIs) were computed.

Results.—A total of 253 (81%) of 311 radiologists from a 30% target population were surveyed. Ninety-one percent (231 of 253; 95% CI: 88%, 94%) of radiologists chose epinephrine as the most important initial medication. No radiologist gave the ideal response, but 41% (94 of 231; 95% CI: 35%, 47%) provided an acceptable administration route, concentration, and dose; 17% ($n = 39$; 95% CI: 12%, 22%) of radiologists provided an overdose. Only 11% (27 of 253; 95% CI: 7%, 15%) of radiologists knew what concentration of epinephrine was available in their drug kit and/or crash cart and what equipment would be required to administer it to a patient.

Conclusion.—Radiologists' knowledge of epinephrine for the management of severe contrast material–induced allergic reactions is deficient.

▶ Let me begin this commentary by simply restating the conclusion of the article: "Radiologists' knowledge of epinephrine for the management of severe contrast material-induced allergic reactions is deficient." This statement, if true, should strike shame and fear into all of us and is worth seriously considering.

The article gives the result of a phone survey of approximately 250 United States and Canadian radiologists who were posed the following straightforward clinical scenario: an adult patient receiving iodinated contrast becomes stridorous, in severe respiratory distress with angioedema and diffuse urticaria, consistent with an allergic reaction. What medicine, dose, concentration, and route of administration do you use? Before reading on, I invite the reader to answer this question for himself/herself.

While a range of answers is acceptable, the single best answer is as follows: give an injection of 0.5 mg of 1:1000 concentrated epinephrine intramuscularly in the lateral aspect of the thigh.

When radiologists provided an incorrect dose of epinephrine, they were more likely to overdose than underdose.

Perhaps even more alarming, 89% of surveyed radiologists did not know what type of epinephrine was available on their emergency cart and/or drug kit nor what equipment is required to administer it.

If you didn't get the correct answer above, you can probably assume most of your colleagues wouldn't either. This article should be a wake-up call for everyone.

A. D. Elster, MD

Quality Assurance

Interobserver Agreement Between On-Call Radiology Residents and Radiology Specialists in the Diagnosis of Pulmonary Embolism using Computed Tomography Pulmonary Angiography

Verweij JW, Hofstee HMA, Golding RP, et al (VU Univ Med Centre, De Boelelaan, Amsterdam)

J Comput Assist Tomogr 33:952-955, 2009

Objective.—The objective of this study was to assess the interobserver agreement between on-call radiology residents and radiology specialists in the evaluation of pulmonary embolism (PE) using computed tomographic pulmonary angiography (CTPA).

Materials and Methods.—Single center, prospective cohort study of 119 consecutive inPatients and outpatients with clinically suspected PE in whom CTPA was performed during out-of-office hours and weekends.

The on-call radiology resident interpreted the CTPA for the presence of PE. An experienced radiology specialist, without knowledge of earlier reading, reviewed the CTPA the following morning. In discordant cases, a consensus reading was conducted. The radiology specialists' assessment and the consensus reading in case of discordance were considered as the gold standard.

Results.—Five CTPA studies were excluded from the analysis. In the remaining cases, radiology residents and radiology specialists agreed on the interpretations of CTPAs (no PE vs PE present) in 106 (93%) of the 114 cases, with a κ value of 0.84 (95% confidence interval, 0.74–0.95). The consensus reading equally disagreed with both radiology resident and radiology specialist in 8 (7%) of the 114 discordant cases. Seven out of 8 discordant cases had suboptimal scan quality.

Conclusions.—The assessment of CTPA in patients with suspected PE by radiology residents seems reliable during out-of-office hours.

▶ At academic centers worldwide, radiology residents typically interpret imaging studies after hours, including computed tomography (CT) pulmonary arteriograms. How well do they perform compared with faculty level interpretations? Research from United States centers suggests they do very well, as the 93% concordance between faculty and residents in the study by Ginsberg et al[1] documents.

This new article is interesting in part because it is from Europe—specifically VU University Medical Centre, a large university teaching hospital in The Netherlands. It also uses better CT technology than previous ones. The Dutch investigators found a discordance between radiology residents and specialists in 8 (7%) of the 114 cases. Interestingly, in almost all the discordant cases, scan quality was suboptimal.

A. D. Elster, MD

Reference

1. Ginsberg MS, King V, Panicek DM. Comparison of interpretations of CT angiograms in the evaluation of suspected pulmonary embolism by on-call radiology fellows and subsequently by radiology faculty. *AJR Am J Roentgenol.* 2004;182: 61-66.

Medico-legal claims against English radiologists: 1995–2006

Halpin SFS (Univ Hosp of Wales, Health Park, CF, UK)

Br J Radiol 82:982-988, 2009

A list of claims against radiologists from 1995–2006 was obtained from the NHS Litigation Authority. It shows a total of 440 claims. The largest number of claims (199) related to delayed or missed diagnoses of cancer, and 73 claims related to breast radiology. There is a trend for a mild increase in the number of claims each year. 30 claims were made after a false-positive diagnosis of cancer. Just under £8.5 million has so far

TABLE 1.—Causes of Claims Against Radiologists from 1995/6 to 2005/6

Cause of claim	No. of claims
Total number of claims	455
Missed/delayed cancer	199
Breast cancer	72
Bronchial cancer	32
All abdominal organs cancer	27
Unspecified cancer	16
Musculoskeletal	124
Missed fracture/dislocation	95
Antenatal ultrasound	17
Brain	29
Vascular	12
Chest	12
Others/miscellaneous	71

been paid in damages, with a further £5 million in legal fees. A claim for multiple missed diagnoses of breast cancer led to a pay-out of £464 000 (£673 000 after legal fees); the largest sum awarded following a delay in the diagnosis of an individual cancer was £300 000. The subtle legal distinction between error and negligence is reviewed here. The reason why breast radiologists are more likely to be sued than any other type of British radiologist is also discussed, along with the implications for UK radiological practice, particularly in light of the recent Chief Medical Officer's report on revalidation. A method is proposed that may protect radiologists from allegations of clinical negligence in the future.

▶ Medical malpractice issues have become an increasing concern for radiologists in the United States, so this article is of interest because it describes a similar phenomenon, which has been occurring in the United Kingdom. The UK experience of malpractice claims from 1995 to 2006 is summarized in Table 1.

The greatest number of claims (199) involved delayed or missed diagnoses of cancer of the following organs: breast (36%), abdomen (13%), and lungs (12%). The second largest category (124 claims) involved musculoskeletal radiology, the great majority of which were for missed diagnosis of fracture. The distribution of these claims is not too dissimilar to those in the United States, although the number of claims is significantly lower. The level of payouts was also very modest by US standards—the largest sum being £464 000 (£673 000 after legal fees). Based on the overall ratio of damages to the legal fees paid during the period (£8.5 million vs £5.0 million), it appears that the contingency fee equivalent is approximately 37%.

We conclude that the United States is not alone in the increase and types of medical malpractice claims being levied against radiologists, although the levels in the United Kingdom are significantly lower.

A. D. Elster, MD

Radiology Education and Training

Female Medical Students' Interest in Radiology Careers

Roubidoux MA, Packer MM, Applegate KE, et al (Univ of Michigan Health System, Ann Arbor; Resurrection Med Ctr, Chicago, IL; Indiana Univ School of Medicine, Indianapolis; et al)

J Am Coll Radiol 6:246-253, 2009

Objective.—Women are underrepresented in radiology. The aim of this study was to measure first-year medical students' level of interest in radiology and their attitudes toward factors that could affect residency specialty choices to further understand how to recruit women into radiology careers.

Materials and Methods.—First-year medical students were administered surveys before and after a 7-week required introductory radiology course. Students rated interest in radiology on a scale ranging from 0 to 10 (low to high). Ten factors that could affect residency choice ("competitive residency," "shorter residency," "role models," "more jobs," "above average income," "flexible work hours," "work is technological," "work is visual," "intellectual challenge," and "more patient contact") were each rated as negative, neutral, or positive. Correlations between level of interest in radiology and the 10 factors were analyzed using Spearman's coefficients.

Results.—The mean levels of interest in radiology were 4.5 for men and 4.0 for women ($P = .38$) among 116 precourse respondents and 5.2 for men and 4.3 for women ($P = .11$) among 80 postcourse respondents. The factors most frequently rated as having a positive impact on residency choice were "flexible work hours," "intellectual challenge," "role models," and "more patient contact." Compared with men, women less frequently rated "work is technological" (20% vs 43%; $P = .0002$) and "work is visual" (50% vs 72%; $P = .03$) as having a positive impact and more frequently rated "more patient contact" (89% vs 77%; $P = .02$) as having a positive impact. For women, the strongest correlation between level of interest in radiology and the 10 factors was for "role models" (correlation coefficient = .30, $P = .03$).

Conclusion.—Interest in radiology did not differ by gender. Opportunities in radiology for flexible hours, intellectual challenge, patient care, and mentoring should be promoted early in medical education to female students to maintain and increase their interest in radiology.

► For the last 15 years fewer female medical students than expected choose careers in radiology.[1] Approximately half of all medical students are women, yet only a little more than a quarter choose to pursue residency training in radiology. This lower number of women in the radiology workforce may partially explain existing shortages in radiological subspecialties long popular among women, including pediatric radiology and mammography.

This study was designed to evaluate interest in radiology among first-year medical students before and after an introductory course. Happily, the overall level of radiology interest among male and female medical students in the first year was not significantly different.

It has been recognized that medical students develop opinions about residency choices during their preclinical years, and misperceptions about radiology that form in the first year persist. If active mentoring and role modeling can be formed in the earliest years, the authors propose that more women may not lose interest as they advance to the later years of medical school.

A. D. Elster, MD

Reference

1. Potterton VK, Ruan S, Sunshine JH, Applegate K, Cypel Y, Forman HP. Why don't female medical students choose diagnostic radiology? A review of the current literature. *J Am Coll Radiol.* 2004;1:583-590.

Research

Effective Doses in Radiology and Diagnostic Nuclear Medicine: A Catalog

Mettler FA Jr, Huda W, Yoshizumi TT, et al (New Mexico Veterans Administration Healthcare System; Med Univ of South Carolina, Charleston; Duke Univ Med Ctr, Durham, NC; et al)

Radiology 248:254-263, 2008

Medical uses of radiation have grown very rapidly over the past decade, and, as of 2007, medical uses represent the largest source of exposure to

TABLE 1.—Adult Effective Doses for Various Diagnostic Radiology Procedures

Examination	Average Effective Dose (mSv)	Values Reported in Literature (mSv)
Skull	0.1	0.03–0.22
Cervical spine	0.2	0.07–0.3
Thoracic spine	1.0	0.6–1.4
Lumbar spine	1.5	0.5–1.8
Posteroanterior and lateral study of chest	0.1	0.05–0.24
Posteroanterior study of chest	0.02	0.007–0.050
Mammography	0.4	0.10–0.60
Abdomen	0.7	0.04–1.1
Pelvis	0.6	0.2–1.2
Hip	0.7	0.18–2.71
Shoulder	0.01	...
Knee	0.005	...
Other extremities	0.001	0.0002–0.1
Dual x-ray absorptiometry (without CT)	0.001	0.001–0.035
Dual x-ray absorptiometry (with CT)	0.04	0.003–0.06
Intravenous urography	3	0.7–3.7
Upper gastrointestinal series	6*	1.5–12
Small-bowel series	5	3.0–7.8
Barium enema	8*	2.0–18.0
Endoscopic retrograde cholangiopancreatography	4.0	...

*Includes fluoroscopy.

TABLE 2.—Adult Effective Doses for Various CT Procedures

Examination	Average Effective Dose (mSv)	Values Reported in Literature (mSv)
Head	2	0.9–4.0
Neck	3	...
Chest	7	4.0–18.0
Chest for pulmonary embolism	15	13–40
Abdomen	8	3.5–25
Pelvis	6	3.3–10
Three-phase liver study	15	...
Spine	6	1.5–10
Coronary angiography	16	5.0–32
Calcium scoring	3	1.0–12
Virtual colonoscopy	10	4.0–13.2

TABLE 3.—Adult Effective Doses for Various Interventional Radiology Procedures

Examination	Average Effective Dose (mSv)*	Values Reported in Literature (mSv)
Head and/or neck angiography	5	0.8–19.6
Coronary angiography (diagnostic)	7	2.0–15.8
Coronary percutaneous transluminal angioplasty, stent placement, or radiofrequency ablation	15	6.9–57
Thoracic angiography of pulmonary artery or aorta	5	4.1–9.0
Abdominal angiography or aortography	12	4.0–48.0
Transjugular intrahepatic portosystemic shunt placement	70	20–180
Pelvic vein embolization	60	44–78

*Values can vary markedly on the basis of the skill of the operator and the difficulty of the procedure.

the U.S. population. Most physicians have difficulty assessing the magnitude of exposure or potential risk. Effective dose provides an approximate indicator of potential detriment from ionizing radiation and should be used as one parameter in evaluating the appropriateness of examinations involving ionizing radiation. The purpose of this review is to provide a compilation of effective doses for radiologic and nuclear medicine procedures. Standard radiographic examinations have average effective doses that vary by over a factor of 1000 (0.01–10 mSv). Computed tomographic examinations tend to be in a more narrow range but have relatively high average effective doses (approximately 2–20 mSv), and average effective doses for interventional procedures usually range from 5–70 mSv. Average effective dose for most nuclear medicine procedures varies between 0.3 and 20 mSv. These doses can be compared with the average annual effective dose from background radiation of about 3 mSv.

▶ This is an extraordinarily useful article that summarizes and compiles the average effective radiation doses of a wide range of diagnostic radiology and

TABLE 5.—Effective Doses for Adults from Various Nuclear Medicine Examinations

Examination*	Effective Dose (mSv)	Administered Activity (MBq)†	Effective Dose (mSv/MBq)‡
Brain (^{99m}Tc-HMPAO–exametazime)	6.9	740	0.0093
Brain (^{99m}Tc-ECD–Neurolite)	5.7	740	0.0077
Brain (^{18}F-FDG)	14.1	740	0.019
Thyroid scan (sodium iodine 123)	1.9	25	0.075 (15% uptake)
Thyroid scan (^{99m}Tc-pertechnetate)	4.8	370	0.013
Parathyroid scan (^{99m}Tc-sestamibi)	6.7	740	0.009
Cardiac stress-rest test (thallium 201 chloride)	40.7	185	0.22
Cardiac rest-stress test (^{99m}Tc-sestamibi 1-day protocol)	9.4	1100	0.0085 (0.0079 stress, 0.0090 rest)
Cardiac rest-stress test (^{99m}Tc-sestamibi 2-day protocol)	12.8	1500	0.0085 (0.0079 stress, 0.0090 rest)
Cardiac rest-stress test (Tc-tetrofosmin)	11.4	1500	0.0076
Cardiac ventriculography (^{99m}Tc-labeled red blood cells)	7.8	1110	0.007
Cardiac (^{18}F-FDG)	14.1	740	0.019
Lung perfusion (^{99m}Tc-MAA)	2.0	185	0.011
Lung ventilation (xenon 133)	0.5	740	0.00074
Lung ventilation (^{99m}Tc-DTPA)	0.2	1300 (40 actually inhaled)	0.0049
Liver-spleen (^{99m}Tc–sulfur colloid)	2.1	222	0.0094
Biliary tract (^{99m}Tc-disofenin)	3.1	185	0.017
Gastrointestinal bleeding (99mTc-labeled red blood cells)	7.8	1110	0.007
Gastrointestinal emptying (^{99m}Tc-labeled solids)	0.4	14.8	0.024
Renal (^{99m}Tc-DTPA)	1.8	370	0.0049
Renal (^{99m}Tc-MAG3)	2.6	370	0.007
Renal (^{99m}Tc-DMSA)	3.3	370	0.0088
Renal (^{99m}Tc-glucoheptonate)	2.0	370	0.0054
Bone (^{99m}Tc-MDP)	6.3	1110	0.0057
Gallium 67 citrate	15	150	0.100
Pentreotide (^{111}In)	12	222	0.054
White blood cells (^{99m}Tc)	8.1	740	0.011
White blood cells (^{111}In)	6.7	18.5	0.360
Tumor (^{18}F-FDG)	14.1	740	0.019

*DMSA = dimercaptosuccinic acid, DTPA = diethylenetriaminepentaacetic acid, ECD = ethyl cysteinate dimer, ^{18}F = fluorine 18, FDG = fluorodeoxyglucose, HMPAO = hexamethylpropyleneamine oxime, ^{111}In = indium 111, MAA = macroaggregated albumin, MAG3 = mercaptoacetyltriglycine, MDP = methylene diphosphonate, ^{99m}Tc = technetium 99m.

†Recommended ranges vary, although most laboratories tend to use the upper end of suggested ranges.

‡From reference 74.

nuclear medicine procedures. It is a summary of approximately 200 measurement studies that have appeared in the world literature since 1980.

Although there are multiple methods by which radiation exposure can be measured, perhaps the most current and most important is the "effective dose," defined by the International Commission on Radiological Protection (ICRP). The ICRP first assigns radiation weighting factors depending on the type of radiation (photons are assigned a value of 1.0), allowing an equivalent dose to be obtained. To estimate the potential injury from cancer and hereditary effects, effective dose is used. The effective dose is calculated by multiplying the average organ equivalent dose by an ICRP-defined tissue weighting factor and summing the results over the whole body. Effective dose is expressed in units of millisieverts (mSv). It is a single dose parameter that reflects the risk of a nonuniform exposure in terms of whole-body exposure.

Effective dose is also age and sex averaged. The most important dose information is provided in Tables 1-3, and 5.

A. D. Elster, MD

Incidental findings in healthy control research subjects using whole-body MRI

Morin SHX, Cobbold JFL, Lim AKP, et al (Imperial College London, Hammersmith Campus, UK; et al)
Eur J Radiol 72:529-533, 2009

Aim.—Magnetic resonance imaging (MRI) is a powerful clinical tool used increasingly in the research setting. We aimed to assess the prevalence of incidental findings in a sequential cohort of healthy volunteers undergoing whole-body MRI as part of a normal control database for imaging research studies.

TABLE 1.—Abnormalities Ordered by Clinical Significance and Frequency (percentage of total scans in which an abnormality of the specified severity is found)

Abnormality	Frequency, % Total Scans	Number (% Abnormalities)
High clinical significance	3.4	5 (10.2)
Ovarian lesion; dermoid cyst		2 (4.1)
Mediastinal lymphadenopathy		1 (2.0)
Psoas mass		1 (2.0)
Splenomegaly		1 (2.0)
Moderate clinical significance	9.5	15 (30.6)
Gallstones		6 (12.2)
Diverticulosis		3 (6.1)
Enlarged uterus/fibroids		3 (6.1)
Kidney agenesis (unilateral)		2 (4.1)
Subdiaphragmatic cystic lesion		1 (2.0)
Low clinical significance	18.9	29 (59.2)
Renal cyst, one or more		21 (42.9)
Liver cyst or haemangioma		5 (10.2)
Splenic cyst/haemangioma		2 (4.1)
Simple bone cyst		1 (2.0)

Clinical significance is graded high, moderate and low. Absolute numbers of abnormalities are tabulated with the percentage of all abnormalities in parentheses.

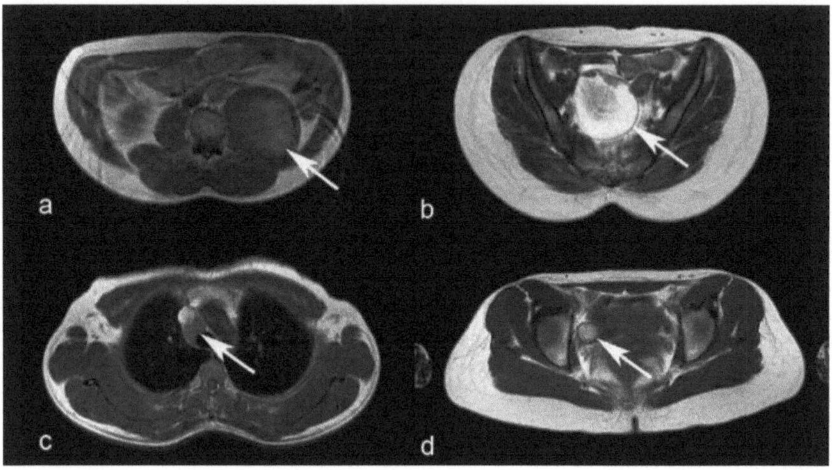

FIGURE 1.—Examples of axial T_1-weighted images showing pathology. Panel (a) shows a large left psoas muscle mass, mild-to-moderately hyperintense, compared to the adjacent muscle. Subsequent biopsy showed this to be a Schwannoma. Panel (b) shows a large hyperintense right-sided pelvic mass. This was found to be amature cystic teratoma and was excised surgically. Panel (c) shows right-sided paratracheal lymphadenopathy of unknown aetiology (subject declined further investigation and was lost to follow-up). Panel (d) shows a small, hyperintense right ovarian lesion, most likely a haemorrhagic cyst, although a small dermoid cyst cannot be ruled out. (Reprinted from Morin SHX, Cobbold JFL, Lim AKP, et al. Incidental findings in healthy control research subjects using whole-body MRI. *Eur J Radiol.* 2009;72:529-533, with permission from Elsevier.)

Materials and Methods.—148 healthy volunteers (median age 36 years, range 21–69 years; 63.5% males, 36.5% females) were enrolled into a prospective observational study at a single hospital-based MRI research unit in London, UK. Individuals with a clinical illness, treated or under investigation were excluded from the study.

Results.—43 (29.1%) scans were abnormal with a total of 49 abnormalities detected. Of these, 20 abnormalities in 19 patients (12.8%) were of clinical significance. The prevalence of incidental findings increased significantly with both increasing age and body mass index (BMI). Obese subjects had a fivefold greater risk of having an incidental abnormality on MRI (OR 5.4, CI 2.1–14.0).

Conclusions.—This study showed that more than one quarter of healthy volunteers have MR-demonstrable abnormalities. There was an increased risk of such findings in obese patients. This has ethical and financial implications for future imaging research, particularly with respect to informed consent and follow-up of those with abnormalities detected during the course of imaging studies.

▶ The detection of incidental findings (sometimes referred to as "incidentalomas") on imaging studies performed for other reasons is perplexing. These lesions cannot simply be ignored, but do they warrant an expensive work-up and, possibly, invasive therapy if identified? The situation is particularly distressing when incidentalomas are discovered in research volunteers who, by definition, are healthy and asymptomatic. It is, therefore, important to have some idea of the frequency and nature of such lesions. This article addresses that important issue at least for body magnetic resonance (MR) imaging.

In this study, approximately 29% of normal volunteered harbored an incidentaloma of which slightly less than half were considered to be of potential clinical significance. The nature of these lesions is reported in Table 1. Representative imaging examples are shown in Fig 1.

The findings are not unexpected based on the experience of clinical radiologists with whole body imaging.[1] Similar findings have been described by others in other body regions, such as the brain[2] and pituitary.[3]

A. D. Elster, MD

References

1. Obuchowski NA, Holden D, Modic MT, et al. Total-body screening: preliminary results of a pilot randomized controlled trial. *J Am Coll Radiol.* 2007;4:604-611.
2. Yue NC, Longstreth WT Jr, Elster AD, Jungreis CA, O'Leary DH, Poirier VC. Clinically serious abnormalities found incidentally at MR imaging of the brain: data from the Cardiovascular Health Study. *Radiology.* 1997;202:41-46.
3. Elster AD. Modern imaging of the pituitary. *Radiology.* 1993;187:1-14.

Patient Dose Reference Levels for Interventional Radiology: A National Approach

Vano E, Sanchez R, Fernandez JM, et al (Hosp Clinico San Carlos, Madrid, Spain; et al)

Cardiovasc Intervent Radiol 32:19-24, 2009

A set of patient dose reference levels (RLs) for fluoroscopically guided interventional procedures was obtained in a survey launched by the National Society of Interventional Radiology (IR), involving 10 public hospitals, as recommended by the European Medical Exposures Directive. A sample of 1391 dose values (kerma area product [KAP]) was collected randomly during clinical procedures for seven of the most frequent procedures. Third quartiles of the KAP distributions were used to set the RLs. A regular quality control of the X-ray systems and a calibration of the dose meters were performed during the survey. The fluoroscopy time and total number of digital subtraction angiography images per procedure were also analyzed. The RL values proposed were 12 Gy cm^2 for fistulography (hemodialysis access; sample of 180 cases), 73 Gy cm^2 for lower limb arteriography (685 cases), 89 Gy cm^2 for renal arteriography (55 cases), 80 Gy cm^2 for biliary drainage (205 cases), 289 Gy cm^2 for hepatic chemoembolization (151 cases), 94 Gy cm^2 for iliac stent (70 cases), and 236 Gy cm^2 for uterine embolization (45 cases). The provisional national RL values are lower than those obtained in a similar survey carried out in the United States from 2002 to 2004. These new values could be used to improve the practice of centers

TABLE 2.—Results for Fluoroscopy Time (Minutes)

Procedure	Max	Min	Mean	SD	Median	3rd Quartile
Fistulography	52	0.3	2.5	4	1.4	2.5
Lower limb arteriography	56.8	0.5	3	3	2.2	3.3
Renal arteriography	27.5	0.3	5.6	6	3.9	8.6
Biliary drainage	79.4	0.6	14.5	12	11.7	19.8
Hepatic chemoembolization	80	2.8	19.8	11	17.8	24.3
Iliac stent	67.1	1.3	9.2	9	6.5	10.9
Uterine embolization	44.5	0.3	23.8	9	23.4	30.3

Note: Sample sizes are given in Table 1.

consistently working with doses higher than the RLs. This national survey also had a positive impact, as it helped increase the awareness of the members of the National Society of IR on a topic as crucial as patient dose values and programs on radiation protection.

▶ How well is your facility doing at minimizing radiation exposure to patients in your interventional radiology area?

This is a relatively straightforward question that might be asked by a Joint Commission representative or even a concerned patient. A suitable answer is difficult to generate, because patient doses are not commonly measured during interventional radiology procedures and national standards are not widely available.

Perhaps the easiest and most commonly measured quantity is fluoroscopy time, which bears a moderately direct relationship to radiation dose. A second metric that can be computed is kerma area product (KAP), a quantity numerically equal to the dose-area product (DAP). KAP is defined as the integral of air kerma (absorbed dose to air) across the entire X-ray beam emitted from the X-ray tube. KAP is a surrogate measurement for the entire amount of energy delivered to the patient by the beam. The units of KAP are Grays per square centimeter (Gy/cm^2).

The article includes some useful tables and graphs for KAP and fluoroscopy time for several common interventional procedures (Table 2). Other sites can compare their time and dose estimates and see how they compare with the Spanish hospitals. And so perhaps, this metric is the best we have to use. The data in this article are from 10 public hospitals in Spain.

A. D. Elster, MD

Socioeconomic Issues

Insight From Patients for Radiologists: Improving Our Reporting Systems

Johnson AJ, Easterling D, Williams LS, et al (Wake Forest Univ School of Medicine, Winston Salem, NC; Indiana Univ School of Medicine, Indianapolis, IN; et al)

J Am Coll Radiol 6:786-794, 2009

Purpose.—The aim of this study was to seek patients' perspectives on radiology reporting systems, so that reporting systems can begin to be reorganized and made more patient-centered by giving patients greater access to their personal health information.

Methods.—Focus group methodology was used to explore which aspects of radiology information are important to patients and to identify their preferred means of access to and format of this information. Subjects for the two groups were outpatients who had recently undergone MR imaging at a single academic medical center. Transcripts were analyzed using thematic content analysis.

Results.—Most subjects were dissatisfied with current reporting systems, citing delays and a lack of detail as the most important problems. Subjects varied with regard to preferences for who should relay results to them, with some expressing a desire for increased direct input from radiologists because they have greater expertise in imaging interpretation. Most subjects wanted results in writing and in detail, with attached lay language explanations, though a few subjects preferred less detail. Subjects were decidedly in favor of having the option to access results immediately via an online system, proposing some potential problems and potentially multiple benefits of such a system.

Conclusions.—Whatever system revisions are attempted to increase the patient-centeredness of care as regards to radiology reporting, patients will need to be able to choose their preferred levels of access and will need to have the option of accessing full details.

▶ For many years, radiologists were strongly discouraged by clinicians to reveal test results directly to patients. With the exception of mammography, direct communication of imaging test results to patients has not been attempted in most areas of radiology. However, there is abundant sociological evidence that patients do not generally receive as much information as they want about their condition from any doctor, particularly with regard to diagnosis, prognosis, and treatment options.[1,2] Moreover, good evidence also exists that such involvement by patients in medical decision making positively affects patient outcomes.[3]

Dr Johnson and her colleagues organized a set of racially and socioeconomically diverse focus groups, finding agreement on several key issues and marked variability on others. A common theme is that most participants were dissatisfied with the current systems of notification of radiologic test results, with delays and lack of details being the most important. Delays in receiving radiologic test results were deemed significant stressors for nearly all participants.

Participants were fairly evenly split with regard to whom they would prefer, a radiologist or their own physician to provide test results to them. Most participants preferred written reports to include much detail, with associated lay explanations of terms, but a minority preferred that written reports should include minimal detail and that their physicians should provide them with further details in person. There was strong preference for a system of optional online access to results, but a minority of patients felt that reporting by telephone would also be acceptable.

The results of this study reflect the desire of most patients to become more empowered to learn the results of their imaging tests sooner and potentially directly from radiologists.

A. D. Elster, MD

References

1. Ley P. Giving information to patients. In: Eiser JR, ed. *Social Psychology and Behavioral Science*. Chichester, UK: John Wiley; 1982:339-373.
2. Ryan J, Sysko J. The contingency of patient preferences for involvement in health decision making. *Health Care Manage Rev.* 2007;32:30-36.
3. Greenfield S, Kaplan S, Ware JE. Expanding patient involvement in care. Effects on patient outcomes. *Ann Intern Med.* 1985;102:520-528.

The Association Between Hospital Outcomes and Diagnostic Imaging: Early Findings

Lee DW, Foster DA (GE Healthcare, Waukesha, WI; Thomson Reuters, Ann Arbor, MI)

J Am Coll Radiol 6:780-785, 2009

Purpose.—Resource use variation across the United States prompts the important question of whether "more is better" when it comes to health care services. The aim of this study was to examine correlations between the use of 4 common imaging modalities (CT, MR, ultrasound, and radiography) and in-hospital mortality and costs.

Methods.—Using clinical and utilization data for 1.1 million inpatient admissions at 102 US hospitals during 2007, two hospital-specific, risk-adjusted imaging utilization measures for each modality were constructed that controlled for patients' demographic and clinical characteristics and for hospital characteristics were constructed for each modality. First, logistic regression was used to estimate the odds that each type of imaging service would be provided during an admission. Second, the mean number of services per admission was estimated using output from a two-part ordinary least squares model. Hospital-specific, risk-adjusted inpatient mortality and total hospital costs were also computed, and correlations between the imaging utilization measures and the mortality and cost outcome measures were then assessed using Pearson's correlation coefficients ($P < .05$). The correlation analyses were weighted by hospital admission volume.

Results.—Hospitals in which patients were more likely to receive imaging services during admissions had lower mortality, even after controlling for potential confounders. Correlation coefficients were −0.2 for all modalities ($P = .02\text{-}.05$). Weaker correlations existed between mean services per admission and mortality, while costs trended insignificantly higher with greater utilization.

Conclusions.—This study lays the foundation for further exploration of the relationship between resource use and the clinical and economic outcomes associated with imaging utilization.

▶ Are we overusing our imaging resources? This is clearly a question that has profound economic indications for our society. Questions have even been raised to suggest that we are beyond the "top of the curve" and that overuse of imaging is actually decreasing patient outcomes by the additional morbidity, mortality, and expense of chasing down meaningless findings on imaging.

This article is a welcome addition to an area of health policy debate that is clearly not yet settled and contains contradictory literature. For example, The Dartmouth Atlas of Health Care[1] showed no association between high utilization of imaging services and outcomes for hip fracture, colorectal cancer, and acute myocardial infarction in a general population sample, but the study by Beinfeld and Gazelle[2] reported that shorter hospital stays correlated with higher spending on imaging services for stroke, appendectomy, lung cancer, upper gastrointestinal procedures, colorectal cancer, and back problems.

This study is of interest, suggesting that hospitals where imaging is more widely used have better patient outcomes than those where imaging is used less. This is admittedly an indirect way to answer the "imaging overuse" question but one that does add positively to the debate.

A. D. Elster, MD

References

1. Dartmouth Institute for Health Policy and Clinical Practice. The Dartmouth Atlas of Health Care. http://www.dartmouthatlas.org. Accessed March 15, 2010.
2. Beinfeld MT, Gazelle GS. Diagnostic imaging costs: are they driving up the costs of hospital care? *Radiology.* 2005;235:934-939.

6 Interventional Radiology

Introduction

The continuing evolution of interventional radiology procedures from adjunctive therapies to first-line treatments of choice testifies to physician and patient preference for minimally invasive options for clinical care. Technology continues to improve and evolve, while innovation is driven by the demand for alternatives to traditional surgery. Clinical specialties that previously dismissed many of these interventional procedures now have fully embraced them, and practitioners are increasingly positioning themselves as the new experts in certain therapies. As I noted in last year's YEAR BOOK OF DIAGNOSTIC RADIOLOGY, this widespread acceptance obviously has resulted in tremendous options for our patients, but it has also caused competition between various clinical specialties and their interventional radiology colleagues. This multidisciplinary acceptance of these minimally invasive therapies is reflected in the array of medical journals from which I have made this year's selections.

As long as the end result is that of improved clinical care, competition in medicine can have a positive impact, as the competitors are driven to continually improve the existing technology, produce better and more predictable results, and offer the patient a variety of treatment options. The challenge for interventional radiologists is to continue to evolve into true clinicians, to maintain superior technical skills, and to pursue innovative scientific research. The variety of topics I have chosen for this year's YEAR BOOK show this specialty's commitment, as well as that of nonradiological physicians, to providing quality options for patient care and conducting scientific research aimed at identifying new therapies and improving existing ones.

I hope this year's readers enjoy the selection of current, interesting articles from the scientific literature.

T. Gregory Walker, MD

Balloon Angioplasty/Stents

AMS INSIGHT—Absorbable Metal Stent Implantation for Treatment of Below-the-Knee Critical Limb Ischemia: 6-Month Analysis

Bosiers M, on Behalf of the AMS INSIGHT Investigators (AZ St-Blasius, Dendermonde, Belgium; et al)

Cardiovasc Intervent Radiol 32:424-435, 2009

Endoluminal treatment of infrapopliteal artery lesions is a matter of controversy. Bioabsorbable stents are discussed as a means to combine mechanical prevention of vessel recoil with the advantages of long-term perspectives. The possibility of not having a permanent metallic implant could permit the occurrence of positive remodeling with lumen enlargement to compensate for the development of new lesions. The present study was designed to investigate the safety of absorbable metal stents (AMSs) in the infrapopliteal arteries based on 1- and 6-month clinical follow-up and efficacy based on 6-month angiographic patency. One hundred seventeen patients with 149 lesions with chronic limb ischemia (CLI) were randomized to implantation of an AMS (60 patients, 74 lesions) or stand-alone percutaneous transluminal angioplasty (PTA; 57 patients, 75 lesions). Seven PTA-group patients "crossed over" to AMS stenting. The study population consisted of patients with symptomatic CLI (Rutherford categories 4 and 5) and de novo stenotic (>50%) or occlusive atherosclerotic disease of the infrapopliteal arteries who presented with a reference diameter of between 3.0 and 3.5 mm and a lesion length of <15 mm. The primary safety endpoint was defined as absence of major amputation and/or death within 30 days after index intervention and the primary efficacy endpoint was the 6-month angiographic patency rate as confirmed by core-lab quantitative vessel analysis. The 30-day complication rate was 5.3% (3/57) and 5.0% (3/60) in patients randomized for PTA alone and PTA followed by AMS implantation, respectively. On an intention-to-treat basis, the 6-month angiographic patency rate for lesions treated with AMS (31.8%) was significantly lower ($p = 0.013$) than the rate for those treated with PTA (58.0%). Although the present study indicates that the AMS technology can be safely applied, it did not demonstrate efficacy in long-term patency over standard PTA in the infrapopliteal vessels (Fig 1, Table 3).

▶ There has been a treatment strategy change for the management of critical limb ischemia (CLI), with the endovascular approach becoming much more accepted as a valid alternative to surgical bypass. The bypass versus angioplasty in severe ischemia of the leg (BASIL) trial participants played a pivotal role in this shift toward this approach for patients with CLI caused by infrainguinal disease, with the first published results of a controlled, randomized study comparing bypass surgery results with those of percutaneous transluminal angioplasty (PTA) in CLI patients. They observed similar outcomes for both treatments, with angioplasty being the less expensive alternative in the

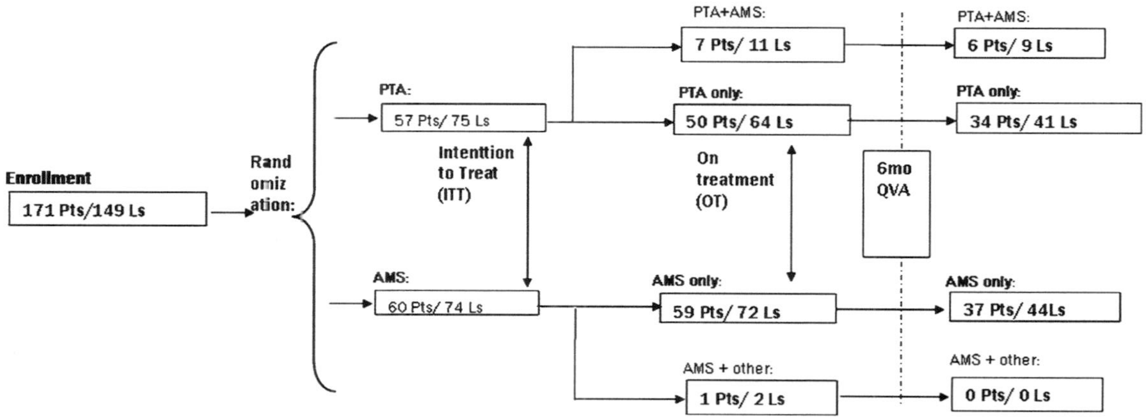

FIGURE 1.—Flowchart for patient distribution after randomization. (With kind permission from Springer Science+Business Media: Bosiers M, on Behalf of the AMS INSIGHT Investigators. AMS INSIGHT-absorbable metal stent implantation for treatment of below-the-knee critical limb ischemia: 6-month analysis. *Cardiovasc Intervent Radiol.* 2009;32:424-435.)

TABLE 3.—Intention-to-treat Primary Efficacy Endpoint: 6-month Follow-up Lesion Patency Based on Quantitative Vessel Analysis (QVA) (Pre- and Postprocedure Lesion QVA Results)[a]

	PTA Preproc	AMS Preproc	PTA Postproc	AMS Postproc	6-mo QVA: PTA	6-mo QVA: AMS
No. patients	57	59	57	59	40	37
No. lesions	74	72	74	72	50	44
Patency	na	na	na	na	29/50 (58%) $p = 0.0134$	14/44 (31.8%)
Non patency	na	na	na	na	21/50 (42%)	30/44 (68.2%)
Binary restenosis, mm					$p = 0.0134$	
Lesion length	12.0 ± 5.0	10.6 ± 4.9				
Min–max	3.5–30.4	3.1–27.6				
Stenosis diameter, mm Mean ± SD	68.7 ± 11.5	69.0 ± 10.7	21.9 ± 12.4	15.1 ± 10.2	47.8 ± 22.7	66.4 ± 27.1
(Min–max)	(51.4–100)	(51.3–100)	(−7.2–53.6)	(−18.4–38.9)	(3.8–100)	(2.5–100)
MLD, mm Mean	0.8 ± 0.3	0.8 ± 0.3	2.1 ± 0.5	2.2 ± 0.4	1.4 ± 0.7	0.9 ± 0.7
(Min–max)	(0.0–1.6)	(0.0–1.5)	(1.0–3.2)	(1.4–3.4)	(0.0–2.9) $p = 0.0009$	(0.0–2.9)
RVD, mm Mean	2.7 ± 0.5	2.6 ± 0.5	2.7 ± 0.5	2.6 ± 0.5	2.7 ± 0.5	2.6 ± 0.5
(Min–max)	(1.4–4.4)	(1.5–4.4)	(1.5–4.6)	(1.5–4.6)	(1.8–4.5)	(1.9–3.9)
Late lumen loss, mm Mean ± SD	na	na	na	na	0.7 ± 0.7	1.4 ± 0.8
(Min–max)					(−0.3–2.9) $p = 0.0001$	(−0.4–2.9)

Note: PTA, percutaneous transluminal angioplasty; preproc, preprocedure; AMS, absorbable metal stent; postproc, postprocedure; MLD, minimal lumen diameter; RVD, reference vessel diameter

[a]For three lesions (one PTA patient and two AMS patients), angiographic procedure data were not available

short-term; this resulted in the statement in the renewed TASC-II guidelines that there is increasing evidence to support a recommendation for PTA in patients with CLI and infrapopliteal artery occlusive disease.

While intravascular stents are commonly used in other peripheral vessels, their use in the infrapopliteal circulation remains controversial. Early thrombosis and delayed restenosis due to intimal hyperplasia can potentially lead to insufficient long-term patency rates and may limit surgical revascularization options are cited as concerns for stent implantation in these small-diameter vessels. There is limited evidence available suggesting acceptable outcomes using either passive coated stents, balloon-expandable drug-eluting stents, or self-expanding nitinol stents. However, target lesion revascularization (TLR) is still required in a significant percentage of patients to enable limb salvage.

Because an artificial implant such as a stent is believed to be the potential trigger for late restenosis, recent stenting technology has focused on the development of temporary implants composed of biocompatible materials, which mechanically support the vessel during the period of high risk for recoil and then completely degrade. These bioabsorbable stents possibly represent a means of combining the mechanical prevention of vessel recoil with the potential advantages of the absence of a permanent implant in the treated vessel. This might potentially maintain patency, allow for late vessel remodeling, and improve reintervention options.

The bioabsorbable magnesium-alloy stent developed by BIOTRONIK AG was the first of its kind for which it was proven that it could be implanted safely in infrapopliteal and human coronary arteries and that it was absorbed in the intended time frame. In the absorbable metal stent (AMS) below-the-knee study, Bosiers et al presented the first clinical results using AMSs for treatment of infrapopliteal lesions in 20 patients with CLI. After 6 months, the resulting values for primary clinical patency and limb salvage indicated a promising performance for the treatment of below-the-knee lesions in CLI patients. Moreover, in the PROGRESS-AMS study the authors proved that bioabsorbable magnesium stents can achieve immediate angiographic results similar to the results with other permanent metal stents and is safely degraded after 4 months in native coronary artery single de novo lesions.

In this article, safety results based on 1- and 6-month clinical follow-up and efficacy data based on 6-month angiographic results for the AMS INSIGHT study in patients with CLI are presented and discussed. The primary aim of the study was to prove the superiority of the AMS stent compared with PTA alone for infrapopliteal indications. This was a prospective, multicenter, randomized study, designed to evaluate the safety and performance of the first-generation AMS for the treatment of infrapopliteal lesions in patients with CLI. The results after AMS implantation are compared with the results after stand-alone PTA. The patient enrollment and randomization data are presented in Fig 1.

The primary safety end point of the study was defined as the absence of clinical complications at 1-month postprocedure. Complications were defined as major amputations or any cause of death. The primary efficacy end point of this study was to analyze and compare the 6-month angiographic patency rate after PTA alone or PTA followed by AMS implantation in patients with stenotic or occlusive atherosclerotic disease of the infrapopliteal arteries. Patency was defined as the absence of a hemodynamically significant restenosis (> 50%) documented by digital subtraction angiography, with core lab confirmation. The secondary end points included (1) immediate angiographic procedural success, which was defined in both therapy groups as < 30% final residual diameter stenosis of the target segment; (2) late lumen loss (LLL) as diagnosed at 6-month angiographic control and defined by the difference between the in-stent minimal lumen diameter (MLD) postprocedure and the MLD at follow-up measured with angiography; (3) the limb salvage rates, defined as lack of major amputations at the different prescheduled follow-up visits until 12 months after index intervention; and (4) the primary patency rates at each visit as determined by color flow Doppler ultrasound (CFDU) and defined as either the absence of a hemodynamically significant restenosis (> 50%), a major amputation, or a TLR.

The intention-to-treat primary efficacy end points are summarized in Table 3. The 6-month angiographic patency rate for lesions treated with AMS (31.8%) was significantly lower ($P = .013$) than the rate for those treated with PTA (58.0%). The study demonstrated that the AMS technology can be safely applied, but unfortunately it did not demonstrate efficacy in long-term patency over standard PTA in the infrapopliteal vessels. The findings were interpreted to indicate that the current-generation stent does not meet efficacy criteria and that the stent needs to be improved and re-evaluated before commercial use can be an option.

T. G. Walker, MD

Novel Treatment of Patients With Lower Extremity Ischemia: Use of Percutaneous Atherectomy in 579 Lesions

McKinsey JF, Goldstein L, Khan HU, et al (Columbia Univ College of Physicians and Surgeons and Weill Med College of Cornell Univ, NY)
Ann Surg 248:519-527, 2008

Objectives.—Endovascular options for the treatment of lower extremity peripheral arterial disease are typified with diminished patency. A novel alternative to standard angioplasty and stent is the excision of the obstructing arterial plaque using a minimally invasive technique, the Silverhawk Atherectomy device.

Methods.—A prospective database was maintained of SilverHawk procedures from 2004 to 2007. A total of 579 lesions were treated in 275 patients (claudication 101 patients (36.7%) and critical limb ischemia (CLI) in 174 patients (63.3%). Noninvasive laboratory evaluation was performed at 1, 3, and 6 months, and yearly. Stand-alone atherectomy was attempted, and adjunct therapy was used for residual stenosis.

Results.—Mean age was 70.0 years (range, 37–102) (62.5% male, 46.2% smokers, 67.6% diabetes mellitus). Lesion characteristics were 199 superficial femoral arteries, 110 popliteal, 218 tibials, and 52 multi-level. Thirty-day perioperative mortality was 1.8%. Mean follow-up was 12.5 months (range, 0.5–48.2). 18-month primary and secondary patency for all lesions was 52.7% ± 2.8 and 75.0% ± 2.4 ($P < 0.0001$). Eighteen-month primary and secondary patency for claudicants was 58.0% ± 4.3 and 82.5% ± 3.5 ($P < 0.0001$) and for CLI was 49.4% ± 3.7 and 69.9% ± 3.2 ($P < 0.0001$), respectively. The reintervention rate was 25.3% in claudicants and 30.1% for CLI. Limb salvage was 100% in claudicants and overall limb salvage was 92.4% per patient at 18 months and only 4.4% required bypass.

Conclusion.—This is the largest reported series with the longest follow-up showing the SilverHawk device as an effective endovascular therapy for patients with claudication and CLI with a low mortality, low complications rate, low amputation rate, and rare need for conversion to surgical bypass. The majority of reintervention was performed with endovascular techniques.

▶ The role of atherectomy as an effective endovascular therapy for the treatment of peripheral vascular disease has been controversial, with many investigators maintaining that there is an unacceptably high risk of peripheral embolization from the procedure and that there is little gain from this technology as compared with other endovascular options.[1] Others view this as a "niche" technology that will never gain widespread acceptance, while some feel that there may be a useful role for atherectomy.[2] This article investigates the role of atherectomy in the treatment of lower extremity ischemia caused by atherosclerotic steno-occlusive disease.

Saphenous vein bypass graft has been considered the gold standard for the management of patients with limb threatening ischemia and severe disabling claudication. To maintain long-term patency 22% to 36% of lower extremity bypass grafts must undergo either percutaneous angioplasty or local surgical revision. Not only do surgical bypasses require reintervention, there are early and late failures associated with bypass grafting and a multitude of possible complications associated with the bypass grafting. When no autologous conduit is available, synthetic bypass grafting alone or with adjuncts, such as hemodynamic hoods, vein cuffs, or arterial venous fistulas have been recommended. All of these interventions have lower patency rates when compared with saphenous vein bypass grafts and potentially a higher complication rate.

The overall patient morbidity and the length of recovery time have created a dramatic shift in the treatment paradigm for patients with lower extremity ischemia. Over the past decade there has been a dramatic change toward a minimally invasive approach for the treatment of patients with symptomatic lower extremity arterial occlusive disease. Initially these patients were treated with balloon angioplasty restricted to iliac lesions and non-total occlusions. The TASC (TransAtlantic Surgical Consensus) recommends endovascular therapy for those short non-occluded lesions (TASC A). Surgical bypass was recommended for more complex lesions involving total occlusions and long-segment lesions (TASC D). The TASC classification was unable to make a definitive recommendation regarding the intermediate lesions (TASC B and C). Subsequent revisions of the TASC classifications were further clarified. The overall recommendations regarding the recommended interventions for TASC A, B, C, and D lesions were unchanged. To further evaluate the role of surgical bypass versus minimally invasive therapy the Bypass versus Angioplasty for Severe Ischemia of the Leg (BASIL) trial evaluated a prospective randomized trial in the United Kingdom evaluating patients with critical limb ischemia and their outcomes from surgical bypass versus angioplasty with adjunctive stenting. Their primary end point was amputation-free survival, and at 6 months there was no difference in the amputation rate between the 2 patient treatment groups, but there was an overall higher hospital cost for treating those patients with surgical bypass.

With the increasing acceptance of percutaneous intervention for treatment of lower extremity occlusive disease, the optimal endovascular therapy has yet to be determined. The initial therapy was angioplasty with bailout stenting for those cases with a residual stenosis greater than 30% or a flow-limiting dissection. This is still the primary endovascular therapy to date for those patients with below-the-knee critical limb ischemia. Those patients that were primarily treated with angioplasty and/or stenting are generally amenable to repeat endovascular procedure without a change in the patency of their outflow vessels.

Debulking technologies such as atherectomy have been reported since the 1990s and have had varying success. The Silverhawk atherectomy device (EV3, Minneapolis, MN) now allows for treating up to the common femoral artery and distally to the pedal vessels on to the foot with debulking of the obstructing atheromatous plaque rather than disruption and dilatation of the affected artery. In this study, it was found that the primary femoropopliteal lesions treated with Silverhawk directional atherectomy device had a primary patency of 73% and secondary patency of 91% at 18 months. Those patients with restenosis or in-stent restenosis had a worse outcome with a primary patency at 18 months of 42% to 49%, respectively, and a secondary patency of 65% to 75%, respectively. The data in our series report comparable outcomes despite the fact that there was a majority of patients with infrapopliteal lesions, as well as patients with in-stent restenosis, restenotic lesions, and de novo lesions. In this series the primary patency at 12 and 18 months was 59.1% and 49.4% with a limb-salvage rate of 87.9% at 18 months for patients with critical limb ischemia and 100% limb salvage in patients with claudication for a combined limb salvage rate of 92.4%. When comparing the results for patients with limb threatening ischemia to that of surgical bypass as well as with angioplasty and stenting the BASIL trial had a 16% overall amputation rate at 6 months between the 2 treatment modalities. The 18-month 7.6% amputation rate using atherectomy in this series certainly compares favorably to the BASIL results. Periprocedural complications of embolization and thrombosis occurred rarely in this study. The authors attribute their low embolization rate to the technique of passing the atherectomy device slowly, packing frequently, and not making long passes of the atherectomy device without packing the device. Additionally, the authors maintain therapeutic anticoagulation intraprocedurally.

There was no difference in patency or limb salvage between the primary atherectomy group and the atherectomy group assisted with percutaneous transluminal angioplasty (PTA) or stenting. Because this was not a randomized trial evaluating the role of primary versus assisted or adjunctive atherectomy, these lesions and results cannot be compared but it can be assumed that those lesions that required additional intervention were more complicated.

Assisted atherectomy in more complicated lesions appears warranted to obtain patency and limb salvage rates comparable to primary atherectomy alone.

The authors feel that one of the significant advantages of atherectomy compared with angioplasty with adjunct stenting is the ability to reintervene without having a foreign body such as the stent in place. With increasing emphasis on the deleterious effects of stents and stent fractures, this could become even a more significant point. For those 29% of the patients that required reintervention in this series, 70% of those were able to be treated with percutaneous intervention with repeat atherectomy or angioplasty.

Only 4.4% of the patients required surgical bypass, 3 of those in which the atherectomy site was actually still patent and used either as an inflow source or a separate tibial target vessel was required to increase perfusion to the foot. Of the patients treated requiring reintervention, 7.6% were treated with amputation either above or below the knee. Of all patients undergoing surgical bypass the distal target vessel was unchanged, when compared with the angiogram before the initial percutaneous atherectomy intervention.

The authors conclude that this study is the first to critically evaluate the long-term patency and limb salvage of patients with symptomatic peripheral arterial disease utilizing directional atherectomy. There was a low instance of complications and low rate of limb loss associated with percutaneous peripheral arterial interventions using atherectomy. The potential for surgical bypass was not changed by performing endovascular intervention first in all cases in this series. They feel that endovascular intervention with Silverhawk directional atherectomy can result in excellent secondary patency and limb salvage for patients with claudication and critical limb ischemia.

T. G. Walker, MD

References

1. Reekers J. Challenging a myth: directional atherectomy. *Cardiovasc Intervent Radiol.* 2009;32:203-204.
2. Vorwerk D. Challenging the myth: directional atherectomy—a reply. *Cardiovasc Intervent Radiol.* 2009;32:205.

Predicting embolic potential during carotid angioplasty and stenting: Analysis of captured particulate debris, ultrasound characteristics, and prior carotid endarterectomy

Malik RK, Landis GS, Sundick S, et al (Mount Sinai School of Medicine, NY)
J Vasc Surg 51:317-322, 2010

Introduction.—Extracranial carotid stenoses exhibit significant variance in embolic potential, with restenotic lesions having a particularly low propensity for embolization. This study sought to identify characteristics associated with increased generation of embolic debris during carotid angioplasty and stenting (CAS).

Methods.—Captured particulate was available for analysis in 56 consecutive patients. Demographics were mean age, 74 years (range, 60-94 years); mean stenosis, 88% (range, 70%-99%); symptomatic, 27%; prior carotid endarterectomy (CEA), 27%; prior radiotherapy, 7%. Plaque echogenicity, heterogenicity, ulceration, and irregularity were assessed with B-mode duplex ultrasound analysis. Gray scale median (GSM) was calculated from normalized B-mode VHS video recordings. Calcification and degree of stenosis were determined angiographically. Captured particulate debris was evaluated for total number; number >200 μm, >500 μm, >1000 μm; mean and median size. Hematoxylin and eosin, trichrome, and von Kossa stains were used for histologic analysis of captured material.

Results.—Restenotic carotid stenoses after prior CEA generated minimal embolic debris compared with primary stenoses. Four of 15 patients (27%) with restenotic lesions demonstrated embolic particles; all debris was <500 μm. All 41 patients with primary stenoses had some embolic debris; particulate size was >200 μm in 91%, >500 μm in 72%, and >1000 μm in 43%. In primary lesions, the number and size of captured particulate correlated with GSM and with the combined ultrasound findings of echogenicity, heterogenicity, and luminal irregularity/ulceration ($P < .02$, 95% confidence interval, 4.5-27.6). None of these ultrasound factors correlated independently with embolic particulate ($P =$ NS). Patients aged >70 years exhibited more total particles (8.1 vs 2.3, $P = .008$) and increased mean particle size (370 vs 157 μm, $P = .02$). No significant correlation was observed between the number and size of captured embolic particulate and any other variable (stenosis percentage, prior radiotherapy, preprocedural symptoms, periprocedural symptoms, and calcification). Histologically, the embolic debris consisted of extensive amorphous, acellular proteinaceous material. Calcium debris in the embolic particulate was associated with heavily and moderately calcified lesions.

Conclusions.—Considerable variation exists in the number and size of embolic particles generated during CAS. Embolic potential is positively correlated with lesion GSM and the combination of lesion echogenicity, heterogenicity, and irregularity. Restenosis after prior CEA is associated with minimal embolic particulate generation, suggesting that embolic protection may not be necessary for CAS of restenotic lesions.

▶ The safety of carotid angioplasty and stenting (CAS) as an alternative to CEA with regard to periprocedural cerebrovascular events is the subject of ongoing investigation. Embolization of atheromatous and thrombotic debris during angioplasty and stent deployment represents one of the significant risks of this procedure.

Multiple factors increase the risk of embolization from carotid lesions. Investigators have shown that anatomic considerations and plaque characteristic are likely the patient-dependent factors that lead to the generation of debris. A more important finding is that the characteristics of the carotid lesion itself have been correlated with embolic potential. Intraplaque hemorrhage and an echolucent appearance on ultrasound imaging have been associated with increased rates of symptomatic presentation in patients undergoing CEA. In addition, restenotic intimal hyperplastic carotid lesions that occur after CEA are associated with lower rates of symptomatic presentation compared with primary atherosclerotic carotid stenosis.

This study evaluated the relationship between the size and number of embolic debris retained in the embolic protection device during CAS procedures with lesion type (primary or restenotic), lesion characteristics, and the patient age. The CAS procedures were for primary atherosclerotic occlusive disease in 41 patients (73%) and for restenosis after previous CEA in 15 (26%). All patients with restenosis presented ≤2 years of the primary procedure, plaque echogenicity, heterogenicity, ulceration, and irregularity were assessed with duplex ultrasound analysis. Calcification and degree of stenosis were determined angiographically. Captured particulate debris was available for analysis in 56 consecutive patients. Embolic debris was recovered from 100% of primary lesions and 27% of restenotic lesions. The mean number of particles recovered in each specimen correlated with lesion type (13.2 for primary vs 0.7 for restenosis). Captured particulate debris was evaluated for total number (number > 200 μm, > 500 μm, > 1,000 μm), mean, and median size. Hematoxylin and eosin, trichrome, and von Kossa stains were used for histologic analysis of the captured material.

The investigators reported that patients with primary atherosclerotic disease had embolic debris 100% of the time compared with 27% of those with restenotic lesions, a difference that was highly statistically significant. Those patients with restenotic lesions demonstrated embolic particles in which all debris was < 500 μm. Of the patients with primary stenoses, the particulate size of the embolic debris was > 200 μm in 91%, > 500 μm in 72%, and > 1,000 μm in 43%. Thus, the patients with primary disease had an increased number of particles (13.2) and a larger mean particle size (382 μm) compared with patients with restenotic lesions, which was again highly statistically significant. This relates to the theory that because primary lesions are more friable and unstable, there is a larger risk of embolization during CAS procedures in de novo lesions compared with the early restenotic lesions.

In primary lesions, the number and size of captured particulate correlated with the combined ultrasound findings of echogenicity, heterogenicity, and luminal irregularity/ulceration (95% confidence interval, 4.5–27.6; $P < .02$). None of these ultrasound factors correlated independently with embolic particulate. Patients aged > 70 years exhibited more total particles (8.1 vs. 2.3; $P = .008$) and increased mean particle size (370 vs. 157 μm; $P = .02$).

The authors concluded that there is considerable variation in the number and size of embolic particles generated during CAS. Embolic potential is positively correlated with the combination of lesion echogenicity, heterogeneity, and irregularity. Restenosis after previous CEA is associated with minimal embolic particulate generation, suggesting that embolic protection may be optional for CAS of restenotic lesions. No significant correlation was observed between the number and size of captured embolic particulate and any other variable, such as percentage of stenosis, previous radiotherapy, preprocedural symptoms, periprocedural symptoms, or calcification.

T. G. Walker, MD

Primary Patency of Wallstents in Malignant Bile Duct Obstruction: Single vs. Two or More Noncoaxial Stents

Maybody M, Brown KT, Brody LA, et al (Memorial Sloan-Kettering Cancer Ctr, NY)

Cardiovasc Intervent Radiol 32:707-713, 2009

The purpose of this study was to determine the primary patency of two or more noncoaxial self-expanding metallic Wallstents (Boston Scientific, Natick, MA) and to compare this with the primary patency of a single stent in malignant bile duct obstruction. From August 2002 to August 2004, 127 patients had stents placed for malignant bile duct obstruction. Forty-five patients were treated with more than one noncoaxial self-expanding metallic stents and 82 patients had a single stent placed. Two patients in the multiple-stent group were lost to follow-up. The primary patency period was calculated from the date of stenting until the first poststenting intervention for stent occlusion, death, or the time of last documented follow-up. The patency of a single stent was significantly different from that of multiple stents ($P = 0.0004$). In the subset of patients with high bile duct obstruction, the patency of a single stent remained significantly different from that of multiple stents ($P = 0.02$). In the single-stent group, there was no difference in patency between patients with high vs. those with low bile duct obstruction ($P = 0.43$). The overall median patency for the multistent group and the single-stent group was 201 and 261 days, respectively. In conclusion, the patency of a single stent placed for malignant low or high bile duct obstruction is similar, and significantly longer than, that of multiple stents placed for malignant high bile duct obstruction. Given the median patency of 201 days, when indicated, percutaneous stenting of multiple bile ducts is an effective palliative measure for patients with malignant high bile duct obstruction.

▶ In this study the authors retrospectively reviewed patency rates of Wallstents placed for malignant bile duct obstruction to determine whether placing 2 or more non-coaxial Wallstents results in a different stent patency rate than placing only a single Wallstent. The only stent used in this study was the self-expanding non-covered metallic Wallstent (Boston Scientific, Natick, MA). In the article the authors define "coaxial" as placement of more than one stent overlapping each other at one or both ends over a single axis. The word "non-coaxial" refers to placement of more than 1 stent in a non-overlapping configuration, so that they do not share a single axis. This is then further defined so that "coaxial" stents that are placed to treat a single bile duct are considered a single stent whereas the term "multiple stents" refers only to a "non-coaxial" stent configuration.

High obstruction is defined as obstruction involving the confluence or higher level, such that one catheter or stent would not definitely drain the entire biliary tree. It includes Bismuth types II–IV. Low obstruction is defined as obstruction below the confluence, such that all the intrahepatic ducts could be drained via 1 catheter or stent and includes Bismuth type I.

In this retrospective study, 131 consecutive patients were identified who had Wallstents placed percutaneously for bile duct obstruction. Forty-six patients were treated with more than one non-coaxial stent; 85 patients had a single stent placed. Patients were excluded for lack of follow-up data or because they were stented for benign disease. Three patients were excluded from the multi-stent group due to lack of follow-up information ($n = 2$) or stent placement for benign disease ($n = 1$). Three patients from the single stent group were excluded because they had stents placed for benign disease. The primary patency of stents in the remaining 125 patients was determined retrospectively by reviewing medical records, interventional radiology reports, relevant interventional and cross-sectional imaging studies, physician progress notes, clinic visits, telephone contacts, and death notifications. The primary patency period was calculated from the date of stenting until the first post-stenting intervention for stent occlusion, death, or the time of last documented follow-up.

TABLE 1.—Overall Statistical Data on Biliary Stent Placement in 131 Patients From August 2002 to August 2004

	Single Stent	Multiple Stents
Total patients (*N* = 131)	85	46
Benign biliary obstruction	3	1
Lack of follow-up	0	2
Patients in study (*N* = 125)	82	43
No. of stents per patient		
1	82	
2		29
>2		14
Median patency relative to no. of stents (days)		
1 stent	261	
2 stents		201
>2 stents		181
Overall		201
Level of obstruction		
High	39	43
Low	43	
Median patency relative to level of obstruction (days)		
High obstruction	249	201
Low obstruction	271	
Overall	261	
Etiology (most common)		
High obstruction		
Cholangiocarcinoma	16 (41%)	9 (21%)
Colorectal CA	7 (18%)	6 (14%)
Gallbladder CA	4 (10%)	18 (42%)
Pancreatic CA	4 (10%)	6 (14%)
Low obstruction		
Gastric CA	3 (7%)	
Pancreatic CA	34 (79%)	

CA cancer

Primary patency was calculated from the date the original stent or stents were placed to the first post-stenting procedure necessary to reestablish stent patency or relieve obstruction, until death of the patient, or until the latest follow-up at the time of data analysis (Table 4) The overall median patency of 261 days in the single-stent group was significantly longer than the 201 days in the multiple-stent group ($P = .0004$) (Fig 4).

All patients with low bile duct obstruction (n = 43) received a single stent. Among the patients with high bile duct obstruction (n = 82) a single stent was placed in 39 patients, with a median patency of 249 days, and multiple stents were placed in 43 patients, with a median patency of 201 days (Table 1). In the subset of patients with high bile duct obstruction, the patency rate of a single stent was significantly longer than that of multiple stents ($P = .02$). Cumulative risk regression was performed to determine if age at the time of biliary stenting had any effect on stent patency, and the result was non-significant. In this series the 30-day mortality rate was 9% (n = 4). Deaths did not occur as a result of stent placement but resulted from progression of disease (3 patients) and pre-existing sepsis with multiorgan failure (1 patient).

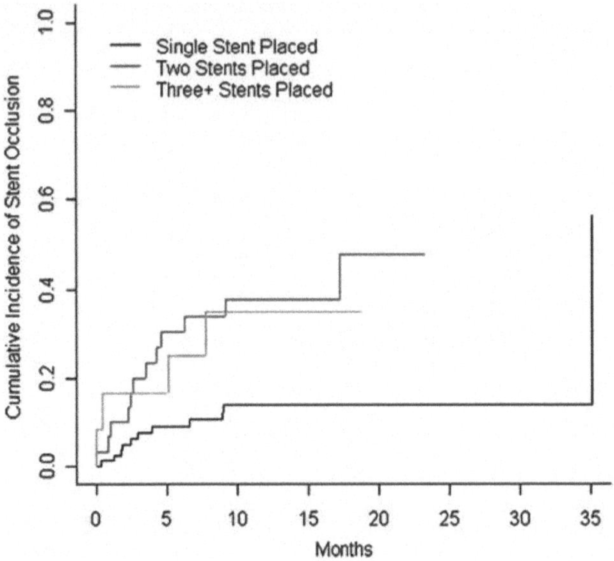

FIGURE 4.—Kaplan–Meier curve of the cumulative incidence of stent occlusion with regard to number of stents. The patency rate for a single stent was significantly different from that for multiple stents ($P = 0.0004$), while the patency rate for two stents was not different from that for three or more stents. (With kind permission from Springer Science+Business Media: Maybody M, Brown KT, Brody LA, et al. Primary patency of wallstents in malignant bile duct obstruction: single vs. two or more non-coaxial stents. *Cardiovasc Intervent Radiol.* 2009;32:707-713.)

The authors conclude that patients requiring only 1 stent to adequately palliate their malignant biliary obstruction have a median stent patency which is significantly longer than that of patients who require multiple non-coaxial stents to do so. When indicated, multiple non-coaxial stents, with a median patency of 7.6 months, may provide adequate palliation for many patients with malignant biliary obstruction.

T. G. Walker, MD

Revascularization versus Medical Therapy for Renal-Artery Stenosis

ASTRAL Investigators (Univ of Birmingham, UK)
N Engl J Med 361:1953-1962, 2009

Background.—Percutaneous revascularization of the renal arteries improves patency in atherosclerotic renovascular disease, yet evidence of a clinical benefit is limited.

Methods.—In a randomized, unblinded trial, we assigned 806 patients with atherosclerotic renovascular disease either to undergo revascularization in addition to receiving medical therapy or to receive medical therapy alone. The primary outcome was renal function, as measured by the reciprocal of the serum creatinine level (a measure that has a linear

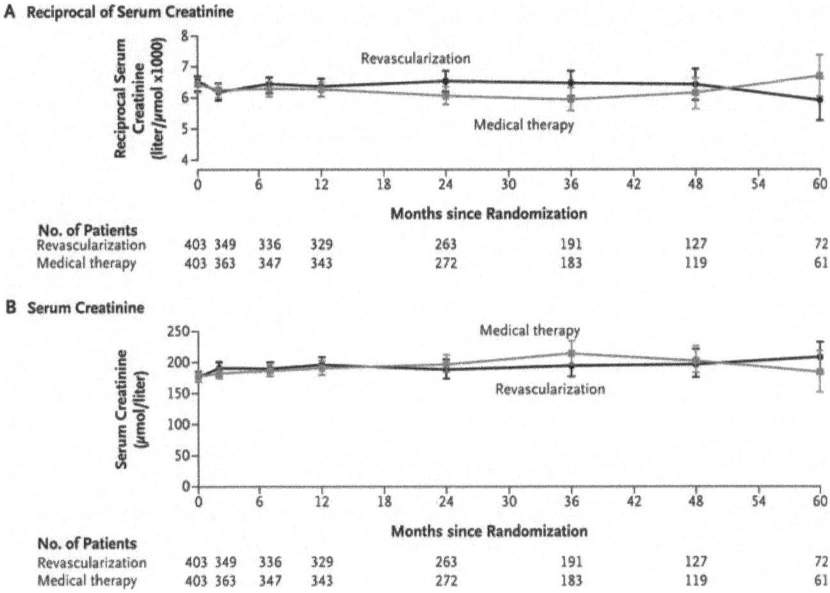

FIGURE 1.—Renal Function in Patients with Renal-Artery Stenosis Treated with Revascularization or Medical Therapy Alone. Shown are mean values for the reciprocal of the serum creatinine level (Panel A) and for the serum creatinine level (Panel B). The second measures for both values were performed 1 to 3 months after baseline; the third measures were performed 6 to 8 months after baseline. The I bars indicate 95% confidence intervals. (Reprinted from The ASTRAL Investigators. Revascularization versus medical therapy for renal-artery stenosis. *N Engl J Med.* 2009;361:1953-1962. Copyright 2009 Massachusetts Medical Society. All rights reserved.)

relationship with creatinine clearance). Secondary outcomes were blood pressure, the time to renal and major cardiovascular events, and mortality. The median follow-up was 34 months.

Results.—During a 5-year period, the rate of progression of renal impairment (as shown by the slope of the reciprocal of the serum creatinine level) was -0.07×10^{-3} liters per micromole per year in the revascularization group, as compared with -0.13×10^{-3} liters per micromole per year in the medical-therapy group, a difference favoring revascularization of 0.06×10^{-3} liters per micromole per year (95% confidence interval [CI], −0.002 to 0.13; P = 0.06). Over the same time, the mean serum creatinine level was 1.6 µmol per liter (95% CI, −8.4 to 5.2 [0.02 mg per deciliter; 95% CI, −0.10 to 0.06]) lower in the revascularization group than in the medical-therapy group. There was no significant between-group difference in systolic blood pressure; the decrease in diastolic blood pressure was smaller in the revascularization group than in the medical-therapy group. The two study groups had similar rates of renal events (hazard ratio in the revascularization group, 0.97; 95% CI, 0.67 to 1.40; P = 0.88), major cardiovascular events (hazard ratio, 0.94; 95% CI, 0.75 to 1.19; P = 0.61), and death (hazard ratio, 0.90; 95% CI, 0.69 to 1.18; P = 0.46). Serious complications associated with revascularization occurred in 23 patients, including 2 deaths and 3 amputations of toes or limbs.

Conclusions.—We found substantial risks but no evidence of a worthwhile clinical benefit from revascularization in patients with atherosclerotic renovascular disease. (Current Controlled Trials number, ISRCTN59586944) (Fig 1).

▶ Renal artery stenosis, a common manifestation of atherosclerosis, is associated with both hypertension and chronic kidney disease, although it is not clear whether these associations are causal. Treatment has traditionally focused on correcting renal-artery stenosis, with endovascular revascularization replacing open surgical techniques as the preferred intervention. The benefit versus the associated potential risks of such interventions remains unclear, however. Three small, randomized, controlled trials showed no significant benefits of angioplasty over medical therapy,[1-3] but these studies were underpowered, even when their results were combined, to detect possible moderate but clinically worthwhile improvements in renal function or blood pressure or a reduction in mortality. The Angioplasty and Stenting for Renal Artery Lesions (ASTRAL) trial was designed to determine whether revascularization together with medical therapy improves renal function and other outcomes, as compared with medical therapy alone, in patients with atherosclerotic renal artery stenosis.

In this multicenter, randomized, unblinded clinical trial, patients were screened for enrollment if clinical findings (eg, uncontrolled or refractory hypertension or unexplained renal dysfunction) suggested a diagnosis of atherosclerotic renovascular disease. Patients with such findings underwent renal-artery imaging (with the use of intra-arterial, computed tomographic, or magnetic resonance angiography), renal ultrasonography, and other relevant laboratory and clinical assessments. They were eligible to participate if they had substantial anatomical atherosclerotic stenosis in at least 1 renal artery that was considered potentially suitable for endovascular revascularization and if the patient's doctor was uncertain that the patient would definitely have a worthwhile clinical benefit from revascularization, based upon available evidence. Patients were ineligible if they required surgical revascularization or were considered to have a high likelihood of requiring revascularization within 6 months, if they had nonatheromatous cardiovascular disease, or if they had undergone previous revascularization for renal-artery stenosis. Patients were randomly assigned either to undergo revascularization in addition to receiving medical therapy or to receive medical therapy alone, in a 1:1 ratio.

Patients in the revascularization group underwent the procedure as soon as possible after randomization. The precise revascularization procedure (angioplasty either alone or with stenting) was determined by local practitioners, and renal protection devices were not used. Patients in both study groups received medical therapy, according to local protocols; this therapy typically consisted of statins, antiplatelet agents, and optimal blood-pressure control.

The primary outcome was the change in renal function, which was assessed by measuring the mean slope of the reciprocal of the serum creatinine level over time. Secondary outcomes included blood pressure, the time to the first renal event, the time to the first major cardiovascular event, and mortality. Renal events were defined as a new onset of acute kidney injury, the initiation of dialysis, renal transplantation, nephrectomy, or death from renal failure.

Fig 1 summarizes the results of renal function in patients with renal artery stenosis treated with revascularization or medical therapy alone. In a per-protocol analysis, there was no significant difference in the primary outcome between the 317 patients who underwent successful revascularization and the 379 patients who received medical-therapy only. Also during the 5-year study period, systolic blood pressure decreased in the 2 study groups, with no significant difference between the groups. As regards renal events, cardiovascular events, and mortality among patients who underwent at least 1 follow-up assessment, 73 renal events occurred in 57 patients in the revascularization group, as compared with 80 events in 58 patients in the medical therapy group ($P = .97$), 238 cardiovascular events were reported in 141 patients in the revascularization group, as compared with 244 events in 145 patients in the medical-therapy group ($P = .96$), and there was no statistical significance between group difference in overall survival (hazard ratio in the revascularization group, 0.90; 95% confidence interval [CI], 0.69 to 1.18; $P = .46$), with 103 deaths in the revascularization group and 106 deaths in the medical therapy group. A total of 38 periprocedural complications (defined as complications occurring within 24 hours after the procedure) were reported in 31 of the 359 patients (9%) who underwent revascularization; 19 of these events (in 17 patients) were considered to be serious complications.

The investigators concluded that revascularization carried substantial risk but was not associated with any benefit with respect to renal function, blood pressure, renal or cardiovascular events, or mortality in patients with atherosclerotic renal-artery stenosis. The upper confidence limits for a benefit from revascularization with respect to renal function were below levels that would be considered clinically relevant. No significant improvements in blood pressure or reductions in renal or cardiovascular events or mortality were seen. The authors noted, however, that there was an important limitation to the trial concerning the population that was studied: as noted, patients were enrolled in the trial only if their own physician was uncertain as to whether revascularization would provide a worthwhile clinical benefit. There is a belief (not evidence-based) that certain groups of patients with severe renal-artery stenosis (eg, those presenting with acute kidney injury or "flash" pulmonary edema) should be treated with revascularization. The investigators felt that by excluding such patients, the trial population remained representative of patients undergoing revascularization in clinical practice.

T. G. Walker, MD

References

1. Plouin PF, Chatellier G, Darne B, Raynaud A. Blood pressure outcome of angioplasty in atherosclerotic renal artery stenosis: a randomized trial. *Hypertension.* 1998;31:823-829.
2. Webster J, Marshall F, Abdalla M, et al. Randomized comparison of percutaneous angioplasty vs. continued medical therapy for hypertensive patients with atheromatous renal artery stenosis. *J Hum Hypertens.* 1998;12:329-335.

3. van Jaarsveld BC, Krijnen P, Pieterman H, et al. The effect of balloon angioplasty on hypertension in atherosclerotic renal artery stenosis. *N Engl J Med.* 2000;342: 1007-1014.

Safety and Short-term Outcomes following Controlled Blunt Microdissection Revascularization of Symptomatic Arterial Occlusions of the Pelvis and Lower Extremities

Thatipelli MR, Misra S, Sanikommu SR, et al (St Elizabeth's Med Ctr, Boston, MA; Mayo Clinic, Rochester, MN)
J Vasc Interv Radiol 20:1541-1547, 2009

Purpose.—To present the safety and short-term outcomes of using controlled blunt microdissection catheter-assisted revascularization of symptomatic chronic total occlusions of the lower extremity.

Materials and Methods.—A retrospective study was performed on 61 patients (46 men) with a mean age of 72.3 years ± 9.4 who underwent 67 procedures in 86 arteries between June 2003 and March 2007 for claudication (38 procedures, 57%), rest pain (19 procedures, 28%), and tissue loss (10 procedures, 15%). Technical success was defined as successful traversal of the occlusion. Duplex ultrasonography (US) was used to assess patency. Clinical patency was defined as at least one category improvement in Rutherford score from baseline and absence of target limb revascularization or major amputation at 6 months.

Results.—Chronic total occlusions were located in aortoiliac (11 arteries, 13%), infrainguinal (72 arteries, 83%), and infrapopliteal (four arteries, 5%) arteries. The mean lesion length was 14.2 cm ± 8. The tibial run-off vessels was 1.9 vessels ± 0.8. The technical success rate of the procedure was 84%. Advanced age ($P = .04$), renal function ($P = .02$), and target lesion length ($P \leq .01$) were predictors of technical failure. The clinical success rate at 6 months was 92%, and the primary patency with duplex US was 87%. Renal function ($P \leq .01$), length of the occlusion ($P \leq .01$), number of stents per procedure ($P \leq .01$), and tibial run-off vessels ($P = .05$) were the predictors of clinical success.

Conclusions.—The controlled blunt microdissection catheter is safe in the revascularization of chronic total occlusions of the lower extremity. The technical success rate was 84% and predicted by age, renal function, and lesion characteristics. Clinical patency at 6 months was 92% and predicted by renal function, lesion characteristics, and run-off.

▶ Surgical bypass remains the gold standard for long femoropopliteal occlusions (TransAtlantic InterSociety Consensus [TASC] C and D classes). Chronic total occlusions are characterized by fibrous and calcified lesions that have poor technical success, clinical success, and long-term patency rates. It is hypothesized that this is also from the dense calcification that can often be present at the proximal and distal ends of the occlusion. As a consequence, minimally invasive devices have been invented to improve the revascularization rates of

TABLE 1.—Demographic and Clinical Characteristics at Baseline

Parameter	Findings	No. of Cases from Which Data were Available
No. of patients	61	
No. of men	46 (75)	61
Mean age (y)	72.3 ± 9.4	61
No. of Caucasians	57 (93)	61
No. of procedures	67	
No of revascularized arterial segments	87	
Cardiovascular risk factors		
Hypertension	60 (98)	61
Diabetes	27 (44)	61
Dyslipidemia	61 (100)	61
Smoking	47 (77)	60
Coronary artery disease	43 (70)	61
Indication for revascularization (no. of procedures)		67
Severe claudication	38 (57)	
Rest pain	19 (28)	
Tissue loss	10 (15)	
Mean resting ABI of the reference limb	0.55 ± 0.22	45
Mean baseline serum creatinine level (mg/dL)	1.34 ± 0.69	67
Baseline GFR (mL/min)*	49.71 ± 17.81	67

Note.—Numbers in parentheses are percentages.
Editor's Note: Please refer to original journal article for full references.
*The GFR was more than 30 in 54 of the 67 procedures (79%) and less than or equal to 30 mL/min in 14 (22).

these vessels in patients in whom surgery is not an option. One such device is the Frontrunner controlled blunt microdissection catheter (Cordis Endovascular, Miami, Florida). Although some endovascular specialists have adopted this device for peripheral vessels, there are few studies regarding its safety, efficacy, and clinical outcomes. This study looks at safety and short-term outcomes following revascularization of symptomatic lower-extremity chronic total occlusions with this device and also examines the predictors of technical and short term clinical success.

This retrospective study evaluated 61 consecutive patients who underwent 67 procedures for 86 chronic totally occluded arteries as the study sample. All patients underwent controlled blunt microdissection catheter–assisted percutaneous revascularization of symptomatic arterial chronic total occlusions of the lower extremities. All study patients were evaluated before the procedure and at follow-up 1, 6, and 12 months after the procedure with noninvasive vascular studies.

The device used is the Frontrunner XP catheter, which is designed to cross arterial occlusions by means of blunt microdissection, which creates a passage through or along the occluded vessel. It is intended to facilitate the intraluminal placement of conventional guide wires beyond the occluded segment before further percutaneous intervention. It is available with shaft lengths of 90 cm and 140 cm and has an outside diameter of 5.4 F terminating in a radiopaque blunt distal assembly. This articulated nitinol dissecting assembly with hinged

TABLE 2.—Angiographic Characteristics

Parameter	No. of Cases	No. of Cases from which Data are Available
Location of chronic total occlusion (arterial segments = 87)		87
Aortoiliac	11 (13)	
Femoropopliteal	72 (83)	
Infrapopliteal	4 (5)	
Access		67
Femoral (retrograde)	59 (88)	
Femoral (antegrade)	3 (5)	
Brachial	5 (7)	
Anticoagulation		66
Heparin	60 (91)	
Angiomax	6 (9)	
Iliac artery tortuousity		51
0–1	32 (63)	
2–3	19 (37)	
Calcification severity		62
0–1	32 (52)	
2–3	30 (48)	
Mean target lesion length	14.13 ± 8	56
<10 cm	25 (45)	
≥10 cm	31 (55)	
TASC grade A	0	65
B	29 (45)	
C	20 (31)	
D	16 (25)	
Procedural complications		67
Local complications (related to the Angioseal [Kensey Nash Corporation, Exton, Pennsylvania])	2	
Acute in-stent thrombosis	3	
Technical success with Frontrunner alone	29 (53)	55
Reentry with Glidewire	18 (33)	
Reentry assisted by reentry catheter	8 (14)	
Outback	2	
Pioneer	6	
Mean no. of stents used per procedure	3 ± 1.7	53
Postprocedural crural run-off	1.9 ± 0.8	40
Good	23 (58)	
Poor	17 (42)	

Note.—Numbers in parentheses are percentages.

jaws can be opened, rotated, and advanced by using the distal control knob to displace, separate, and fracture atheromatous plaque in the occluded artery.

Diagnostic angiography was performed in all patients and the reference chronic total occlusion, length, and calcification were defined by using the TASC classification system. A hydrophilic-coated straight or angled-tip guide wire was used to assist the controlled blunt microdissection catheter traverse the occlusion either intraluminally or subintimally. Once the occlusion was successfully traversed, the catheter was removed over a 0.18- or 0.035-inch non-hydrophilic guide wire; subsequently, a stent was deployed after

TABLE 3.—Summary of Results

Parameter	No. of Cases (%)	No. of Cases from Which Data are Available
Technical failure of the device	11 (16)	67
Technical success of the device*	56 (84)	67
6-mo outcomes		
Clinical patency	50/54 (92)	54
Primary patency (by duplex)	35/41 (87)	41
Secondary patency (by duplex)	37/41 (92)	41
Hemodynamic success	14/15 (93)	15
Survival at 6 mo	53/56 (95)	56
Outcome of clinical failures at 6 mo†		54
Percutaneous revascularization	2	
Below-knee amputation	1	

Note.—Numbers in parentheses are percentages.
*Technical success of the device was defined as successful traversal of the occlusion by using the Frontrunner +/– reentry device usage.
†One patient underwent femorofemoral crossover bypass 8 months after index percutaneous revascularization.

TABLE 4.—Predictors of Technical and 6-month Clinical Success

Technical Success	*P* Value	Clinical Success	*P* Value
Age	.04	Age	.12
Sex	.99	Sex	.15
Hypertension	.53	Hypertension	.69
Diabetes	.65	Diabetes	.43
Dyslipidemia	NA	Dyslipidemia	NA
Smoking	.07	Smoking	.54
Coronary artery disease	.38	Coronary artery disease	.84
Baseline resting ABI of index limb	.93	Baseline resting ABI of index limb	.17
Baseline Rutherford clinical score of index limb	.71	Baseline Rutherford clinical score of index limb	.53
Baseline serum creatinine level	.02	Baseline serum creatinine level	.56
Access	.55	Baseline GFR	.01
Target lesion location	.48	Access	.59
Target lesion length	<.01	Target lesion location	.39
		Target lesion length	.01
		No. of stents used per procedure	<.01
		Postprocedural crural run-off	.05

Note.—NA = not applicable; 100% of study patients had dyslipidemia.

pre-dilating the vessel. The diameter and length of the stent used was determined according to the length of the occlusion by using the TASC classification and the diameter of the proximal non-diseased vessel. In the case of inadvertent subintimal passage of the catheter or Glidewire, a lumen reentry catheter (Outback, [Cordis Endovascular] or Pioneer [Medtronic AVE, Minneapolis, Minnesota]) was used to gain access into the true lumen. A variety of different self-expandable nitinol stents were used in most procedures except those in the common iliac arteries, where balloon- expandable stents were used because of their high radial strength. Patients were monitored overnight and discharged the following day on dual antiplatelet therapy consisting of 75 mg of clopidogrel bisulfate by mouth for 1 month and 325 mg of aspirin indefinitely.

Tables 1 and 2 respectively detail the demographics/clinical characteristics at baseline and the angiographic characteristics of the study group. In 29 of 67 procedures (53%), use of the controlled blunt dissection catheter alone was successful. In 18 procedures (32%), the Glidewire was needed and in 8 procedures (14%) either the Outback (n = 2) or Pioneer (n = 6) reentry catheter was used. In 8 procedures, the Frontrunner was successful in true lumen reentry after failed initial attempts with the Glidewire. The mean number of stents used per procedure was 3 ± 1.7. At completion, all arterial segments distal to the reference artery were found to be widely patent with brisk antegrade blood flow. The mean number of tibial artery run-off vessels after the procedure was 1.9 ± 0.8, with 23 of 40 (57%) having good and 17 of 40 (43%) having poor run-off. The technical success rate of the device was 84% (56 out of 67). There were 12 technical failures, and these were secondary to the inability to advance either the controlled blunt microdissection catheter or the Glidewire through the proximal cap (n = 3), body of occlusion (n = 2), or reentry into the true lumen (n = 6). The clinical patency rate at 6 months was 92% (data were available for 50 out of 54 patients). Among the 4 clinical failures, 2 patients had percutaneous revascularizations, 1 patient had below-knee amputation within 6 months of revascularization, and 1 patient had femoral-femoral crossover bypass grafting 9 months after the index procedure for occluded iliac artery. The duplex ultrasound (US) guided primary patency rate was 87% (data available = 68%), and the hemodynamic success rate was 97% (data available = 70%). The secondary patency rate at 6 months was 92% (data available = 66%). These results are summarized in Table 3, and the predictors of clinical success are noted in Table 4.

This single-center study shows results for safety, technical, short-term clinical, and hemodynamic success rates of controlled blunt microdissection catheter-assisted (Frontrunner XP) revascularization of symptomatic lower-extremity chronic total occlusions. The technical success rate with use of this catheter was 84% in the study, and technical success was predicted by age, renal function, and lesion characteristics. Clinical patency at 6 months was 92% and was predicted by renal function, lesion characteristics, and run-off. The role of endovascular revascularization for symptomatic chronic total occlusion has undergone a paradigm change during the past decade, moving from surgical revascularization as the gold standard to minimally invasive therapies. The strength of endovascular therapy is that it has decreased morbidity and mortality while helping preserve the venous conduits, although the long-term patency rates remain inferior to those of surgical bypass. Most patients with chronic total occlusions presenting with critical limb ischemia have prohibitive mortality risk from surgical bypass due to concomitant cardiovascular conditions. Also, the current experience is that the treatment of critical limb ischemia supports the fact that the presence of long term patency of the treated vessel following successful revascularization neither correlates nor is essential for symptomatic resolution or limb salvage. The reported technical success rates of various conventional devices in the revascularization of lower limb chronic total occlusions range from 26% to 100%. The wide range of rates highlights the fact that several variables influence the technical success, including lesion location, length, calcification, type of device used, clinical stage at intervention, and operator's experience.

Given the increasing choice of devices that are available to assist in the endovascular treatment of chronic total occlusions, it is important to have objective data regarding their use. This study examines data for the Frontrunner controlled blunt microdissection catheter and shows that it is safe and associated with a low 30-day mortality rate when used for revascularization of lower extremity symptomatic chronic total occlusions. Age, baseline renal function, and target lesion characteristics are predictive of technical success. A technically successful revascularization with controlled blunt microdissection is associated with a short-term clinical patency rate of 92% and a primary patency rate of 87%. Baseline renal function, lesion length, number of stents deployed, and crural run-off are predictive of the 6-month clinical patency rate.

T. G. Walker, MD

Subintimal Angioplasty: Predictors of Long-term Success

Keeling AN, Khalidi K, Leong S, et al (Beaumont Hosp, Dublin, Ireland; et al)

J Vasc Interv Radiol 20:1013-1022, 2009

Purpose.—To determine the clinical outcomes and success rates after percutaneous subintimal angioplasty (SIA) in patients with lower-limb occlusive lesions causing intermittent claudication (IC) or critical limb ischemia (CLI) at midterm to long-term follow-up. The secondary aim was to elicit factors predictive of a successful outcome.

Materials and Methods.—Between January 1999 and June 2006, 75 consecutive patients (45 men; age range, 46–91 years; CLI in 79%) underwent SIA of iliac and infrainguinal (84%) occlusions. Outcomes were determined on an intent-to-treat basis. The composite endpoint of interest was major adverse clinical outcome (MACO) of the treated limb at follow-up, which was defined as the development of IC, CLI, or need for subsequent endovascular or surgical revascularization. Actuarial freedom from MACO was assessed via Kaplan-Meier curves and multivariable Cox proportional-hazards regression.

Results.—SIA was performed on 75 lesions, with an initial procedure success rate of 83% ($n = 62$). Procedure failure was caused by heavily calcified lesions ($n = 5$) and failure of reentry ($n = 8$). A total of 56.3% of patients with claudication were free from ipsilateral claudication at follow-up (mean, 32 months; range, 1–64 months), and those with CLI had a 79.7% limb salvage rate at a mean follow-up of 30.7 months (range, 0.5–91 months). On Cox regression analysis, the following variables were identified as independent predictors of MACO within the limb treated with SIA: ABI after SIA (hazard ratio, 0.21; 95% CI, 0.05–0.89; $P = .035$) and number of patent runoff vessels (ie, ≤ 1 vs ≥ 2; hazard ratio, 0.29; 95% CI, 0.15–0.59; $P = .001$).

TABLE 2.—Characteristics of Lesions Treated with SIA

Variable	Value
Mean ABI (range)	
Before SIA	0.50 (0.15–1.1)
After SIA	0.74 (0.3–1.21)
Site	
Aortoiliac	12 (16)
Femoropopliteal	63 (84)
Side	
Right	38 (50.7)
Left	37 (49.3)
Severity	
Occlusion	100 (100)
Stenosis	0
Mean (range) length (mm)	64 (20–120)
Lesion calcification	47 (62.7)
Balloon diameter (mm)	
Mean	5.5
Range	3–8
Technical outcome*	
Success	62 (82.7)
Failure	13 (17.3)

Note.—Values in parentheses are percentages unless specified otherwise.
*Percentages of percutaneous transluminal angioplasty cases are shown.

Conclusions.—SIA is a feasible therapeutic option for occlusive atherosclerotic lesions in IC and CLI and is the evolving preferred strategy in CLI and perhaps IC with long-segment occlusions.

▶ Subintimal angioplasty (SIA) or percutaneous intentional extraluminal revascularization, in which there is intentional subintimal dissection with subsequent generation of a new lumen, is frequently used in treating occluded lower-limb arteries. Studies have reported successful clinical outcomes and favorable patency rates for this method of recanalizing occluded vascular segments. Primary patency rates at 12-month follow-up vary among published reports, from 22% to as high as 92%. However, the average 1-year primary patency rate of 50% is lower than the figures of 80% to 90% quoted for surgical bypass. The long-term clinical outcome, measured by limb salvage rates and improvement in claudication distance is the most important outcome measure from the patient's perspective.

There is 1 report that limb salvage rates after SIA compare favorably with those of surgical bypass. For SIA to be a viable alternative to surgical bypass, long-term patency and limb salvage rates need to be evaluated and compared with those associated with surgery. SIA has less associated morbidity and less mortality and is economically more viable than surgical bypass. Also, if bypass is ultimately required, an earlier SIA procedure generally does not interfere with the surgical procedure or the collateral supply.

At present, there are few publications determining the midterm to long-term outcomes after SIA. This study was performed to determine the midterm to long-term clinical outcomes and success rates after SIA in lower-limb occlusive

TABLE 3.—Association between Variables and MACO

Variable	MACO	No MACO	*P* Value
Female sex	11 (30.6)	19 (48.7)	.109
Symptoms: CLI vs IC	29 (80.6)	30 (76.9)	.701
SIA level: aortoiliac segment	7 (19.4)	5 (12.8)	.434
Right-sided SIA	16 (44.4)	22 (56.4)	.300
Lesion calcification	26 (72.2)	21 (53.8)	.100
Complication of SIA	4 (11.1)	1 (2.6)	.138
Contralateral symptoms	22 (61.1)	14 (35.9)	.029*
Contralateral CLI vs IC	8 (36.4)	5 (35.7)	.968
Previous PTA	5 (13.9)	8 (20.5)	.449
Previous bypass	2 (5.6)	2 (5.1)	.934
Previous amputation	0	4 (10.3)	.116
Ulcers	20 (55.6)	18 (46.2)	.416
Rest pain	21 (58.3)	23 (59.0)	.955
Gangrene	5 (13.9)	5 (12.8)	.892
Hypertension	26 (72.2)	29 (74.4)	.834
Hypercholesterolemia	23 (63.9)	27 (69.2)	.624
Family history	15 (41.7)	14 (35.9)	.608
Diabetes mellitus	13 (36.1)	16 (41.0)	.662
Smoking	28 (77.8)	27 (69.2)	.403
Atherosclerosis at remote site	22 (61.1)	24 (61.5)	.970
Lesion length ≥10 cm	1 (2.8)	5 (12.8)	.20
≥2 Patent distal runoff vessels	14 (38.9)	32 (82.1)	<.0001*
Clopidogrel	17 (47.2)	16 (41.0)	.589
Aspirin	34 (94.4)	35 (89.7)	.453
Warfarin	6 (16.7)	7 (17.9)	.883
Statin	30 (83.3)	27 (69.2)	.153
Blood pressure control on follow-up	23 (63.9)	26 (66.7)	.801
Cholesterol control on follow-up	28 (77.8)	27 (69.2)	.403
Continued smoking in follow-up	10 (27.8)	16 (41.0)	.228
Exercise program	4 (11.1)	11 (28.2)	.064
Age >70 y	16 (44.4)	23 (59.0)	.208

Note.—PTA = percutaneous transluminal angioplasty.
*Significant at $P < .05$.

lesions, causing intermittent claudication (IC) or critical limb ischemia (CLI). This was a retrospective study of all consecutive patients undergoing SIA in the interventional radiology department at a single institution over a 7.5-year period. Follow-up data were available for a maximum period of 8.5 years and a minimum period of 18 months. Patients with treated lower limb atherosclerotic occlusive lesions at all sites were included. Patients who had coexisting stenotic lesions treated at the same sitting with percutaneous transluminal angioplasty were also included. All patients had a conventional digital subtraction angiogram or magnetic resonance angiogram obtained for diagnosis and characterization of lesions before SIA.

All patients had a full vascular assessment by vascular surgery, and a full clinical history and physical examination were performed. Laboratory indices, including risk factor profile assessment and serum creatinine levels were measured. Ankle-brachial index (ABI) was evaluated in a dedicated vascular laboratory. On the basis of this clinical assessment, patients were then labeled as having IC or CLI, with IC being defined as lower-limb pain (ie, calf, thigh, or

TABLE 4.—Cox Regression Multivariable Analysis for MACO on Treated Limb

Predictor	HR (95% CI)	*P* Value
ABI after SIA	0.21 (0.05–0.89)	.035*
Contralateral symptoms	1.25 (0.63–2.50)	.522
Degree of patent runoff (2–3 vessels)	0.29 (0.15–0.59)	.001*

Note.—HR = hazard ratio.
ABI = ankle brachial index.
*Significant at $P < .05$.

buttock) brought on by exertion and relieved by rest, and CLI being defined as rest pain, arterial ulceration, and/or gangrene.

Seventy-eight patients had a lower-limb arterial occlusion treated with SIA. After exclusions, 75 patients were included for the purposes of this study. The SIA procedure followed a well-documented technique, with creation of a subintimal channel and neolumen formation. No reentry devices were used during the course of this study. Stents were reserved for "bailout" procedures. Technical success was defined by residual stenosis of 30% or less, fast flow of contrast agent during completion angiography, and preservation of distal runoff. Technical failure occurred when failure to enter/progress within the subintimal channel occurred, there was a failed distal reentry, or there was a residual stenosis greater than 30% within the treated arterial segment. All patients were routinely prescribed oral aspirin 75 mg once daily for life if not already taking same, and all patients were prescribed oral clopidogrel 75 mg once daily for 3 months provided no contraindications to either medication were present.

The composite end point of interest was major adverse clinical outcome (MACO) of the treated limb at follow-up, which was defined as the development of one or more of the following:

IC, CLI, or need for subsequent endovascular or surgical revascularization. Clinical follow-up data were collected at each clinic visit.

The characteristics of the lesions treated with SIA are summarized in Table 2. The cases of SIA failure (17.3%; n = 13) occurred in occluded arteries in which there was a failure to enter or progress within the subintimal channel as a result of heavily calcified vessels (n = 5) or a failure to reenter the true arterial lumen from the newly created subintimal channel distal to the occlusion (n = 8). There were no major complications. ABIs significantly improved from a mean of 0.50 before SIA (range, 0.15-1.1) to 0.74 (range, 0.30-1.21) after SIA.

All 75 patients—irrespective of SIA technical success or failure—were followed up, as all patients were included on an intent-to-treat basis. In total, 35 patients (46.7%) among the entire study population had persistent or recurrent symptoms after SIA at a mean of 8 months (range, 0.5-48 months). For the intent-to-treat recanalizations, 56.3% (n = 9) of patients with IC were free from IC at mean follow-up of 32 months (range, 1-64 months). On an intent-to-treat basis, among patients with initial CLI, 52.5% of patients (n = 31) were symptom-free at a mean follow-up of 30.7 months (range, 0.5-91 months), with limb salvage occurring in 79.7% (n = 47) and clinical symptomatic improvement in 64.4% (n = 38). Clinical outcome measures for the successful

SIA procedures were a limb salvage rate of 83.7% (n = 41), clinical symptomatic improvement rate of 73.5% (n = 36), and symptom-free incidence of 59.2% (n = 29) in the CLI group, with an ipsilateral claudication–free rate of 61.5% (n = 8) in the IC group.

Significant predictors of adverse outcome within the limb treated with SIA included the following variables: contralateral symptoms ($P = .029$), no more than one patent distal runoff vessel ($P = .0001$), and ABI after SIA ($P = .018$). There was a trend toward significance for participation in an exercise program in the follow-up period ($P = .064$), but this was not considered as a clinically significant variable (Table 3).

The following variables were identified as independent protectors against a major adverse outcome within the limb treated with SIA: improved ABI after SIA and number of patent runoff vessels (ie, ≥ 2 patent vessels). Contralateral symptoms failed to show any statistical significance (Table 4).

Thus, over a 7.5-year study period, in a population with peripheral vascular disease resulting in IC and CLI, this study reports a symptom-free rate of 52.5%, a limb salvage rate of 79.7%, and a clinical symptom improvement rate of 64.4% in patients with CLI, and a claudication-free rate of 56.3% in patients with IC after SIA of occluded peripheral arterial segments on an intent-to treat basis, inclusive of secondary interventions. Improved ABI after SIA and the number of patent runoff vessels (ie, 2-3 patent vessels) were identified as significant independent protectors against a MACO within the limb treated with SIA. The authors note that reentry can be a major limitation of SIA and is usually cited as the most common cause of failure, occurring in 62% of failures in this study.

The main limitations of this study are its retrospective nature and the fact that primary patency was not documented for all patients. However, the primary patency rate does not always translate into a good clinical outcome, and limb salvage rates are more important clinical markers of procedural success.

The authors conclude that lower-limb SIA is a feasible therapeutic option for occlusive atherosclerotic lesions in IC and CLI, with favorable clinical outcomes at midterm follow-up. The number of patent runoff vessels and the ABI are significant predictors of adverse outcome. They feel that a large multicenter randomized controlled trial is now necessary to directly compare SIA outcomes with those of surgical bypass.

T. G. Walker, MD

Embolization

SPECT/CT with ^{99m}Tc-MAA in Radioembolization with ^{90}Y Microspheres in Patients with Hepatocellular Cancer

Hamami ME, Poeppel TD, Müller S, et al (Univ at Duisburg-Essen, Germany)
J Nucl Med 50:688-692, 2009

Radioembolization with ^{90}Y microspheres is a novel treatment for hepatic tumors. Generally, hepatic arteriography and ^{99m}Tc-macroaggregated albumin (MAA) scanning are performed before selective internal

TABLE 1.—Detection of Gastrointestinal Shunting with Planar Imaging, SPECT, and SPECT/CT

Imaging Method	Accuracy	Sensitivity	Specificity
Planar	49/68 (72%)	4/16 (25%)	45/52 (87%)
SPECT	54/68 (79%)	9/16 (56%)	45/52 (87%)
SPECT/CT	65/68 (96%)	16/16 (100%)	49/52 (94%)

In 58 patients, 68 hepatic arteriographies and ^{99m}Tc-MAA scans with planar imaging, SPECT, and SPECT/CT were performed. Accuracies, sensitivities, and specificities were calculated to evaluate potential additive value of SPECT/CT compared with planar imaging or conventional SPECT. Difference between sensitivity of planar and SPECT/CT was statistically significant (χ^2 test, $P = 0.03$).

radiation therapy to detect extrahepatic shunting to the lung or the gastrointestinal tract. Whereas previous studies have used only planar or SPECT scans, the present study used ^{99m}Tc-MAA SPECT/CT scintigraphy (SPECT with integrated low-dose CT) to evaluate whether SPECT/CT and additional diagnostic contrast-enhanced CT before radioembolization with ^{90}Y microspheres are superior to SPECT or planar imaging alone for detection of gastrointestinal shunting.

Methods.—In a prospective study, we enrolled 58 patients (mean age, 66 y; SD, 12 y; 10 women and 48 men) with hepatocellular carcinoma who underwent hepatic arteriography and scintigraphy with ^{99m}Tc-MAA using planar imaging, SPECT, and SPECT with integrated low-dose CT of the upper abdomen (acquired with a hybrid SPECT/CT camera). The ability of the different imaging modalities to detect extrahepatic MAA shunting was compared. Patient follow-up of a mean of 180 d served as the standard of reference.

Results.—Gastrointestinal shunting was revealed by planar imaging in 4, by SPECT in 9, and by SPECT/CT in 16 of the 68 examinations. For planar imaging, the sensitivity for detection of gastrointestinal shunting was 25%, the specificity 87%, and the accuracy 72%. For SPECT without CT, the sensitivity was 56%, the specificity 87%, and the accuracy 79%. SPECT with CT fusion had a sensitivity of 100%, a specificity of 94%, and an accuracy of 96%. In 3 patients, MAA deposits in the portal vein could accurately be attributed to tumor thrombus only with additional information from contrast-enhanced CT. The follow-up did not show any gastrointestinal complications.

Conclusion.—SPECT with integrated low-dose CT using ^{99m}Tc-MAA is beneficial in radioembolization with ^{90}Y microspheres because it increases the sensitivity and specificity of ^{99m}Tc-MAA SPECT when detecting extrahepatic arterial shunting. The overall low risk of gastrointestinal complications in radioembolization may therefore be further reduced by SPECT/CT (Table 1).

▶ Radioembolization with ^{90}Y microspheres via hepatic arterial administration as a treatment for patients with primary and metastatic liver cancer is currently approved in the United States for the treatment of hepatocellular carcinoma

(HCC) (TheraSphere; MDS Nordion) and colorectal cancer (SIR-Spheres; Sirtex Medical). The microspheres are injected into the arterial supply of the liver, where they preferentially flow into hypervascular tumor areas resulting in a significantly higher irradiation of tumor tissue than of normal liver parenchyma. Rare but severe complications that may occur after administration of ^{90}Y include nontarget irradiation leading to gastrointestinal ulceration, pancreatitis, cholecystitis, or radiation pneumonitis. Aggressive preprocedural prophylactic embolization of any arterial branches that could potentially allow for nontarget irradiation is recommended because ^{90}Y-induced ulcers may be refractory to medical therapy.

Generally, hepatic arteriography and radionuclide ^{99m}Tc-macroaggregated albumin (MAA) scanning are performed before radioembolization with ^{90}Y microspheres to detect extrahepatic shunting to the lung or the gastrointestinal tract. Although previous studies have used only planar imaging or single photon emission computed tomography (SPECT) of the upper abdomen to assess MAA distribution, the present study used ^{99m}Tc-MAA SPECT/CT scintigraphy (SPECT with integrated low-dose CT). SPECT/CT, a dual-modality imaging system that provides functional (SPECT) and anatomic (CT) images in the same scanning session, is now widely used in many fields of cancer imaging, dosimetry for radionuclide therapy, and the monitoring of anticancer drug distribution.

This study evaluated whether SPECT/CT and additional diagnostic, contrast-enhanced CT before selective internal radiation therapy may be superior to SPECT or planar imaging alone when assessing treatment with ^{90}Y microspheres in patients for gastrointestinal shunts before therapy. Fifty-eight consecutive patients with unresectable HCC who underwent radioembolization with TheraSphere were prospectively enrolled in the study. Two of the 58 patients underwent separate MAA scans of the right and left liver lobes on different days. In 9 patients, MAA scans were repeated because of gastrointestinal shunting. Thus, 68 MAA scans were performed. All patients underwent ^{99m}Tc-MAA planar imaging, SPECT, and SPECT/CT of the liver.

Depending on the anatomy of the patient, 150 MBq of ^{99m}Tc-MAA were injected with a microcatheter either into the proper hepatic artery or the common hepatic artery after coil embolization of all visible nonhepatic arterial flow. If injection into the common or proper hepatic artery was not possible, the left and right hepatic arteries were injected separately. In variant anatomies, accessory or replaced vessels were injected separately. After MAA injection, the patient was transferred to the nuclear medicine department for imaging.

Anterior and posterior planar images of the whole body were obtained within 30 min after the MAA injection using a dual detector γ-camera with a mounted 2-row CT. SPECT/CT images of the upper abdomen were acquired immediately after the delayed planar images. Acquisition parameters for SPECT were a 128×128 matrix with 128 frames (25 s/frame). The scan parameters for CT were 130 kV, 17 mAs, 5-mm slices, and image reconstruction with a medium smooth kernel. SPECT images were corrected for attenuation and scatter. The reconstructed data were visualized in sagittal, coronal, and axial slices. Fusion images were generated from the coregistered SPECT and low-dose CT images.

A combination of clinical and radiologic follow-up served as the standard of reference for extrahepatic gastrointestinal shunting. This follow-up included physical examination and laboratory tests on days 2, 7, 14, 21, 28, 90, and 180 after the intervention. Abdominal contrast-enhanced ultrasound, CT, and magnetic resonance imaging (MRI) were performed on days 30, 90, and 180 postintervention. On the basis of the standard of reference, potential gastrointestinal shunts detected on planar imaging, SPECT, and SPECT/CT were rated as true-positive, true-negative, false-positive, or false-negative. Sensitivities, specificities, and accuracies were calculated for all imaging modalities.

Table 1 details the results for the detection of gastrointestinal shunting with planar imaging, SPECT, and SPECT/CT. Extrahepatic shunting was revealed by planar imaging in 4, SPECT in 9, and SPECT/CT in 16 of the 68 examinations (true-positives). The number of false-positive results was 7 for planar imaging (13%), 7 for SPECT (13%), and 3 for SPECT/CT (6%). The number of false-negative results was 12 for planar imaging (75%), 7 for SPECT (44%), and 0 for SPECT/CT (100%). On the basis of these findings, the sensitivity in detecting extrahepatic shunting with planar imaging was 25%, the specificity was 87%, and the accuracy was 72%. For SPECT, the sensitivity was 56%, the specificity was 87%, and the accuracy was 79%. SPECT/CT had a sensitivity of 100%, a specificity of 94%, and an accuracy of 96%.

In 3 patients, MAA deposition in the liver hilum was misinterpreted as extrahepatic on SPECT/CT. Comparison of the SPECT/CT data of these 3 patients with previously acquired diagnostic contrast-enhanced CT showed that these MAA depositions were tumor uptake in portal vein tumor thrombi. In 10 patients, extrahepatic MAA deposition was detected with SPECT/CT. This deposition was localized within the gastrointestinal tract in all patients. In 6 of these patients, the underlying vessel was found on repeated angiography and coil embolized. In 2 others, the underlying vessel was detected but could not be embolized because of its small diameter. These patients were treated from a more distal catheter position. In the remaining 2 patients, no underlying artery was found on repeated angiography, but MAA uptake persisted. These patients were not treated. Thereby, nontarget seeding was successfully avoided. No treated patient experienced gastrointestinal complications. None of the imaging modalities performed on days 30, 90, and 180 after the treatment revealed signs of cholecystitis or pancreatitis.

These data show that SPECT with integrated low-dose CT before radioembolization with ^{90}Y microspheres is superior to SPECT or planar imaging alone in the detection of gastrointestinal MAA deposition. Additional diagnostic contrast-enhanced CT further increases specificity in patients with portal vein thrombosis by differentiating tumor-associated MAA deposition in the portal vein from that in extrahepatic locations. The investigators found approximately 40% more patients with gastrointestinal MAA accumulation using SPECT/CT than with SPECT alone (16 vs 9 of 68 examinations).

The authors conclude that SPECT with integrated low-dose CT using ^{99m}Tc-MAA is beneficial in radioembolization with ^{90}Y microspheres because it increases the sensitivity and specificity of ^{99m}Tc-MAA SPECT when detecting extrahepatic arterial shunting. The overall low risk of gastrointestinal complications in radioembolization may, therefore, be further reduced by SPECT/CT.

T. G. Walker, MD

Uterine Artery Embolization versus Laparoscopic Occlusion of Uterine Arteries for Leiomyomas: Long-term Results of a Randomized Comparative Trial

Hald K, Noreng HJ, Istre O, et al (Oslo Univ Hosp, Norway)

J Vasc Interv Radiol 20:1303-1310, 2009

Purpose.—To compare long-term symptom recurrence and 6-month magnetic resonance (MR) imaging results after two different treatments for uterine leiomyomas.

Materials and Methods.—Sixty-six women with uterine leiomyomas were randomized to undergo uterine artery embolization (UAE) or laparoscopic bilateral occlusion of uterine arteries. Contrast-enhanced MR imaging was performed before treatment and after 6 months.

Results.—Fifty-eight patients received treatment. Median follow-up time was 48 months (range, 8–73 months). Clinical failure and symptom recurrence occurred in 14 patients after laparoscopy (48%) and in five after UAE (17%; $P = .02$, log-rank test). Hysterectomy was performed in two patients after UAE (7%) and in eight after laparoscopy (28%; $P = .041$). Six-month MR imaging results were available for 26 patients treated with UAE and 22 treated with laparoscopy. The mean uterine volume was reduced by 51% (range, 16%–86%) after UAE treatment, compared with 33% (range, 6%–77%) after laparoscopy ($P = .001$). Complete leiomyoma infarction was seen in all 26 patients in the UAE group and in only five patients in the laparoscopy group ($P < .001$). Eleven patients experienced symptom recurrence later than 6 months. Uterine volume reduction at 6 months was 24% in this group, compared with

TABLE 2.—Clinical Failure and Symptom Recurrence during the Study

Outcome	UAE ($n = 29$)	Laparoscopy ($n = 29$)	P Value
Clinical failure at 6 months	2	6	.13‡
Long-term symptom recurrence	3	8	.035§
Total clinical failure and recurrence	5 (17)	14 (48)	.02§
Pathologic findings other than leiomyoma	2 (7)*	1 (3.4)†	1.00‖
Hysterectomy	2 (7)	8 (28)	.041§

Note.—Values are presented as count (%).
*One endometrial cancer (hysterectomy) and one atypical endometrial hyperplasia (refused hysterectomy).
†Ovarian cancer (hysterectomy).
‡χ^2 test.
§Log-rank test.
‖Fisher exact test.

TABLE 3.—MR Results at 6 Months after UAE or Laparoscopic Occlusion of Uterine Arteries

Result at 6 Months	UAE *n*	UAE Value	Laparoscopy *n*	Laparoscopy Value	*P* Value
Uterine volume reduction (%)	26	51.3 ± 15.4	22	33.0 ± 18.4	.001†
Dominant leiomyoma volume reduction (%)	26	62.8 ± 27.0	22	50 ± 22.1	.083†
Change T2 signal intensity (%)*	23	−63.1 (−229.2–91.6)	19	−26.5 (−68.2–86.3)	.002‡
Contrast enhancement index in dominant leiomyoma	23	1.3 (−16.8–18.7)	20	74.6 (−21.4–255.5)	.058‡
Reduction of contrast enhancement index in dominant leiomyoma (%)	22	99.4 (86.0–115.38)	20	41.7 (−68.9–113.9)	.039‡
Pts. with complete infarction of all leiomyoma tissue at 6 months	26	26 (100)	22	5 (23)	<.001§

Note.—Values presented as mean ± SD or median (range).
*Leiomyoma/skeletal muscle.
†Student *t* test.
‡Mann-Whitney *U* test.
§Fisher exact test.

48% in the 37 patients with no recurrence ($P = .004$). Incomplete infarction of leiomyomas was seen in eight of the 11 cases of recurrence (73%) versus nine of 37 cases without recurrence (24%; $P = .009$).

Conclusions.—Recurrence rate was significantly lower after UAE than after laparoscopic treatment. Larger volume reduction and more complete devascularization of leiomyomas were found after UAE treatment and among patients with no recurrence.

▶ Uterine artery embolization (UAE) has been established as an effective and safe treatment of symptomatic leiomyomas in women who wish to retain their uterus. Limitations of UAE treatment have been reported in prospective studies, with recurrence of symptoms and leiomyoma growth in 16% to 28% of patients within 4 to 7 years after treatment and hysterectomy rates of 12% and 14% during 4- and 5-year follow-up. Studies with short-term follow-up have suggested laparoscopic occlusion of uterine vessels as an equally effective alternative to UAE, but there is little knowledge about the long-term results of this technique. Complete leiomyoma infarction seems important to avoid tumor regrowth and risk of recurrent symptoms.

This prospective randomized study was designed to compare UAE versus laparoscopic occlusion of the uterine arteries. Fifty-eight patients received treatment, equally divided into the 2 groups, and the median follow-up time was 48 months.

The clinical failure and symptom recurrence during the study is detailed in Table 2 and the MR results at 6 months after UAE or laparoscopic occlusion of the uterine arteries is detailed in Table 3. The results of this study suggest that UAE is superior to laparoscopic occlusion of uterine arteries with regard to symptom recurrence rate, reduction of uterine volume, and number of patients with completely infracted leiomyomas at 6-month MR imaging examination. In the group treated with UAE, the symptom recurrence rate was found to be similar to what has been previously reported in several series. The symptom recurrence rate after laparoscopic occlusion of uterine arteries was much higher in this study than the rate of 9% that has been described by other investigators. The MR-measured 63% reduction of leiomyoma volume 6 months after UAE agrees with the findings of other studies.

The reason for better results after UAE could be explained by several factors. During the laparoscopic procedure, the occlusion is created more proximal than the occlusion in UAE. Collateral vessels might exist distal to the occlusion from the cervicovaginal area, from the iliac and femoral arteries, and from vessels in the round ligament. However, the anastomoses that often exist between the uterus and the ovaries are probably most important. Another explanation for unfavorable results after laparoscopic occlusion might be technical surgical failure. The uterine artery has numerous anatomic variations. During the laparoscopic procedure, there is not the same visualization of the vascular anatomy as during a UAE procedure.

One limitation to this study is the relatively small number of patients. Another limitation is the lack of MR imaging examinations later than 6 months. However, based upon this study, the authors recommend UAE instead of laparoscopic occlusion of uterine arteries to most women with symptoms caused by uterine leiomyomas in cases in which hysterectomy or myomectomy is not an option. The study suggests that UAE is superior to laparoscopic occlusion of uterine arteries, with a lower recurrence rate, larger number of patients with completely infarcted leiomyomas, and greater decrease in uterine size. The findings support the notion that incompletely infarcted leiomyomas and lower post-treatment volume reductions are associated with higher recurrence rates.

T. G. Walker, MD

Yttrium-90 Radioembolization for Liver Malignancies: Prognostic Factors Associated with Survival

Dunfee BL, Riaz A, Lewandowski RJ, et al (Aria Health Hosps, Philadelphia, PA; Northwestern Memorial Hosp and Robert H Lurie Comprehensive Cancer Ctr, Chicago, IL)

J Vasc Interv Radiol 21:90-95, 2010

Purpose.—To identify key prognostic clinical and imaging variables in patients undergoing yttrium-90 radioembolization (^{90}Y) for liver malignancies.

Materials and Methods.—Patients with liver malignancies that progressed despite standard-of-care therapy were treated with ^{90}Y from 2002 to 2006. Baseline functional status, laboratory values, and diagnostic imaging were assessed before therapy. Imaging follow-up was performed 1 month after treatment and subsequently at 3-month intervals. Patients were followed for survival from the time of their first ^{90}Y treatment.

Results.—Patients with follow-up imaging after radioembolization ($N = 130$) were included in this analysis. Primary malignancies included colon, neuroendocrine, and others. The following clinical variables had a significant effect on survival on multivariate analysis: Eastern Cooperative Oncology Group (ECOG) performance status (PS) greater than 0 (hazard ratio [HR], 7.98; 95% CI, 3.98–16), hepatic tumor burden of 51%-75% (HR, 2.46; 95% CI, 1.01–6.02), bilirubin level greater than 1.3 mg/dL (HR, 2.60; 95% CI, 1.27–5.34), hepatic metastases from breast cancer (HR, 2.51; 95% CI, 1.13–5.61), response on imaging based on World Health Organization (WHO) criteria (HR, 0.48; 95% CI, 0.24–0.94), and lymphocyte depression (HR, 0.56; 95% CI, 0.31–0.96). Among patients with colorectal cancer metastases to the liver, the HR for survival on univariate analysis for responders compared with nonresponders (per WHO criteria) was 0.26 (95% CI, 0.10–0.69).

Conclusions.—Cancer-related symptoms (ie, ECOG PS > 0), hepatic tumor burden greater than 50%, increased bilirubin levels, and hepatic metastases from breast cancer were found to be negative prognostic factors. Tumor response to therapy and lymphocyte depression were associated with favorable prognosis. Additionally, WHO response was identified to be a favorable prognostic factor in patients with colorectal cancer metastases. These findings may be useful when counseling patients regarding prognosis of their hepatic disease.

▶ Liver metastases are the most common form of hepatic malignancies. Various methods for the treatment of hepatic metastases include resection, systemic chemotherapy, radiofrequency ablation, and external-beam radiation. Recently, yttrium-90 microspheres have been extensively used for the treatment of primary and metastatic hepatic disease. Although the standard treatment for metastatic disease to the liver is chemotherapy and surgical resection, there is currently no reported survival benefit for incomplete tumor debulking compared with conservative therapy. Many clinical and tumor features have been previously examined to determine which patients benefit most from resection. Such factors include perihepatic lymph node involvement and locally recurrent disease or extrahepatic metastases. Other determinants that represent relative contraindications for resection vary among studies; some have included large tumor size and multiplicity, increased tumor markers, presence of nondiploid tumors, and margins of resection less than 1 cm.

Multiple oncologic factors have been studied in determining the prognostic outcome of patients with metastatic disease. The presence of metastases to the liver predicts a worse outcome in nearly all tumors with the site of the primary tumor being an additional factor. Performance status (PS) has also been proven useful in predicting the long-term prognosis. The most commonly used system for grading performance was proposed by the Eastern Cooperative Oncology Group (ECOG). Many studies have demonstrated that a higher ECOG PS is related to a higher short-term mortality rate. In addition, laboratory values such as bilirubin levels may also assist in predicting the effects of the metastatic deposit on the functional status of the liver. Many other prognostic factors previously investigated include sex, age, lactic dehydrogenase level, number of involved segments, and presence of extrahepatic metastases.

Tumor markers, such as carcinoembryonic antigen and cancer antigen 19-9 have been proposed, but the data are not consistent. Additional prognostic indicators can be determined by the response of metastatic lesions after various medical and surgical treatments. Improved survival benefit in patients with metastatic colorectal carcinoma results if the disease does not progress while the patient is receiving neoadjuvant chemotherapy.

The purpose of this study was to determine prognostic clinical variables for patients who underwent catheter-directed yttrium-90 therapy in a large cohort of patients with liver metastases. The goal was to determine the prognostic factors that had the most influence on outcome of patients after treatment.

This was an open label cohort study of 137 patients with liver malignancies refractory to standard-of-care therapy. One hundred thirty patients with follow-up imaging were studied for this analysis.

TABLE 1.—Exploratory Analysis of Prognostic Benefit of Variables

Variable	Total ($N = 130$)	At Study Completion Alive ($n = 67$)	Dead ($n = 63$)	P Value
Age (y)				
<65	54	31 (57)	23 (43)	.34
≥65	76	36 (47)	40 (53)	
ECOG PS				
>0	48	8 (17)	40 (83)	<.001
0	82	59 (72)	23 (18)	
Lobar involvement				
Bilobar	105	51 (49)	54 (51)	.24
Unilobar	25	16 (89)	9 (20)	
Hepatic tumor burden (%)				
>75	0	0	0	.03
51–75	6	0 (0)	6 (100)	
26–50	18	7 (39)	11 (61)	
0–25	106	60 (57)	46 (43)	
Extrahepatic metastases				
Yes	66	27 (41)	39 (59)	.02
No	64	40 (63)	24 (37)	
Increased bilirubin (>1.3 mg/dL)				
Yes	11	0 (0)	11 (100)	<.001
No	119	67 (56)	52 (44)	
Higher dose utilization (≥104 Gy)				
Yes	88	46 (53)	42 (47)	.95
No	42	21 (50)	21 (50)	
Tumor response (WHO Criteria)				
Yes	48	32 (67)	16 (33)	.01
No	82	35 (43)	47 (57)	
Tumor				
Breast carcinoma	20	12 (60)	8 (40)	.046
Neuroendocrine	17	13 (76)	4 (24)	
Colon carcinoma	50	26 (52)	24 (48)	
Cholangiocarcinoma	7	4 (57)	3 (43)	
Other carcinomas	36	12 (33)	24 (67)	
Lymphocyte depression				
Yes	67	32 (48)	35 (52)	1.00
No	63	31 (49)	32 (51)	

Note.—Values in parentheses are percentages.

TABLE 2.—Univariate and Multivariate Analyses

Variables	Univariate Analyses HR (95% CI)	*P* Value	Multivariate Analysis HR (95% CI)	*P* Value
Age (y)				
<65	Reference	.1552	—	—
≥65	1.45 (0.87–2.43)		—	
Sex				
Female	Reference	.56	—	—
Male	1.16 (0.70–1.90)		—	
ECOG PS				
0	Reference	<.0001	Reference	<.0001
>0	10.0 (5.47–18.5)		7.98 (3.98–16.0)	
Lobar involvement				
Unilobar	Reference	.0860	—	—
Bilobar	1.87 (0.92–3.80)		—	
Hepatic tumor burden (%)				
51–75	4.50 (1.91–10.6)	.0006	2.46 (1.01–6.02)	.048
26–50	1.35 (0.70–2.59)	.3680	Reference	
0–25	Reference			
Extrahepatic metastases				
Yes	2.30 (1.38–3.83)	.0015	—	—
No	Reference		—	
Increased bilirubin (mg/dL)				
>1.3	2.91 (1.51–5.61)	.0014	2.60 (1.27–5.34)	.0092
≤1.3	Reference		Reference	
Dose utilization (Gy)				
>104	1.10 (0.65–1.86)	.7286	—	—
≤104	Reference		—	
Positive tumor response to treatment				
Yes	0.30 (0.17–0.55)	.0001	0.48 (0.24–0.94)	.034
No	Reference		Reference	
Tumor				
Breast carcinoma	1.59 (0.75–3.38)	.23	2.52 (1.13–5.61)	
Neuroendocrine	0.33 (0.12–0.92)	.03		
Colon carcinoma	0.84 (0.50–1.39)	.49	Reference	.26
Cholangiocarcinoma	0.84 (0.26–2.70)	.78		
Other carcinomas	Reference			
Lymphocyte depression				
Yes	0.62 (0.38–1.02)	.06	0.55 (0.31–0.96)	.036
No	Reference		Reference	

The data were collected prospectively. The PS of each patient was documented according to the ECOG grading criteria. Laboratory values including bilirubin levels were obtained for all patients.

Each patient had preprocedural imaging, such as computed tomography (CT) and/or magnetic resonance (MR) imaging, depending on the primary neoplasm.

The following inclusion criteria were met for each patient: (i) progressive liver disease despite standard-of-care therapy; (ii) disease not amenable to surgery; (iii) ECOG PS of 0-2; (iv) noncompromised pulmonary function; (v) ability to undergo angiography and selective visceral catheterization; (vi) adequate hematologic parameters and renal function (creatinine <2.0 mg/dL); (vii) adequate liver function (ie, bilirubin <2.0 mg/dL); and (viii) limited extrahepatic disease. Exclusion criteria included (i) evidence of any uncorrectable flow to the gastrointestinal tract observed on angiography or technetium-99m macroaggregated albumin (^{99m}Tc-MAA) scan; (ii) radiographic/pathologic confirmation of hepatocellular carcinoma; and (iii) estimated radiation dose greater than 30 Gy (16.5 mCi) delivered to the lungs in a single administration or 50 Gy on multiple administrations.

Imaging follow-up was performed 1 month after treatment and subsequently every 3 months. Response was assessed according to World Health Organization (WHO) guidelines. Tumor response on CT or MR imaging was determined for a maximum of 4 measurable lesions (>1 cm) in terms of the product of the greatest cross-sectional tumor length and the length of the corresponding orthogonal projection. Corresponding lesions from baseline and posttreatment scans were compared for size changes. Complete response was defined by a change in the sum of the cross products to 0, partial response as a decrease in the sum of cross products by at least 50%, stable disease as a decrease in the sum of cross products by less than 50% or an increase less than 25%, and progression as an increase in the sum of cross products by at least 25%. Posttreatment studies were compared and assessed simultaneously with retreatment images. The only laboratory parameter used to assess baseline liver function was baseline bilirubin level.

In the study, 130 underwent 218 intra-arterial administrations of ^{90}Y microspheres.

Seven patients did not have follow-up imaging and hence were not included in the regression analyses. The entire analysis was, therefore, based on 130 patients. The distribution of the primary malignancies is summarized in Table 1. All patients were considered to have unresectable disease. A total of 63 patients (48%) had died, and 67 (52%) were alive at the time of study completion (Table 1).

Table 2 presents the univariate and multivariate analysis for the study. The patients in this study population were not deemed candidates for surgical resection; therefore, the goal of brachytherapy was to prevent progression of liver tumors while maintaining residual liver function. The most common variable factors that might influence prognosis were chosen for analysis. Of the various factors analyzed, poor ECOG PS, increased bilirubin level (ie, >1.3 mg/dL), hepatic tumor burden greater than 50%, and metastases from breast cancer were variables that demonstrated statistical significance on the multivariate analysis. Patients with a positive tumor response on imaging and lymphocyte depression after ^{90}Y treatment were found to have a favorable prognosis. However, unlike the other prognostic indicators, WHO response and lymphocyte suppression may be used only in the decision of whether to continue with additional intra-arterial therapies. The presence of extrahepatic metastases was found to be a significant prognostic factor on univariate analysis. However, it did not show significance on multivariate analyses.

Patients with secondary liver tumors differ from patients with hepatocellular carcinoma. Patients with hepatocellular carcinoma present with compromised liver function as a result of cirrhosis and the tumor. This leads to clinical aberrations, such as decrease in PS, and biochemical aberrations, such as an increase in bilirubin. The decrease in liver function in patients with secondary lesions is most often caused by only the presence of tumor. Hence, the PS and baseline bilirubin values present important and clinically relevant prognostic factors in patients with secondary liver tumors.

The WHO response has been thought to correlate with survival. The data presented in this study show that it is a significant prognostic factor for survival on univariate and multivariate analyses. Colon carcinoma represented the largest number of primary tumors in this study population.

These patients were further analyzed with respect to their tumor response after treatment. A significantly improved survival was observed in patients in whom WHO response was demonstrated.

The authors conclude that various prognostic clinical and imaging factors are important for patients referred for catheter-directed intra-arterial ^{90}Y therapy for hepatic malignancies. Poor ECOG PS, hepatic tumor burden greater than 50%, and increased bilirubin level (ie, >1.3 mg/dL) were determined to be negative prognostic indicators on univariate and multivariate analyses. Compromised PS was the most powerful negative prognosticator in this population with liver malignancies. Tumor response was a favorable prognostic factor for survival.

T. G. Walker, MD

Embolization of the Gastroduodenal Artery Before Selective Internal Radiotherapy: A Prospectively Randomized Trial Comparing Platinum-Fibered Microcoils with the Amplatzer Vascular Plug II

Pech M, Kraetsch A, Wieners G, et al (Univ of Magdeburg, Germany)
Cardiovasc Intervent Radiol 32:455-461, 2009

The Amplatzer Vascular Plug II (AVP II) is a novel device for transcatheter vessel occlusion, for which only limited comparative data exist. Embolotherapy of the gastroduodenal artery (GDA) is essential before internal radiotherapy (SIRT) in order to prevent radiation-induced peptic ulcerations due to migration of yttrium-90 microspheres. The purpose of this study was to compare the vascular anatomical limitations, procedure time, effectiveness, and safety of embolization of the GDA with coils versus the AVP II. Fifty patients stratified for SIRT were prospectively randomized for embolization of the GDA with either coils or the AVP II. The angle between the aorta and the celiac trunk, diameter of the GDA, fluoroscopy time and total time for embolization, number of embolization devices, complications, and durability of vessel occlusion at follow-up angiography for SIRT were recorded. A t-test was used for statistical analysis. Embolizations with either coils or the AVP II were technically feasible in all but two patients scheduled for embolization of the GDA with the AVP II. In both cases the plug could not be positioned due to the small celiac trunk outlet angles of 17° and 21°. The mean diameter of the GDA was 3.7 mm (range, 2.2–4.8 mm) for both groups. The procedures differed significantly in fluoroscopy time (7.8 min for coils vs. 2.6 min for the AVP II; $P < 0.001$) and embolization time (23.1 min for coils vs. 8.8 min for the AVP II; $P < 0.001$). A mean of 6.0 ± 3.2 coils were used for GDA embolization, while no more than one AVP II was needed for

TABLE 1.—Patient Characteristics

	Randomized to Vascular Plug			Randomized to Pushable Coil			*P*		
	Males	Females	All	Males	Females	All	Males	Females	All
Number	17	8	25	16	9	25	–		
Age, year (mean ± SD)	57.1 ± 12.9	55.5 ± 9.2	56.8 ± 11.4	68.3 ± 7.7	56.7 ± 15.2	64.12 ± 12.1	0.004	0.854	0.073
GD artery, diameter (mm)									
Mean ± SD	3.7 ± 0.6			3.6 ± 0.7			0.604		
Range	2.7–4.8			2.2–4.6					
Celiac trunk, angle (deg)									
Mean ± SD	43 ± 14			45 ± 19			0.424		
Range	21–82			17–101					
Transferred to other group	2	0	2	None			–		
	Treated with vascular plug			Treated with pushable coil			*P*		
Number	15	8	23	18	9	27	–		
Age, year (mean ± SD	56.5 ± 7.6	55.5 ± 9.2	56.1 ± 11.6	67.8 ± 7.6	56.7 ± 15.2	64.1 ± 11.7	0.004	0.854	0.073
GD artery, diameter (mm)									
Mean ± SD	3.7 ± 0.5			3.5 ± 0.7			0.244		
Range	2.7–4.8			2.2–4.6					
Celiac trunk, angle (deg)							0.173		
Mean ± SD	41 ± 15			47 ± 18					
Range	17–82			29–101					

Note: GDA, gastroduodenal artery. No significant changes after transfer of two patients from primary randomisation of vascular plug group to the pushable coil group

TABLE 2.—Gastroduodenal Artery Embolization with Either the Amplatzer Vascular Plug II (AVP II) or Platinum-Fiber Coils

	Treated with AVP II	Treated with platinum-fiber coils	Statistical difference
Duration of procedure (min)	5.7 ± 3.3	19.0 ± 9.5	$P < 0.001$
Embolization time (min)	3.6 ± 2.5	14.1 ± 8.1	$P < 0.001$
Fluoroscopy time (min)	1.7 ± 1.2	7.5 ± 6.0	$P < 0.001$
No. of devices needed for GDA occlusion	1.05 ± 0.2[a]	6.0 ± 3.2	$P < 0.001$
Amount of contrast agent (ml)	20.4 ± 10.9	20.0 ± 18.5	n.s.
Material cost per patient (€)	898	1268	−29%

Note: n.s., not significant

[a]One plug implant for each patient, except for a single patient who received two plugs at separate locations (see text).

Legend for Fig. 1

successful vessel occlusion ($P < 0.001$). One coil migration occurred during coil embolization, whereas no procedural complication was encountered with the use of the AVP II. Vessel reperfusion was noted in only one patient, in whom coil embolization was performed. In conclusion, embolization of the GDA with the AVP II is safe, easy, rapid, and highly effective; only an extremely sharp-angled celiac trunk outlet represented an anatomical limitation for device deployment.

▶ Transcatheter embolization of the gastroduodenal artery (GDA) is usually required before selective internal radiotherapy (SIRT) delivery through the hepatic artery for primary or metastatic hepatic neoplasm. This protective artificial embolization of the GDA is performed to confine the radioactive material (in SIRT) to the live and thereby prevent any "non-target" embolization and the potential associated complications that may ensue. Other clinical scenarios in which it may be necessary to embolize the GDA include control of hemorrhage or treatment of a visceral artery aneurysm. Methods for such embolization have included surgical ligature and acrylic cement; the most widespread method today is embolization by insertion of a platinum vasoocclusion coil through a catheter.

Since failure to confine the radioactive agent to the target area can lead to severe adverse side effects, the success and efficiency of embolization are critical for successful clinical outcomes. The placement of the occluding device(s) is therefore an important consideration. The best position is as close to the main hepatic artery as possible, to ensure a sufficient retrograde collateral blood supply from the circulation region of the superior mesenteric artery to as many branches of the GDA as possible, to prevent ischemic complications. Consequently, coil embolization may not be ideal in this application, as there is a significant risk of dislocation of the last coil before the main hepatic artery; if a part of the coil dissociates and flows with the bloodstream into the liver, then it may need to be recovered in a complex procedure, and failure of recovery can result in closure of the main hepatic artery and consequent liver dysfunction; this can also render the planned therapy difficult or even impossible.

There is therefore a need for an efficient embolization technique that gives reliable, complete, and safe blockage of the GDA as a prerequisite for selective internal radiotherapy (SIRT) or for use in other indications.

Fifty patients stratified for SIRT were enrolled in this single-blinded prospective randomized trial, for embolization of the GDA with either coils or the Amplatzer Vascular Plug II (AVP II). The latter is a recently introduced vascular occlusion device that requires a relatively large-lumen delivery catheter. In this study a 5-French guiding catheter was used for this purpose and the diameter of the AVP II was chosen to be approximately 30% to 50% greater than that of the GDA. By contrast in those patients randomized for embolization with coils the GDA was catheterized coaxially with a microcatheter and vessel occlusion was achieved with conventional platinum-fiber microcoils.

Tables 1 and 2 detail the patient demographics/characteristics and the data related to the embolization procedure respectively. The procedure duration and time required for embolization were both significantly shorter with the use of the AVP II as compared with the coils. The cost analysis (Table 2) suggested a margin of ~29% in favor of the plug. The plug method was not successful in cases where the angle between the celiac artery and the aorta was too acute. The use of the AVP II has some drawbacks. First, because it is a cylindrical device, it requires a short portion of blood vessel with a constant diameter. For a tapering vessel, coils may work better. Second, it cannot be introduced over a guidewire but must be introduced through a large-lumen guiding catheter that has to be advanced to the distal point of the embolization site. An important safety feature is that the device is secured with a long cable, which offers the possibility of repositioning the AVP II if the device is not at the desired deployment position. Once embolization is successful the AVP II may be released by unscrewing the cable.

The authors conclude that the AVP II was associated with a shorter procedure duration, shorter embolization times, and less exposure of the medical personnel to radiation. The material cost for embolization was approximately 30% lower with the plug than with the coil method, and after a short learning period it should be possible to reduce the amount of contrast-enhancing agent required for the procedure. However, the use of the plug is restricted by the relative inflexibility of the catheter used to position it.

T. G. Walker, MD

Endografts

Defining high-risk patients for endovascular aneurysm repair

Egorova N, Giacovelli JK, Gelijns A, et al (Mount Sinai School of Medicine, NY; et al)
J Vasc Surg 50:1271-1279, 2009

Background.—Endovascular aneurysm repair (EVAR) is commonly used as a minimally invasive technique for repairing infrarenal aortic aneurysms. There have been recent concerns that a subset of high-risk patients experience unfavorable outcomes with this intervention. To determine whether such a high-risk cohort exists and to identify the characteristics of these patients, we analyzed the outcomes of Medicare patients treated with EVAR from 2000-2006.

Methods.—We identified 66,943 patients who underwent EVAR from Inpatient Medicare database. The overall 30-day mortality was 1.6%. A risk model for perioperative mortality was developed by randomly selecting 44,630 patients; the other one third of the dataset was used to validate the model. The model was deemed reliable (Hosmer-Lemeshow statistics were $P = .25$ for the development, $P = .24$ for the validation model) and accurate ($c = 0.735$ and $c = 0.731$ for the development and the validation model, respectively).

Results.—In our scoring system, where scores ranged between 1 and 7, the following were identified as significant baseline factors that predict mortality: renal failure with dialysis (score = 7); renal failure without dialysis (score = 3); clinically significant lower extremity ischemia (score = 5); patient age ≥85 years (score = 3), 75-84 years (score = 2), 70-74 years (score = 1); heart failure (score = 3); chronic liver disease (score = 3); female gender (score = 2); neurological disorders (score = 2); chronic pulmonary disease (score = 2); surgeon experience in EVAR <3 procedures (score = 1); and hospital annual volume in EVAR <7 procedures (score = 1). The majority of Medicare patients who were treated (96.6%, n = 64,651) had a score of 9 or less, which correlated with a mortality <5%. Only 3.4% of patients had a mortality ≥5% and 0.8% of patients (n = 509) had a score of 13 or higher, which correlated with a mortality >10%.

Conclusion.—We conclude that there is a high-risk cohort of patients that should not be treated with EVAR because of prohibitively high mortality; however, this cohort is small. Our scoring system, which is based on patient and institutional factors, provides criteria that can be easily used by clinicians to quantify perioperative risk for EVAR candidates.

► It appears that endovascular repair of abdominal aortic aneurysm (EVAR) will soon become the predominant method of abdominal aortic aneurysm (AAA) repair. Because it is minimally invasive, EVAR potentially holds great advantage for high-risk patients with multiple comorbidities. The procedure may not require general anesthesia or intensive care unit (ICU) admission postoperatively. Additionally, EVAR requires only femoral artery exposure, eliminating the need for a laparotomy and its complications. There is decreased blood loss compared with open repair, and the major perioperative intravenous fluid shifts observed with open repair are avoided. It has been previously demonstrated that the perioperative morbidity and mortality associated with endovascular repair approaches one fourth that of traditional open surgery.

These advantages suggest that a broad spectrum of AAA patients should be appropriate candidates for the endovascular approach. However, the concept that EVAR is applicable to all patients regardless of the severity of their comorbidities has recently been challenged. In the EVAR 2 trial, a cohort of patients

TABLE 1.—Patient Demographics and 30-Day Mortality by Demographic Groups (n = 66,943 Patients)

Variable	No. of Patients	% of Cohort	Mortality (%)	*P* value
Age, years				
65-69	12,046	17.99	0.9	(reference group)
70-74	16,994	25.39	1.2	0.04
75-79	18,624	27.82	1.6	<0.0001
80-84	13,311	19.88	2.0	<0.0001
≥85	5,968	8.92	3.2	<0.0001
Male	55,485	82.88	1.4	(reference group)
Female	11,458	17.12	2.5	<0.0001
Whites	63,492	94.84	1.6	(reference group)
Blacks	1,847	2.76	1.9	0.29
Hispanics	354	0.53	2.0	0.55
Native Americans	118	0.17	1.7	0.92
Other races	830	1.24	0.7	0.05

TABLE 2.—Comorbidities and their Association with 30-Day Mortality (N = 66,943 Patients)

Comorbidity	No. of Patients	% of Cohort	Mortality (%)	*P* value	Odds Ratio
Renal failure w/dialysis	718	1.07	11.8	<.0001	9.01 [7.12-11.39]
Renal failure w/o dialysis	2,554	3.82	3.8	<.0001	2.64 [2.13-3.26]
PAD	4,855	7.25	3.1	<.0001	2.17 [1.82-2.58]
LE ischemia	1,414	2.11	6.2	<.0001	4.42 [3.52-5.53]
Vascular intestine	141	0.21	2.8	0.23	1.82 [0.67-4.93]
Renal atherosclerosis	3,409	5.09	1.9	0.09	1.24 [0.97-1.60]
Heart failure	9,644	14.41	3.5	<.0001	2.88 [2.53-3.28]
Neurological disorders	7,494	11.19	2.4	<.0001	1.63 [1.39-1.92]
Cerebrovascular and/or paralysis	5,282	7.89	2.1	0.0016	1.38 [1.13-1.68]
Other neurological	2,596	3.88	3.3	<.0001	2.23 [1.79-2.80]
Liver disease	728	1.09	3.2	0.0006	2.05 [1.35-3.13]
Cardiac arrhythmia	16,840	25.16	2.3	<.0001	1.78 [1.57-2.01]
Rheumatoid arthritis	1,311	1.96	2.3	0.04	1.47 [1.02-2.12]
Valvular disease	6,364	9.51	2.2	<.0001	1.44 [1.20-1.72]
Chronic pulmonary	24,854	37.13	2.0	<.0001	1.49 [1.32-1.68]
Atherosclerosis	5,378	8.03	1.7	0.49	1.08 [0.87-1.34]
Cancer	5,135	7.67	1.7	0.57	1.07 [0.85-1.33]
Diabetes	11,013	16.45	1.6	0.80	0.98 [0.83-1.15]
Coronary disease	36,664	54.77	1.5	0.03	0.88 [0.78-0.99]

LE, Lower extremity; *PAD*, Peripheral arterial disease.

was identified as "unfit for open repair," and these patients were randomized to EVAR versus medical treatment. The investigators found that there was no significant difference in all-cause mortality between medical treatment and EVAR, with EVAR patients having a perioperative mortality of 9%. These authors concluded that for high-risk AAA patients, no surgical intervention is warranted. There remain unanswered questions related to high-risk patients

TABLE 5.—Risk Scores for 30-Day Mortality for Evar Patients

Risk Factor	Score
Renal failure w/dialysis	7
LE ischemia	5
Age ≥ 85 years	4
Liver disease	3
CHF	3
Renal failure w/o dialysis	3
Age 80-84 years	2
Female	2
Neurological	2
Chronic pulmonary	1
Surgeon EVAR experience <3	1
Hospital annual volume <7	1
Age 75-79 years	1

CHF, Congestive heart failure; *EVAR*, endovascular aneurysm repair; *LE*, lower extremity.

TABLE 6.—Predicted Mortality Based on Scoring System

Total Risk Score	Predicted 30-day Mortality (%)	No. of Patients
0	0.5	9907
1	0.7	7516
2	0.9	12005
3	1.1	9281
4	1.4	7656
5	1.7	6532
6	2.2	4715
7	2.8	3403
8	3.5	2274
9	4.4	1462
10	5.5	892
11	6.8	543
12	8.5	348
13	10.6	213
14	13.0	99
15	15.6	87
16	19.4	46
17	23.4	35
18	27.9	7
19	32.9	7
20	38.4	8

and EVAR: how large is the subset of patients who are high risk for EVAR and what are the preoperative characteristics that can identify them?

Subgroups of patients, who are at high or even prohibitively high risk for conventional open AAA repair, have been identified through various clinical studies. Variables that have been commonly defined as preoperative risk factors for mortality include: increased age, congestive heart failure (CHF), myocardial ischemia, and renal and pulmonary dysfunction. By comparison, such risk factors have not been identified for patients undergoing EVAR. Specification of these risk factors is essential to delineate those patients at prohibitive risk for EVAR who would benefit from medical therapy alone. The authors analyzed the outcomes of Medicare patients treated with EVAR between 2000 and 2006 to better understand the mortality associated with EVAR and, more importantly, to define patients at excessively high risk for this "minimally invasive" procedure. Only patients with elective admissions were included in this study. To ensure that all comorbidities identified at previous hospitalizations were not missed, a longer time period (1995-2006) was used to define the comorbidities of patients who were ultimately treated with EVAR.

As it is often difficult to differentiate between preexisting comorbidities and postoperative complications (eg, stroke) in large datasets, a diagnosis was included as a comorbidity if (1) it was present on a previous hospital admission or (2) if it appeared during the index hospitalization and was coded as a chronic or "acute on chronic" disorder. The following comorbidities were assessed (primary and all secondary diagnoses): cardiac disease (coronary artery disease, CHF, valvular heart disease, cardiac arrhythmias), diabetes, chronic pulmonary disease, peripheral arterial disease (clinically significant lower extremity ischemia, vascular insufficiency of the intestine, and renal atherosclerosis), renal disease, neurological disorders (cerebrovascular, paralysis, and other neurological diseases), cancer, rheumatoid arthritis, and liver disease.

A total of 66,943 patients aged 65 years or older were identified who underwent EVAR between 2000 and 2006. In terms of demographics, 8.9% of the study population was 85 years of age or older, 82.9% were males, and 94.8% were Caucasians (Table 1). The overall 30-day mortality was 1.6%. The 30-day mortality after EVAR among females was higher than among males (2.5% vs. 1.4%; $P < .0001$). Perioperative mortality also increased with patient age; this became statistically significant for patients 70 years of age or older (Table 1).

A number of baseline comorbidities were associated with 30-day mortality as shown in Table 2. Common risk factors included chronic pulmonary diseases (37.1% of cohort, mortality 2.0%, $P < .0001$), cardiac arrhythmia (25.2%, mortality 2.3%, $P < .0001$), and heart failure (14.4%, mortality 3.5%, $P < .0001$). Patients with renal failure with dialysis represented only 1.1% of the cohort; however, their risk of dying after EVAR was highest (11.8%, $P < .0001$). Another less common risk factor strongly associated with mortality was clinically significant lower extremity ischemia (2.1% of cohort, mortality 6.2%, $P < .0001$).

The authors also evaluated the relationship between hospital annual volume and surgeon cumulative experience with EVAR and perioperative mortality. Mortality declined from 2.3% to 1.4%, with growing hospital annual volume from less than 7 procedures versus volume greater than 73 EVARs. Thirty-day mortality, when EVAR was performed by surgeons with total experience of <2 procedures, was 2.4%, whereas the mortality was in the range of 1.3% to 1.6% for surgeons with a cumulative EVAR experience ≥3 procedures.

A multivariate regression model was used to assess baseline comorbidities and predict 30-day mortality after EVAR. Risk scores were assigned for every statistically significant risk factor, as depicted in Table 5. Risk scores ranged from a minimum of 1 point for chronic pulmonary disorders to a maximum of 7 points for renal failure with dialysis. The total risk score was obtained by summing individual risk points. The relationship between predicted 30-day mortality after EVAR and patients' total risk score is presented in Table 6. The estimated mortality ranged from 0.5% to 38.4% for risk scores that ranged from 0 to 20.

This analysis provides both information about the factors that define patients at high risk for EVAR and insight into the size of the cohort that has a prohibitively high mortality. Of the almost 67 000 patients evaluated in this study, a risk of perioperative mortality of 9% or greater (the EVAR 2 outcome) was found in only 1.3% of the treated population. The authors also identified preoperative characteristics that can determine this small but high-risk cohort. They found that renal failure, lower extremity vascular disease, liver disease, neurological disorders, female gender, age, hospital volume, surgeon experience, heart failure, and chronic pulmonary diseases all increased the potential of death within 30 days following EVAR. They found risk factors in descending order of importance to be: renal failure with dialysis, lower extremity ischemia, age ≥85 years, liver disease, CHF, renal failure without dialysis, female gender, a neurological disorder, chronic obstructive pulmonary disease, and low hospital volume and surgeon experience with EVAR. The patients at highest risk are those with renal failure; also of great significance is the strong association between renal failure and calcified and diseased iliac arteries, as such pelvic arterial disease directly affects access for introduction of the endograft. The consequence can be arterial rupture, occlusion, and the need for a conduit or a prolonged intervention. Thus, renal failure may be a surrogate for complex arterial anatomy. Patients with lower extremity vascular disease are at increased risk presumably for the same reason as those with renal failure. Lower extremity vascular disease is also a marker for generalized atherosclerosis, including myocardial insufficiency.

The authors have composed a scoring system that can be used to assist interventionalists and patients. Assessing the surgical risk of a patient with multiple comorbidities can be remarkably difficult, yet it is these particular patients who benefit most from an accurate preoperative evaluation, as they are likely to have increased early and late mortality as a consequence of their associated illness. The described scoring system can be a useful aid to preoperatively identify patients unfit for even minimally invasive treatment of their aneurysm.

T. G. Walker, MD

Endovascular Abdominal Aortic Aneurysm Repair: Nonenhanced Volumetric CT for Follow-up

Bley TA, Chase PJ, Reeder SB, et al (Univ of Wisconsin-Madison)
Radiology 253:253-262, 2009

Purpose.—To evaluate the clinical usefulness of volumetric analysis at nonenhanced computed tomography (CT) as the sole method with which to follow up endovascular abdominal aortic aneurysm repair (EVAR) and to identify endoleaks causing more than 2% volumetric increase from the previous volume determination.

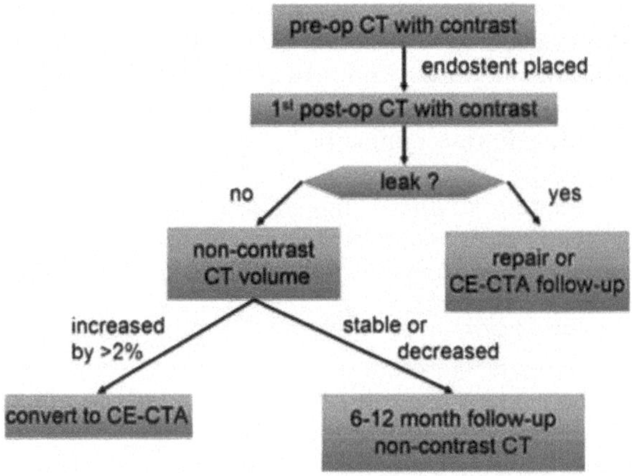

FIGURE 1.—Flowchart shows the imaging protocol for follow-up of EVAR with nonenhanced CT and contrast-enhanced CT angiography (CE-CTA). (Reprinted from Bley TA, Chase PJ, Reeder SB, et al. Endovascular abdominal aortic aneurysm repair: nonenhanced volumetric CT for follow-up. *Radiology.* 2009;253:253-262, Copyright by the Radiological Society of North America.)

TABLE 2.—Significance of Volumetric Changes

Leak Type	Mean Difference (%)	*P* Value
No leak vs types I and III	13.23	<.0001
No leak vs type II	8.55	<.0001
No leak vs types I, II, and III	9.55	<.0001

Note.—No leak means negative for clinically relevant endoleaks and resembles a volumetric change of 2% or smaller. *P* <.001 indicates a significant difference.

Materials and Methods.—The study had institutional review board approval. Images were reviewed retrospectively in a HIPAA-compliant manner for 230 CT studies in 70 patients (11 women, 59 men; mean age, 74 years) who underwent EVAR. The scannning protocol consisted of three steps: (*a*) contrast material–enhanced CT angiography before endovascular stent placement, (*b*) contrast-enhanced CT angiography 0–3 months after repair to depict immediate complications, and (*c*) nonenhanced CT at 3, 6, and 12 months after repair. At each follow-up visit, immediate aortic volume analysis was performed. If the interval volumetric change was 2% or less, no further imaging was performed. If the volume increased by more than 2% on the nonenhanced CT image, contrast-enhanced CT angiography was performed immediately to identify the suspected endoleak. Confidence intervals (CIs) were obtained by using bootstrapping to account for repeated measurements in the same patients.

Results.—Mean volume decrease was −3.2% (95% CI: −4.7%, −1.9%) in intervals without occurrence of a clinically relevant endoleak ($n = 183$). Types I and III high-pressure endoleaks ($n = 10$) showed a 10.0% (95% CI: 5.0%, 18.2%) interval volumetric increase. Type II low-pressure endoleaks ($n = 37$) showed a 5.4% (95% CI: 4.6%, 6.2%) interval volumetric increase. Endoleaks associated with minimal aortic volume increase of less than 2% did not require any intervention. This protocol reduced radiation exposure by approximately 57%–82% in an average-sized patient.

Conclusion.—Serial volumetric analysis of aortic aneurysm with nonenhanced CT serves as an adequate screening test for endoleak, causing volumetric increase of more than 2% from the volume seen at the previous examination (Fig 1, Table 2).

▶ Endovascular abdominal aortic aneurysm repair (EVAR) has been established as a method with which to completely exclude the abdominal aortic aneurysm (AAA) sac from high-pressure arterial blood flow. An unwanted complication after EVAR is the persistence of blood flow within the excluded aneurysm sac (outside the lumen of the endograft, termed an endoleak). The magnitude of the leak determines whether the size of the aneurysm sac remains the same or continues to increase because the endoleak exerts continued high pressure on the aortic aneurysm wall.

With current endografts, endoleak prevalence is reported in approximately 4% to 11% of EVAR procedures. On the basis of the underlying pathophysiology, endoleaks are currently classified into 5 types. They can be detected invasively with angiography or noninvasively with ultrasonography (US), computed tomography (CT), or magnetic resonance (MR) imaging. Of these modalities, CT has become the most widely accepted. It is an efficient and accurate modality with which to assess aneurysm morphology, define the position and integrity of the endograft, and identify endoleaks. Endoleaks that cause an increase in aneurysm volume often require treatment because these patients are at continued risk for aortic rupture. Because endoleaks can develop at any time after stent insertion, lifetime surveillance is mandatory. Repaired aneurysms that remain stable or decrease in size usually do not require intervention and may not be at increased risk for rupture regardless of small endoleaks.

Volumetric assessment is more sensitive than diameter measurement in the detection of changes in aneurysm size. Reproducible volumes can be obtained relatively quickly with CT, and the degree of variability is less than 2% for a well-performed 3-dimensional reformation.

In most studies, endografts have been evaluated by using multiphasic contrast material—enhanced CT angiography, which may include nonenhanced, arterial, and delayed-phase image acquisition. With the goal of radiation reduction, the usefulness of the different phases has been discussed in the literature. Some authors favor eliminating the arterial phase, some advocate eliminating the delayed phase, while others cite specific advantages in acquiring both arterial and delayed-phase images. However, impaired renal function is a common comorbidity in these patients, and use of iodinated contrast material may be contraindicated. Recently, the usefulness of dual energy CT performed during the delayed phase and reconstruction of virtually nonenhanced images has been reported. Decisions regarding management of endoleaks are often based on changes in the size of the aneurysm. Disadvantages of lifetime follow-up algorithms with contrast-enhanced CT angiography include repeated radiation exposure, cost, and numerous administrations of contrast material in a patient population with a higher incidence of renal impairment.

The purpose of this study was to evaluate the validity of performing only follow-up nonenhanced CT with volumetric analysis of infrarenal AAA as a means of detecting substantial endoleaks.

This evaluation was accomplished initially by comparing nonenhanced CT results with contrast-enhanced CT angiography results for the depiction of endoleak. In patients with stable or decreasing aneurysm volumes, this method would eliminate the need for contrast material administration and greatly reduce radiation exposure. Contrast material administration and multiphasic contrast-enhanced CT angiography would be used only in patients with increasing aneurysm volumes and, thus, higher risk for endoleak.

In this retrospective study, the medical records and CT scans of patients who underwent EVAR for infrarenal AAA between January 2000 and August 2006 were reviewed. The preoperative and first postoperative scans at 0 to 3-month follow-up after EVAR were obtained by means of multiphasic contrast-enhanced CT angiography, and infrarenal AAA volumes were calculated. In patients with an identified endoleak, subsequent scans were obtained with routine contrast-enhanced CT angiography and volume calculation. In patients without identifiable endoleak on the first postoperative scan, subsequent scans were obtained without contrast material. Immediate aortic volume was calculated for the current study and the most recent previous scan by a 3-dimensional technologist (10 years experience) while the patient waited. If the current volume change on nonenhanced scans was less than a 2% increase compared with what the previous study showed, the patient was discharged without further scanning or contrast material administration. If the volume increased by more than 2%, the patient was brought back to the CT scanner immediately, and full contrast-enhanced CT angiography was performed to check for an endoleak. Fig 1 summarizes the design of the imaging protocol. The cutoff value of 2% was chosen because small endoleaks that produce a small volume change may not require surgical intervention, and we have shown that a 2% cutoff is higher than the intraobserver variability for repeated volume measurements obtained by an experienced observer.

The patients who underwent scanning during the period from 2000 through 2004 routinely underwent serial nonenhanced CT and contrast-enhanced CT angiography for detection of endoleaks. Subsequently, on the basis of the successful results of our initial evaluation of the authors' protocol, they changed the scanning protocol in 2004, to that detailed in Fig 1. All scans were retrospectively rereviewed, and each volumetric calculation was performed at a workstation (Volume Viewer 2 Advantage; GE Healthcare) by the same dedicated 3-dimensional CT technologist by using a semiautomated protocol and the nonenhanced CT data set.

In a patient undergoing a follow-up noncontrast CT evaluation, volumetric analysis of this study and the most recent previous examination was performed on the same day and by the same technologist to eliminate interobserver variability. Volumes were calculated in a standardized fashion from the level of the renal arteries to the common iliac bifurcation. Percentage of volume change was calculated for one study versus its most recent comparison study. Two hundred and thirty CT studies in 70 patients were available for statistical analysis. In 65 patients, follow-up CT was performed without contrast material. In 6 (9%) of these, contrast-enhanced CT angiography was then performed owing to more than 2% increased aneurysm volume calculation. All 6 of these studies took place more than 12 months after EVAR of an infrarenal AAA. Three of the 6 demonstrated a type II endoleak. In the other 3 cases, no endoleak (Types I-IV) was detected and, therefore, endotension (type V) was presumed to be the cause.

Comparative volume observations in 183 of the 230 CT studies did not suggest a volume change greater than 2% in patients without symptoms. On the basis of the study of technical margin of error, this was arbitrarily defined as excluding a relevant endoleak. Forty-seven CT observations revealed a volumetric increase of 2% or more. On the basis of sequential scans, the authors found a type I endoleak in 4 patients, a type II endoleak in 17 patients, and a type III endoleak in 2 patients. No type IV endoleaks were identified in this study population. On average, endoleaks increased the residual aortic sac volume by 6.4%. Higher volumetric increases were found in the high-pressure endoleaks (types I and III, $n = 10$) than in those in 5 low-pressure endoleaks (type II, $n = 37$): 10.0% and 4% volumetric increases, respectively. As anticipated, a volumetric decrease (−3.2%) was found in CT observations in which there was no relevant endoleak. The differences in volumetric change between no clinically relevant endoleak and high-pressure endoleak types I and III, as well as low-pressure endoleak type II, were significant, as shown in Table 2.

With the authors' protocol of nonenhanced CT for volumetric analysis and limited contrast-enhanced CT angiography (Fig 1), radiation exposure is reduced by approximately 69% to 82% in an average-sized patient compared with radiation exposure in the traditional protocol of combined nonenhanced CT and contrast-enhanced CT angiography with delayed-phase postcontrast CT protocol. Radiation exposure is reduced by 57% to 72% compared with that in the combined nonenhanced CT and contrast-enhanced CT angiography protocol. With the decreasing incidence of endoleaks in newer improved stents, this algorithm may have an even greater effect and is now the standard of care at the authors' institution.

The authors conclude that patients who have undergone EVAR for an infrarenal AAA can be followed up without intravenous contrast material on the basis of serial volumetric analysis of the aneurysm sac. They do not recommend using nonenhanced CT alone in the initial 3 months after repair because volumetric changes may not yet be large enough to be detected, and the prevalence of endoleaks is highest in the immediate postoperative period. For later follow-up, their protocol allows identification of endoleaks causing a volumetric increase of more than 2% from the volume on the previous scan and also facilitates detection of endotension. This protocol has the important benefits of reduced scanning time and radiation exposure and potentially reduced cost and risk of nephropathy from contrast material.

T. G. Walker, MD

Endovascular Aortic Aneurysm Repair with the Endurant Stent-graft: Early and 1-year Results from a European Multicenter Experience

Torsello G, Troisi N, Tessarek J, et al (St Franziskus Hosp, Münster, Germany; Univ Hosp, Münster, Germany; Univ of Florence, Italy)

J Vasc Interv Radiol 21:73-80, 2010

Purpose.—To assess preliminary results of the Endurant stent-graft, which was developed to treat patients with abdominal aortic aneurysms (AAAs) and challenging aortoiliac anatomy.

Materials and Methods.—From November 2007 to October 2008, 45 patients with AAAs were treated with the Endurant stent-graft in three European vascular centers. Thirty-eight of the 45 patients (84%) had a hostile anatomy of the proximal neck. Early (30 days), early midterm

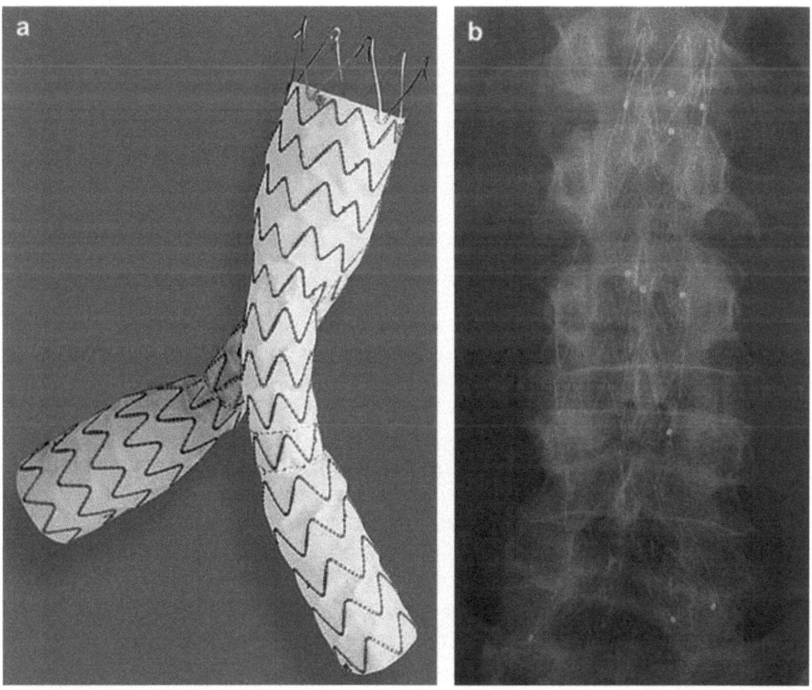

FIGURE 1.—(**a**) Endurant stent-graft. (**b**) Control radiograph obtained at discharge shows all the radiopaque markers. (Courtesy of Torsello G, Troisi N, Tessarek J, et al. Endovascular aortic aneurysm repair with the endurant stent-graft: early and 1-year results from a European multicenter experience. *J Vasc Interv Radiol.* 2010;21:73-80.)

TABLE 1.—Preoperative Demographic Data and Comorbidities

Parameter	No. of Patients ($n = 45$)
Older age (>80 y)	10 (22)
Smoking	34 (76)
Hypertension	38 (84)
Hypercholesterolemia	26 (58)
Diabetes mellitus	12 (27)
Coronary artery disease	25 (56)
History of myocardial infarction	12 (27)
Chronic obstructive pulmonary disease	19 (42)
Cerebrovascular disease	9 (20)
Peripheral artery disease	12 (27)
Renal chronic failure (serum creatinine level ≥1.5 mg/dL)	10 (22)
Previous laparotomy	9 (20)

Note.—Numbers in parentheses are percentages.

TABLE 2.—Preoperative Diagnostic Assessment

Parameter	Mean
Neck diameter at renal level (mm)	24.7 (19–31)
Neck diameter at the beginning of aneurysmal sac (mm)	26.3 (20–33)
Aneurysm diameter (mm)	58 (41–76)
Aortic bifurcation diameter (mm)	32.8 (20–54)
Right common iliac artery diameter (mm)	15.8 (9–30)
Left common iliac artery diameter (mm)	16.7 (11–32)
Right external iliac artery diameter (mm)	8.2 (6–13)
Left external iliac artery diameter (mm)	8.3 (6–14)
Neck length (mm)	16.2 (5–40)
Proximal neck angle	55.5° (20°–90°)
Right iliac angle	43.9° (10°–100°)
Left iliac angle	45.9° (15°–90°)

Note.—Numbers in parentheses are the range.

(6 months), and late midterm (1 year) results were analyzed in terms of technical success, clinical success, morbidity, and mortality.

Results.—Intraoperative immediate technical and clinical success was achieved in all cases. At completion angiography, a type II endoleak was detected in seven of the 45 patients (16%) and an iliac limb stenosis was diagnosed and successfully treated in five (11%). At discharge, computed tomography demonstrated a type I endoleak in one of the 45 patients (2.2%). At 30 days, a graft limb thrombosis was diagnosed (2.2%) and successfully treated. The 30-day overall technical and clinical success rates were 97.8% (44/45 cases) and 95.6% (43/45 cases), respectively. During follow-up (mean duration, 8 months; range, 5–16 months),

TABLE 3.—Anatomic Features of the Proximal Neck

Feature	No. of Patients ($n = 45$)
Short neck (≤10 mm)	14 (31)
Neck bulge	0
Tapered neck	23 (51)
Angulated neck (≥60°)*	26 (58)
Neck thrombus	11 (24)
No. of features for each patient	
0	7 (16)
1	9 (20)
2	23 (51)
3	5 (11)
4	1 (2.2)
5	0

Note.—Numbers in parentheses are percentages.
*The mean distance between the lowest renal artery and the apex of the infrarenal aortic angle was 7.3 mm (range, 4–12 mm).

the type I endoleak was successfully treated with an aortic extension. Estimated freedom from type I or III endoleak and repeat intervention at 1 year was 97.8% (44/45 cases) and 93.3% (42/45 cases), respectively.

Conclusions.—The initial experience shows that the Endurant stent-graft appears to be effective in endovascular repair of AAAs in patients with hostile aortoiliac anatomy. This graft permits a broader group of patients to be treated with endovascular aneurysm repair; however, further studies are needed to evaluate the long-term results.

► Endovascular repair of abdominal aortic aneurysm (EVAR) has become a valuable therapeutic alternative in treating patients with abdominal aortic aneurysms (AAA). Multiple randomized trials have demonstrated that EVAR is a less invasive alternative to open AAA repair, and decreases perioperative morbidity and mortality, blood loss and transfusions, and the use of the intensive care unit and the length of hospital stay. In addition, endovascular techniques make it possible to treat patients whose comorbidities make conventional open repair difficult or risky. For these reasons, many physicians believe that high-risk patients with AAA should be preferentially treated with EVAR. However, several anatomic criteria (eg, short/large/angulated neck, small/large iliac diameters, iliac tortuosity/occlusion, accessory renal arteries) can make an individual an unsuitable candidate for EVAR, which is often the case in patients at high surgical risk.

Severe infrarenal aortic neck angulation is strongly associated with subsequent proximal type I endoleak, and dilatation of the proximal aortic neck represents a clear risk factor for endograft migration. In patients with angulated or dilated proximal aortic necks, proximal sealing of the endograft still remains a challenge. Endovascular repair by using a flexible stent-graft seems to be feasible in patients with significantly angulated necks.

Access-related difficulties, mainly due to tortuous, small, and calcified iliac vessels, also continue to represent a major problem for EVAR. Several strategies have been reported in the literature to overcome access problems and limit arterial access complications; however, the main factor to allow the safe passage of the graft through the iliac vessels is generally considered to be the diameter and trackability of the delivery system. Furthermore, complex aortic anatomy often increases the perioperative and long-term rates of device-related complications. To adapt the prosthesis to the anatomy of the patient, a large number of commercially available stent-grafts have been developed. Devices are generally constructed from a polyester-type material and nitinol or stainless steel; however, none of them have provided durable and successful clinical outcomes in challenging, hostile anatomies as described.

The aim of this study was to assess the safety and effectiveness in the early and midterm period of a new generation device (Endurant stentgraft; Medtronic Cardiovascular, Santa Rosa, California) developed to treat patients with AAAs who have hostile aorto-iliac anatomy. Forty-five consecutive patients who underwent elective surgery for AAA in 3 European centers were included in this study. Inclusion criteria were as follows: (a) AAA with a diameter of 50 mm or larger, 40 mm or larger with documented AAA growth of more than 5 mm within the previous 6 months; (b) adequate iliac/femoral access with a diameter at the access site of at least 6 mm bilaterally; (c) proximal neck diameter of 19 ± 32 mm; and (d) common iliac diameters of 8 ± 25 mm. Hostile neck anatomy was defined as the presence of at least one of the following features: (a) a length between the most caudal renal artery and the beginning of the aneurysm sac of less than or equal to 10 mm ("short neck"); (b) a focal enlargement of the neck of at least 3 mm within the first 15 mm after the most caudal renal artery ("neck bulge"); (c) a gradual dilation of at least 2 mm of the neck within the first 10 mm ("tapered neck"); (d) infrarenal AAA neck angulation of at least 60° within the seal zone of the proximal neck ("angulated neck"); and (e) thrombus covering more than 50% of the circumference of the neck ("neck thrombus").

Technical success was defined as the successful passage of the delivery system to the landing zone, accurate placement of the device in the target zone, proper positioning of the contralateral limb, and complete withdrawal of the delivery system after the deployment without significant limb stenosis and with freedom from conversion and absence of type I or III endoleak, as evidenced at CT angiography performed at discharge. Clinical success was defined as freedom from the following: aneurysm rupture, expansion (> 5 mm), conversion to open repair, type I or III endoleak, graft infection or thrombosis, and aneurysm related death.

The demographic data of the patients and the details of the preoperative diagnostic assessment that includes anatomic features of the AAAs are respectively summarized in Tables 1 and 2. Thirty-eight of the 45 patients (84%) had a hostile anatomy of the proximal neck. All the anatomic features of the proximal neck are listed in Table 3. Mean oversizing of the main body respect to the proximal neck diameter was 19.5% (range, 8% ± 26%).

The Endurant stent-graft is a new-generation device for the treatment of AAAs. It is comprised of polyester graft material externally supported by nitinol M-shaped body stents that are attached to the graft material by sutures and allow for conformability and flexibility. The bare stent design of the suprarenal stent with anchoring hooks allows the stent-graft to be fixed above the renal arteries. Radiopaque markers are placed at the proximal and distal ends of all stent-graft segments, as well as the flow divider (Fig 1). The Endurant stent-graft is loaded into a low-profile delivery system, and the outer diameter ranges from 18 F to 20 F for the main body and from 14 F to 16 F for the extenders. The delivery system features a tip-capture mechanism to ensure the accurate controlled release deployment of the suprarenal stent.

In this multicenter experience, use of the Endurant stent-graft permitted the treatment of patients with very short proximal necks; in fact, more than half of the patients had a neck shorter than 15 mm (24 patients, 53%), and 14 (31%) had a proximal neck length of 10 mm or less. At 30 days, only 1 case of type I endoleak (2.2%) was recorded, and this was successfully treated with an aortic extension. The low profile of the Endurant delivery system allowed for good results with total percutaneous access of the femoral arteries without any recorded iliac injury during delivery of the graft. At 30 days, one graft thrombosis (2.2%) occurred, and these results are similar to those obtained with other stent-grafts available on the market.

The authors conclude that this preliminary experience shows that Endurant stentgraft appears to be efficacious and safe in the endovascular management of AAAs in patients with hostile anatomy of the proximal aortic neck and iliac arteries. The 2 main characteristics of the endograft (suprarenal fixation with controlled deployment of the top stent and lower profile of the delivery system) allow a broader group of patients to be treated with EVAR, even if further studies are needed to evaluate the long-term results of this new device.

T. G. Walker, MD

Patient Care

Catheter-directed Therapy for the Treatment of Massive Pulmonary Embolism: Systematic Review and Meta-analysis of Modern Techniques

Kuo WT, Gould MK, Louie JD, et al (Stanford Univ Med Ctr, CA; Veterans Affairs Palo Alto Health Care System, CA)

J Vasc Interv Radiol 20:1431-1440, 2009

Purpose.—Systemic thrombolysis for the treatment of acute pulmonary embolism (PE) carries an estimated 20% risk of major hemorrhage, including a 3%–5% risk of hemorrhagic stroke. The authors used evidence-based methods to evaluate the safety and effectiveness of modern catheter-directed therapy (CDT) as an alternative treatment for massive PE.

Materials and Methods.—The systematic review was initiated by electronic literature searches (MEDLINE, EMBASE) for studies published from January 1990 through September 2008. Inclusion criteria were applied to select patients with acute massive PE treated with modern CDT. Modern techniques were defined as the use of low-profile devices (<10 F), mechanical fragmentation and/or aspiration of emboli including rheolytic thrombectomy, and intraclot thrombolytic injection if a local drug was infused. Relevant non-English language articles were translated into English. Paired reviewers assessed study quality and abstracted data. Meta-analysis was performed by using random effects models to calculate pooled estimates for complications and clinical success rates across studies. Clinical success was defined as stabilization of hemodynamics, resolution of hypoxia, and survival to hospital discharge.

Results.—Five hundred ninety-four patients from 35 studies (six prospective, 29 retrospective) met the criteria for inclusion. The pooled clinical success rate from CDT was 86.5% (95% confidence interval [CI]: 82.1%, 90.2%). Pooled risks of minor and major procedural complications were 7.9% (95% CI: 5.0%, 11.3%) and 2.4% (95% CI: 1.9%, 4.3%), respectively. Data on the use of systemic thrombolysis before CDT were available in 571 patients; 546 of those patients (95%) were treated with CDT as the first adjunct to heparin without previous intravenous thrombolysis.

Conclusions.—Modern CDT is a relatively safe and effective treatment for acute massive PE. At experienced centers, CDT should be considered as a first-line treatment for patients with massive PE.

▶ Massive pulmonary embolism (PE), defined as hemodynamic shock from acute PE, is a common life-threatening condition. An estimated 530 000 cases of symptomatic PE occur annually in the United States, and approximately 300 000 people die every year from acute PE. The mortality rate can reach more than 58% in patients with acute PE with hemodynamic shock, and most of these deaths occur within 1 hour of presentation. The standard medical treatment of such patients with massive PE is systemic intravenous thrombolysis; however, some patients cannot receive systemic thrombolysis due to contraindications, and there may be insufficient time in the acute setting to infuse full dose intravenous thrombolysis. Furthermore, the use of systemic thrombolysis for treatment of acute PE carries up to a 20% risk of major hemorrhage, including a 3% to 5% risk of hemorrhagic stroke. The safety and efficacy of catheter-directed therapy (CDT) as opposed to systemic thrombolytic therapy has not been firmly established. When systemic thrombolysis is contraindicated, it is also unclear which alternative treatment should be pursued. In addition, many patients treated with systemic thrombolysis do not improve hemodynamically and may require urgent endovascular mechanical thrombectomy with local pharmacologic intervention. The current systematic review and meta-analysis was performed by the authors to evaluate the safety and effectiveness of modern CDT as a potential alternative or early treatment option in patients with massive PE.

T. G. Walker, MD

Fluoroscopically Guided Percutaneous Jejunostomy with Use of a 21-gauge Needle: A Prospective Study in 51 Patients

Hu H-T, Shin JH, Song H-Y, et al (Henan Tumor Hosp, Zhengzhou, China; Univ of Ulsan College of Medicine, Seoul, Korea)

J Vasc Intervent Radiol 20:1583-1587, 2009

Purpose.—To assess the safety and clinical efficacy of fluoroscopically guided percutaneous jejunostomy with use of a 21-gauge needle and a single anchor technique in 51 patients.

Materials and Methods.—From November 2006 to January 2009, 51 consecutive patients (42 men and nine women; mean age, 63.7 years) underwent percutaneous jejunostomy under fluoroscopic guidance. A 7.5-F multifunctional coil catheter was used to insufflate the jejunum. The distended jejunum was punctured with a 21-gauge needle, with the inserted coil catheter as the target. A single anchor was used. The technical success, number of punctures, procedure time, complications, and follow-up data including 30-day mortality rate were evaluated.

Results.—The technical success rate was 100%, and the single anchor technique was used in all but one patient, in whom three anchors were used. The mean number of punctures was 1.7 (range, 1–4), and the mean procedure time was 14.8 minutes (range, 7–29 min). Peritonitis was a major complication in two patients (3.9%), who were treated by changing the catheters from 14 F to 16 F and performing percutaneous drainage procedures. Three minor complications were encountered: superficial cellulitis ($n = 2$) and severe puncture site pain ($n = 1$). The 30-day mortality rate was 5.9% (three of 51), although none of the deaths could be attributed to the jejunostomy procedures.

Conclusions.—Fluoroscopically guided percutaneous jejunostomy with use of a 21-gauge needle and the single anchor technique seems to be safe and effective, with high technical success and low complication rates.

▶ Malnutrition, a common problem in hospitalized patients, can be caused by malignancy, benign stricture, aspiration, or diseases of the central nervous system, such as stroke or traumatic brain injury. When gastric feeding is insufficient as a result of poor gastric emptying, gastric outlet obstruction, gastroesophageal reflux, or aspiration, feeding jejunostomy or gastrojejunostomy is an alternative. Jejunostomy was initially performed surgically, but laparoscopic, endoscopic, and radiologic techniques have also been described. Fluoroscopically guided percutaneous jejunostomy lacks wide acceptance, probably because of its inherent technical difficulties, such as puncturing the mobile and compliant jejunum and maintaining the position of the jejunum during catheter insertion. Recently the usefulness of a 21-gauge needle technique with the use of a vascular or multifunctional coil catheter for percutaneous jejunostomy has been reported. The main advantage of this technique is the easy and safe puncture of the jejunum related to the smaller needle diameter as compared with the 17-gauge needles in previous reports.

The present prospective study was designed to assess the safety and clinical efficacy of this underused procedure in 51 patients. These patients all underwent percutaneous jejunostomy under fluoroscopic guidance. The technique involved transoral or transnasal catheterization of the proximal jejunum using a specially designed "coil catheter" and an exchange length guidewire, followed by administration of water soluble contrast and 200 ml to 300 ml of air for jejunal opacification and insufflation. The catheter tip was used as a target for percutaneous puncture with a 21-gauge Chiba needle, followed by introduction of a 0.018-inch guide wire, which was advanced as far as possible into the jejunal lumen to allow the insertion of a 6-F Neff catheter. A single suture anchor was then deployed through the Neff catheter with use of a 0.035-inch hydrophilic guide wire. The Neff catheter was then removed, leaving the guide wire in place. After serial dilation of the tract a 14-F locking-loop catheter was inserted and the jejunum was secured to the anterior abdominal wall by suturing the skin with a thread connected to the anchor.

All procedures were technically successful and there were no procedural complications. The mean number of punctures was 1.7 (range, 1–4), and the mean procedure time was 14.8 minutes. The major complication rate was 3.9%, consisting of peritonitis, which was treated by changing the catheters from 14 F to 16 F and performing percutaneous drainage procedures. The 30-day mortality rate of 5.9% was unrelated to the procedures.

Various methods may be used to obtain enteral access for artificial nutrition, including surgical or laparoscopic jejunostomy or percutaneous endoscopic or fluoroscopic jejunostomy. Surgical or laparoscopic approaches require general anesthesia, and the rate of serious complications has been reported to be 1% to 17%. Therefore, most surgical or laparoscopic jejunostomy procedures have been replaced by percutaneous endoscopic or fluoroscopic jejunostomy because of the greater patient comfort and less frequent displacement. Percutaneous fluoroscopic jejunostomy has become an increasingly used technique for small bowel enteral feeding because there are several potential advantages of radiologic versus endoscopic placement. These advantages include the ability to traverse tight esophageal strictures or gastric outlet obstructions with small-bore catheters and the ability to opacify the jejunum with contrast medium, both of which reduce the possibility of inadvertent puncture of other organs. There is also a lower likelihood of hemorrhage or long-term fistula on eventual withdrawal of the catheter because of the small catheter size. The percutaneous fluoroscopic jejunostomy procedure has been slow to gain widespread acceptance, perhaps because of the technical difficulties of targeting and puncturing the mobile and compliant jejunum. In this study, the 100% technical success rate may be related to the use of a 21 rather than a 17-gauge needle for achieving enteral access. The authors assume that the smaller needle is sharper and produces less tissue resistance during penetration than a 17-gauge needle, which tends to slip off the bowel wall, especially when the intestine is poorly distended or ascites is present. Additionally, use of a 21-gauge needle allows multiple punctures with lower complication rates because of the needle's smaller diameter. Another factor contributing to the high success rate was felt to be the injection of approximately 200 mL to

300 mL of air to inflate the jejunum, makes the bowel segment more stable, enabling the needle to puncture the jejunal wall more easily.

In this study, the time required for the procedure was satisfactory, as the average procedure time was 14.8 minutes, which is much less than the 40 minutes to 45 minutes noted in reported studies that used endoscopic or fluoroscopic guidance. The major complication rate of 3.9% and the minor complication rate of 5.9% were both lower than the respective 10.5% to 13% and 8.7% to 27.3% rates noted in previous reports. The low complication rates in the present study can be attributed to the safe and easy punctures with the 21-gauge needle and the adequate diameter of the 14-F feeding tube.

The authors conclude that fluoroscopically guided percutaneous jejunostomy with a 21-gauge needle and a single anchor technique seems to be safe and effective and is associated with high technical success and low complication rates.

T. G. Walker, MD

Quality Improvement Guidelines for Recording Patient Radiation Dose in the Medical Record

Miller DL, Balter S, Wagner LK, et al (Natl Naval Med Ctr, Bethesda, MD; Lenox Hill Hosp, NY; Univ of Texas Houston Med School; et al)

J Vasc Interv Radiol 20:S200-S207, 2009

Background.—The Society of Interventional Radiology (SIR) has presented their Standards of Practice for patient radiation dose recording. The US Food and Drug Administration (FDA) offers recommendations on this topic, but state-level regulations are not uniform. The SIR guidelines were written to be included in a quality-improvement program to manage radiation dose from fluoroscopically guided invasive and interventional procedures, although computed tomographic (CT) fluoroscopy is excluded. Definitions of all important terms were included.

General Principles.—Patient radiation dose is recorded primarily to determine the likelihood of radiation-induced effects in any individual patient. Quality-assurance purposes and patient safety are secondary purposes for such recordings. The four standard means for measuring dose are fluoroscopy time and number of fluorographic images, measurement of the total radiation energy entering the patient (DAP), air kerma value at a specific point (CD), and measurement of the highest radiation dose at any point on that patient's skin (PSD). There is some degree of uncertainty in all patient dose measurements. Sources of this uncertainty include the effects of patient size, beam orientation, and technical configuration of the fluoroscope on the conversion of fluoroscopy time to patient dose; instrument response to changes in beam energy, dose rate, and collimator size; the effects of backscatter from the patient; and reconstruction of dose data from fluoroscopy time and number of fluorographic frames.

Recommendations.—Patient radiation doses should be obtained for all procedures known to be associated with relatively high patient radiation

TABLE 2.—Threshold Rates for Recording of Dose Metrics

Procedure Type	Threshold (%)
High-dose procedures (all embolization procedures, TIPS, arterial angioplasty or stent placement at any site in the abdomen or pelvis)	99
Other fluoroscopically-guided interventional procedures (excluding CT fluoroscopy)	95

doses. This includes all embolization procedures, transjugular intrahepatic portosystemic shunt (TIPS) procedures, and abdominal or pelvic arterial angioplasty or stent placement procedures. Medium-dose procedures that are likely to be repeated should also include dose measurements. Although other interventional radiology procedures are less likely to produce high patient doses, it is desirable to record these as well. Getting into this habit makes it probable that the process will not be overlooked for high-dose procedures and supplies information useful for quality-assurance monitoring. The placement of the data is not specified but suggestions include the immediate procedure note, the procedure worksheet, and/or the final report.

Optimal measurement uses the PSD and includes real-time skin dose mapping. If skin dose measurements are not available, other dose metrics are recommended. PSD is the most valuable predictor of deterministic injury; DAP functions similarly for stochastic injury. If PSD is difficult to measure, CD may be used, recognizing that it does not correlate well with PSD in individual cases. Monitoring fluoroscopy alone underestimates the radiation-induced skin effect risk. Fluoroscopy time and number of fluorographic images are a better guide to patient dose; they do not measure dose in and of themselves, but can be used, along with height and weight of the patient, to record patient radiation dose if nothing else is available.

Conclusions.—Predicting the risk of adverse radiation-induced effects depends on knowing the dose received during a procedure. Measurements also help in quality assurance and provide valuable feedback to the operator. The guidelines offered by SIR can standardize the methods used to measure radiation doses.

▶ Fluoroscopically guided procedures are an essential part of the contemporary practice of medicine. In general, the risk of patient injury as a result of radiation exposure during these procedures is low. Some fluoroscopically guided procedures are associated with a risk of radiation injury to the skin. The frequency of injury is unknown. A comprehensive review published in 2001 reported data on radiation-induced skin injuries in 73 patients of which 64% were results of coronary angiography and intervention, 16% were results of cardiac radiofrequency catheter ablation, 10% were results of transjugular intrahepatic portosystemic shunt (TIPS) creation, 4% were results of neuroradiologic interventions, and the type of procedure was not specified for 4 patients. Some reported skin injuries have been associated with renal angioplasty, multiple hepatic/biliary procedures, or embolization.

In a Public Health Advisory of September 30, 1994, the FDA recommended that "information permitting estimation of the absorbed dose to the skin be recorded in the patient's medical record." The International Commission on Radiological Protection (ICRP) has also recommended recording patient radiation dose in the medical record for certain procedures. Monitoring and recording patient dose data for all procedures can be valuable for quality-assurance purposes as well as patient safety. Feedback to the operator may help to optimize radiation doses overall.

The guidelines in this article were written for inclusion in quality-improvement programs used to manage radiation dose from fluoroscopically guided invasive and interventional procedures excluding computed tomographic (CT) fluoroscopy.

Patient radiation dose may be measured and recorded in different ways. There are 4 relatively standard methods for measuring dose during interventional fluoroscopic procedures. These methods differ in usefulness and availability. In the United States, one method is widely available, whereas the others vary in availability from relatively common to extremely uncommon. None of these methods are applicable to dose measurements for CT fluoroscopy. Dose measurement and recording for CT fluoroscopy are not discussed in this article. The simplest and most widely available measurements are fluoroscopy time and number of fluorographic images. These are analogues of dose; that is, they do not measure dose directly, and, by themselves, they are insufficient to determine patient dose. They are the least useful measures of patient dose. To estimate patient dose from fluoroscopy time and number of fluorographic images, the fluoroscopic dose rate and the dose per image must also be measured or estimated. The next most commonly available dose measurement is dose-area-product (DAP), which is a measure of the total radiation energy entering the patient. It is a good indicator of stochastic risk and correlates with operator and staff dose. DAP meters may be integrated into the fluoroscopic unit or installed as add-on devices.

DAP is not an ideal indicator of deterministic risk. The principal deterministic risk to the patient is radiation-induced skin injury. The likelihood and severity of radiation injury at any point on the skin are related to the dose delivered to that portion of skin. DAP is a surrogate measure of skin dose. It does not correlate well with skin dose. A large dose delivered to a small skin area yields the same DAP as a small dose delivered to a large skin area. Air kerma is the energy released per unit mass of a small volume of air when it is irradiated by an x-ray beam. For diagnostic x-rays, air kerma is the same as the absorbed dose delivered to the volume of air in the absence of scatter. Air kerma is measured in Gy. The cumulative dose (CD) is the air kerma accumulated at a specific point in space relative to the fluoroscopic gantry (the interventional reference point) during a procedure. CD does not include tissue backscatter and is measured in Gy. CD is sometimes referred to as cumulative air kerma. The peak skin dose (PSD) is the highest dose at any portion of a patient's skin during a procedure.

The optimal method is measurement of PSD. Ideally, this would include real-time skin-dose mapping as a means for managing patient radiation dose. In the absence of skin dose measurements, other dose metrics should be used. PSD and DAP are the most useful predictors for deterministic and stochastic injury, respectively. CD is an acceptable substitute if PSD is difficult to measure, but it does not correlate well with PSD in individual cases. Fluoroscopy time alone does not correlate with PSD. Monitoring fluoroscopy time alone also underestimates the risk of radiation-induced skin effects. Fluoroscopy time and number of fluorographic images used together can provide a better guide to patient dose but are not themselves measures of dose. They do not provide sufficient information for dose calculations and are, therefore, suboptimal dose metrics. However, if none of the other metrics can be measured, fluoroscopy time and number of fluorographic images, along with the patient's height and weight, can be used for recording patient radiation dose until other means are available.

Certain procedures are known to be associated with relatively high patient radiation doses. Patient radiation dose data should be recorded for all instances of these procedures. To simplify the categorization of high-dose procedures, the Society of Interventional Radiology (SIR) recommends that all embolization procedures, TIPS procedures, and arterial angioplasty or stent placement procedures anywhere in the abdomen or pelvis should be considered potential high-dose procedures. Patient radiation dose data should also be recorded for medium-dose procedures that are likely to be repeated. Medium-dose procedures may be defined as procedures with a CD of 1 to 3 Gy.

Recording patient dose data for all procedures makes it less likely that the process will be omitted inadvertently for high-dose procedures. In addition, monitoring and recording patient dose data can be valuable for quality-assurance purposes as well as for patient safety. Feedback to the operator may help to optimize radiation doses overall. Adequate recording of dose metrics is defined as documentation in the patient record of at least one of the following for all interventional procedures requiring fluoroscopy (in descending order of desirability): skin dose mapping, PSD, CD, DAP, or fluoroscopic time/number of fluorographic images (Table 2).

This article provides a very comprehensive review of radiation dose metrics, which is obviously of significant importance in providing optimal patient care during fluoroscopically guided procedures.

T. G. Walker, MD

Research

Atherosclerosis and Matrix Metalloproteinases: Experimental Molecular MR Imaging in Vivo

Amirbekian V, Aguinaldo JGS, Amirbekian S, et al (Mount Sinai School of Medicine, NY; et al)

Radiology 251:429-438, 2009

Purpose.—To evaluate the capability of P947, a magnetic resonance (MR) imaging contrast agent that molecularly targets matrix metalloproteinases (MMPs), to aid detection and imaging of MMPs in atherosclerotic lesions in vivo; its specificity compared with that of P1135; expression and distribution of MMPs in atherosclerotic vessels; and in vivo distribution and molecular localization of fluorescent europium (Eu) P947.

Materials and Methods.—The Animal Care and Use Committee approved all experiments. P947 was synthesized by attaching a gadolinium chelate (1,4,7,10-tetraazacyclododecane-N,N′, N″,N‴-tetraacetic acid) to a peptide that specifically binds MMPs. Scrambled form of P947 (P1135) was synthesized by replacing the targeting moiety of P947 with a scrambled peptide lacking the ability to bind MMPs. P947, P1135, and gadoterate meglumine were injected into atherosclerotic apolipoprotein E-deficient and wild-type mice. The aortic MR imaging enhancement produced by the contrast agents was measured at different times and was compared by using one-way analysis of variance. MMP expression was investigated in the aortas by using MMP immunostaining and in situ MMP zymography. A fluorescent form of P947 (Eu-P947) was synthesized to compare the in vivo distribution of the contrast agent (Eu-P947) with specific MMP immunofluorescent staining.

Results.—MMP-targeted P947 facilitated a 93% increase ($P < .001$) in MR image signal intensity (contrast-to-noise ratio [CNR], 17.7 compared with 7.7; $P < .001$) of atherosclerotic lesions in vivo. Nontargeted P1135 (scrambled P947) provided 33% MR image enhancement (CNR, 10.8), whereas gadoterate meglumine provided 5% (CNR, 6.9). Confocal laser scanning microscopy demonstrated colocalization between fluorescent Eu-P947 and MMPs in atherosclerotic plaques. Eu-P947 was particularly present in the fibrous cap region of plaques.

Conclusion.—P947 improved MR imaging for atherosclerosis through MMP-specific targeting. The results were validated and provide support for further assessment of P947 as a potential tool for the identification of unstable atherosclerosis.

▶ Despite great advances in cardiovascular medicine, atherosclerosis remains the foremost cause of death in developed industrialized societies and has become the leading cause of death globally. We do not currently possess clinical tools capable of accurately evaluating atherosclerosis directly or at early stages of disease when interventions may have the greatest effect on altering disease progression. The rapidly evolving field of molecular imaging will ultimately allow for accurate assessment of the molecular, cellular, and compositional components of atherosclerotic lesions. This information will be clinically vital to identify vulnerable patients and facilitate personalized medical management. Molecular magnetic resonance (MR) imaging is especially appealing as a noninvasive modality to evaluate atherosclerosis because it lacks ionizing radiation and possesses the superb spatial resolution necessary to image small-scale pathologic findings such as atherosclerotic lesions. Matrix metalloproteinases (MMPs) are implicated in the progression, destabilization, and rupture of atherosclerotic lesions. Abnormal expression and regulation of MMPs in atherosclerotic plaques may lead to destabilization, resulting in plaque rupture with subsequent thrombosis, occlusion, and detrimental clinical consequences, such as myocardial infarction or stroke. MMP overexpression has been demonstrated in the vulnerable shoulder regions of human atheroma. Macrophage-derived foam cells that are characteristic of unstable plaques have been shown to be major sources of MMPs in human and experimental atherosclerotic lesions. Therefore, MMP expression not only is a part of the pathogenesis of atherosclerosis, but it may be predictive of plaque instability. Such a potentially pivotal role in the pathogenesis of atherosclerosis suggests that MMPs are a rewarding target for molecular imaging.

The authors investigated the molecular MR imaging contrast agent P947, which specifically targets MMPs. For the purpose of demonstrating the specificity of P947 for MMPs in vivo, 2 other contrast agents were used as controls: P1135 (Guerbet), which is an MR imaging analog of P947 for which the peptide sequence has been scrambled, leading to a loss of affinity for MMPs; and P1191 (Guerbet), which is a europium (Eu) derivative of P947 (Eu-P947) with fluorescence emission properties. The goals of their investigation were as follows: (a) to study and validate the in vivo capability of P947 (targeted to MMPs) to assess atherosclerosis in apolipoprotein E (ApoE) knockout (ApoE deficient) mice by using MR imaging; (b) to evaluate the specificity of P947 for MMPs by comparing its in vivo MR imaging effectiveness to that of P1135 (untargeted, scrambled form of P947 lacking specific binding to MMPs); (c) to demonstrate the expression and examine the distribution of MMPs in the atherosclerotic vessels of the mice we used for the in vivo MR imaging experiments; and (d) to investigate the in vivo distribution of fluorescent Eu-P947 (also known as P1191) in atherosclerosis and to study the in vivo molecular localization of Eu-P947 in relation to specific MMP fluorescent immunostaining in atherosclerotic lesions (examination of possible colocalization between fluorescent Eu-P947 and specific MMPs).

In vivo MR imaging was performed with a 9.4-T, 89-mm bore MR imaging system operating at a proton frequency of 400 MHz. After preinjection baseline MR imaging, ApoE-deficient atherosclerotic mice were injected with either MMP-targeted P947 ($n = 16$) or the untargeted scrambled control P1135 ($n = 5$) at 100 μmol of gadolinium per kilogram. For the control group, ApoE-deficient mice ($n = 6$) were injected (at 100 μmol of gadolinium per kilogram) with the standard nonspecific gadolinium-based MR contrast agent gadoterate meglumine. In addition, as another control group, WT mice were injected with equivalent doses of P947 ($n = 4$). MR imaging was performed at 1, 2, 3, and 22 hours after injection. Slices were precisely anatomically matched to the sections obtained on the preinjection baseline image.

After the 22-hour postinjection image, the animals were sacrificed. The aortas were carefully isolated by using microscopic dissection and were fixed for standard pathologic analysis. The histologic slides from the atherosclerotic aortas were then matched to the MR image sections for comparison and correlation. To quantitatively analyze the MR imaging results, signal intensity(SI) measurements were obtained by using regions of interest on the aortic wall, with 4 points in 4 Cartesian quadrants of the aorta on each section.

Before injection of the contrast agent, MR imaging was performed in all mice at baseline. Before injection of the contrast agent, MR imaging of ApoE-deficient mice showed a heterogeneous distribution of aortic wall thickening. Aortic wall thickening correlates strongly with the presence of atherosclerotic lesions. As anticipated, baseline MR imaging of WT mice showed no regions of aortic wall thickening. After injection of P947, designed to specifically target MMPs, the authors observed very strong and significant heterogeneous enhancement of the aortic wall in ApoE-deficient mice (Fig 1a in the original article). The change in the MR image SI of the aorta was measured by using the NER and percentage NER. In particular, by using ApoE-deficient mice, at 1hour after injection, P947 provided an average aortic MR image signal enhancement of 127% normalized enhancement ratio (NER) when standardized to normal adjacent muscle. In comparison, the percentage NER was 33% (mean NER, 1.33 ± 0.12) when we used P1135 and 4.6% (mean NER, 1.05 ± 0.06) with gadoterate meglumine, both also standardized to muscle.

As demonstrated in Figs 1 and 2 in the original article, these results translate into a three-fold increase in the MR image SI in the atherosclerotic aortic wall when P947 was used as compared with P1135, a permuted derivative of P947 with the amino acid sequence scrambled such that specific binding to MMPs is lost, supporting an MMP-specific image enhancement mechanism. In addition, the SI increase with P947 was 15-fold higher when compared with the standard nonspecific gadolinium-based MR imaging contrast agent gadoterate meglumine ($P < .001$) (Fig 2 in the original article). In nonatherosclerotic WT mice, at 1 hour after injection, P947 provided minimal change of the average aortic MR imaging SI that was homogeneous in nature and was not significant from baseline (Fig 2 in the original article). At 2 and 3 hours after injection, no significant changes were seen ($P > .05$) of the aortic MR image SI values in WT mice (Fig 2 in the original article). After the MR imaging experiments were completed, the mice were sacrificed for various types of pathologic analyses of the atherosclerotic aortas (all animals). Pathologic analysis of aortic tissue sections was performed to validate expression of MMPs in atherosclerotic plaques in ApoE-deficient mice used for the MR imaging experiments in this study.

A major result of this investigation was that MMP-targeted P947 facilitated significant MR image enhancement (increase of 93%) of the atherosclerotic vessel wall of ApoE-deficient mice in vivo. These results suggest that the bulk of atherosclerosis enhancement seen with P947 is caused by specific binding, accumulation, or retention of P947, making P947 substantially better than any of the MR imaging contrast agents that were tested in this study.

Additionally, the MR imaging results of this study correlated well with the standard pathologic findings in the regions of vessels that were imaged.

MR imaging offers superb anatomic resolution that is orders of magnitude higher than that of scintigraphy, including PET or SPECT. Molecular MR imaging, such as P947-enhanced MR imaging, will offer molecular information combined with superior anatomic resolution, which is vital for imaging small-scale pathologic findings such as atherosclerosis. This study showed that P947 targeted to MMPs facilitated specific in vivo assessment of atherosclerosis at a molecular pathologic level. Knowledge about MMP content and activity may aid in the prediction of the vulnerability of plaque to rupture and thrombosis. Therefore, with proper development and investigation, imaging with P947 (or similar MMP imaging approaches) may contribute substantially to preclinical and clinical evaluation of atherosclerosis.

T. G. Walker, MD

Tumor Ablation

Small Hepatocellular Carcinoma: Is Radiofrequency Ablation Combined with Transcatheter Arterial Chemoembolization More Effective than Radiofrequency Ablation Alone for Treatment?

Shibata T, Isoda H, Hirokawa Y, et al (Kyoto Univ Graduate School of Medicine, Shogoin, Sakyoku, Japan)

Radiology 252:905-913, 2009

Purpose.—To prospectively compare combined radiofrequency ablation and transcatheter arterial chemoembolization (TACE) with radiofrequency ablation alone for the treatment of small (≤3 cm) hepatocellular carcinomas (HCCs).

Materials and Methods.—The ethics committee of the study institution approved the study protocol. Written informed consent was obtained from all patients at enrollment. Eighty-nine patients with 93 HCC nodules 0.8–3.0 cm (mean ± standard deviation, 1.7 cm ± 0.5) in diameter were randomly treated with combined radiofrequency ablation and TACE (46 patients with 49 nodules) or with radiofrequency ablation alone (43 patients with 44 nodules). The patients were followed up for 9–68 months (mean ± standard deviation, 30.4 months ± 14.0). Rates of local tumor progression, overall survival, local progression-free survival, and event-free survival were evaluated by using the Kaplan-Meier method.

Results.—The 1-, 2-, 3-, and 4-year rates of local tumor progression, overall survival, local progression-free survival, and recurrence-free survival were as follows: Local tumor progression rates were 14.4%, 17.6%, 17.6%, and 17.6%, respectively, in the combined treatment group and 11.4%, 14.4%, 14.4%, and 14.4%, respectively, in the radiofrequency ablation group ($P = .797$). Overall survival rates were 100%, 100%, 84.8%, and 72.7%, respectively, in the combined treatment group and 100%, 88.8%, 84.5%, and 74.0%, respectively, in the radiofrequency ablation group ($P = .515$). Local progression-free survival rates were 84.6%, 81.1%, 69.7%, and 55.8%, respectively, in the combined treatment group and 88.4%, 74.1%, 74.1%, and 61.7%, respectively, in the radiofrequency ablation group (P=.934). Event-free survival rates were 71.3%, 59.9%, 48.8%, and 36.6%, respectively, in the combined treatment group and 74.3%, 52.4%, 29.7%, and 29.7%, respectively, in the radiofrequency ablation group (P=.365).

Conclusion.—Combined radiofrequency ablation plus TACE and radiofrequency ablation alone have equivalent effectiveness for the treatment of small (≤3 cm) HCCs, so the combination treatment may not be necessary.

▶ Current options for the curative treatment of hepatocellular carcinoma (HCC) consist of surgical resection, liver transplantation, and image-guided ablation therapy. Surgical resection is the first-line treatment for small HCCs, but a small proportion of patients with HCC are not candidates for surgical resection because of poor hepatic reserve or multifocal distribution of tumor nodules. Liver transplantation is another curative treatment for end-stage liver disease, but it is limited because of the donor shortage. Recently, radiofrequency ablation (RFA) has been widely performed for the treatment of small HCCs, because it is less invasive than other therapies and is reported to be effective.

Local tumor control is an important factor in achieving good survival rates among patients treated with RFA, but local tumor progression is not infrequent. RFA combined with transcatheter arterial chemoembolization (TACE) has been reported to be effective for local control of medium-sized HCCs and small HCCs. However, there have not been any investigations to prospectively assess whether RFA combined with TACE is more effective than RFA alone for the treatment of small (≤3 cm) HCCs. This study was designed to prospectively compare the effectiveness of combined RFA and TACE with that of RFA alone in the treatment of these small HCCs.

Eighty-nine consecutive patients with HCC participated in this study. Inclusion criteria were as follows: (1) 3 or fewer HCC nodules 3 cm in diameter or smaller, (2) Child-Pugh class A or B cirrhosis, (3) no imaging evidence of tumor invasion into the major portal or hepatic vein branches, (4) no extrahepatic metastases, and (5) a platelet count of more than 40 000 cells/mm^3. The patients were randomly assigned to the combined treatment group (46 patients with 49 HCC nodules) or the RF group (43 patients with 44 HCC nodules). In all patients, ultrasonography (US) and dynamic computed tomography (CT) were performed as pretreatment imaging studies.

The US diagnosis of HCC was based on the presence of lesions with different echogenicity—hypoechoic, hyperechoic, isoechoic, or a mixed pattern—compared with that of the surrounding liver parenchyma.

The CT diagnosis of HCC was based on the presence of an enhancing lesion on early-phase CT scans and hypoattenuation on late-phase CT scans. The European Association for the Study of the Liver has demonstrated that HCC can be diagnosed on the basis of coincident findings of at least 2 modalities—among US, CT, and magnetic resonance imaging—that show characteristic features in a focal lesion. Of 93 nodules, 83 (89%) were diagnosed as

HCC on the basis of characteristic findings at both US and dynamic CT. In 10 (11%) of 93 nodules, which were low attenuating on early-phase CT scans, the histologic diagnosis was obtained with US-guided needle biopsy.

One author with 22 years of experience in interventional radiology performed RFA and another author with 15 years of experience in interventional radiology performed TACE. The treatment course and follow-up are shown in Fig 1 in the original article.

In the 46 patients in the combined treatment group, 46 sessions of combined RFA and TACE were performed. In the 43 patients in the RF group, 46 sessions of RFA were performed. A single RF session was performed in 40 patients, and 2 RF sessions were performed in 3 patients because the residual enhanced lesion was noted at dynamic CT performed 1 to 2 days after the first session. In all 93 nodules in both groups, a nonenhancing area with a diameter greater than that of the treated nodule was noted at dynamic CT performed a month after the treatment session. Effectiveness of the primary technique was considered 100% in both groups. The major complication rate of 2% was equal in both groups.

During follow-up, local tumor progression occurred in 8 nodules (8 patients) in the combined treatment group and in 6 nodules (6 patients) in the RFA group. The rates of local tumor progression at 1, 2, 3, and 4 years were 14.4%, 17.6%, 17.6%, and 17.6%, respectively in the combined treatment group and 11.4%, 14.4%, 14.4%, and 14.4%, respectively in the RFA group. No significant difference was observed between the 2 groups. During follow-up, 5 patients in the combined treatment group and 6 patients in the RFA group died. The overall survival rates at 1, 2, 3, and 4 years were 100%, 100%, 84.8%, and 72.7%, respectively, in the combined treatment group and 100%, 88.8%, 84.5%, and 74.0%, respectively in the RFA group. No significant difference was observed between the 2 groups.

The rates of local progression-free survival at 1, 2, 3, and 4 years were 84.6%, 81.1%, 69.7%, and 55.8%, respectively in the combined treatment group and 88.4%, 74.1%, 74.1%, and 61.7%, respectively in the RFA group. No significant difference was observed between the 2 groups.

The rates of event-free survival at 1, 2, 3, and 4 years were 71.3%, 59.9%, 48.8%, and 36.6%, respectively in the combined treatment group and 74.3%, 52.4%, 29.7%, and 29.7%, respectively in the RFA group. No significant difference was observed between the 2 groups. No significant prognostic factors, including treatment allocation, were identified by using Cox proportional hazards regression models.

This study had several limitations. First, because needle biopsy was performed in only 10 (11%) of 93 nodules, neither histologic staging of the HCCs nor tumor differentiation could be performed. Poorly differentiated HCCs might be resistant to locoregional treatments, such as RFA, TACE, and combined RFA and TACE, and the rates of poorly differentiated HCCs in both groups were unknown in this study. In addition, the duration of follow-up may have been too short to allow for evaluating long-term prognoses.

However, the authors conclude that patients with HCCs 3 cm or smaller who are treated with combined RFA and TACE have local tumor progression rates, overall survival rates, and local progression-free survival rates that are equivalent to those of patients who undergo RFA alone. Hence, the addition of TACE to RFA for HCCs 3 cm or smaller does not appear to be warranted.

T. G. Walker, MD

Vascular Imaging

Detection of common carotid artery stenosis using duplex ultrasonography: A validation study with computed tomographic angiography

Slovut DP, Romero JM, Hannon KM, et al (North Shore Med Ctr, Salem; Massachusetts General Hosp, Boston)

J Vasc Surg 51:65-70, 2010

Background.—Severe stenosis of the common carotid artery (CCA), while uncommon, is associated with increased risk of transient ischemic attack and stroke. To date, no validated duplex ultrasound criteria have been established for grading the severity of CCA stenosis. The goal of this study was to use receiver-operating curve (ROC) analysis with computed tomographic angiography as the reference standard to establish duplex ultrasound criteria for diagnosing ≥50% CCA stenosis.

Methods.—The study cohort included 64 patients (42 men, 22 women) with a mean age of 65 ± 12 years (range, 16-89 years) who had CCA peak systolic velocity (PSV) ≥150 cm/sec and underwent computed tomographic angiography (CTA) of the cervical and intracerebral vessels within 1 month of the duplex examination. One study was excluded because the CTA was technically inadequate, whereas another was excluded because the patient underwent bilateral CCA stenting. The CCA ipsilateral to any of the following was excluded from the analysis: innominate artery occlusion (n = 1), previous stenting of the ICA or CCA (n = 7), carotid endarterectomy (n = 1), or carotid-to-carotid bypass (n = 1). Thus, the data set included 62 patients and 115 vessels. Bland-Altman analysis was used to examine the agreement between two measures of luminal reduction measured by CTA: percent diameter stenosis and percent area stenosis. Receiver operating characteristic (ROC) analysis was used to determine optimal PSV and EDV thresholds for diagnosing ≥50% CCA stenosis.

Results.—Severity of CCA stenosis was <50% in 76 vessels, 50%-59% in eight, 60%-69% in eight, 70%-79% in nine, 80%-89% in three, 90%-99% in five, and occluded in six. Duplex ultrasonography identified six of six (100%) patients with 100% CCA occlusion by CTA. Bland-Altman analysis showed poor agreement between percent stenosis determined by vessel diameter compared with percent stenosis determined by reduction in lumen area. Therefore, subsequent analysis was performed using percent stenosis by area. ROC analysis of different PSV thresholds for detecting stenosis ≥50% showed that >182 cm/sec was the most accurate with a sensitivity of 64% and specificity of 88% ($P < .0001$). Sensitivity, specificity, and accuracy of carotid duplex were higher when the stenosis was located in the mid or distal aspects of the CCA (sensitivity 76%, specificity 89%, area under curve 0.84, $P < .001$) than in the intrathoracic and proximal segment of the artery ($P = \text{NS}$). ROC analysis of different EDV thresholds for detecting CCA stenosis ≥50% showed that >30 cm/sec was the most accurate with a sensitivity of 54% and a specificity of 74% ($P < .0239$).

Conclusions.—Duplex ultrasonography is highly sensitive, specific, and accurate for detecting CCA lesions in the mid and distal CCA. Use of peak systolic velocity may lead to improved detection of CCA disease and initiation of appropriate therapy to reduce the risk of stroke.

▶ Duplex ultrasonography (DUS) is the primary initial modality for detecting internal carotid artery (ICA) stenosis. Using pulsed-wave Doppler velocity measurements and correlating these to invasive contrast arteriography, carotid DUS has excellent sensitivity and specificity for grading the severity of ICA stenosis. However, a major shortcoming of DUS is an absence of validated Doppler criteria for grading the severity of common carotid artery (CCA) stenosis. Although some vascular laboratories attempt to diagnose CCA stenosis based on alterations in carotid waveforms, this analysis may be unreliable as it may be based on differences in the gain applied and is not strictly quantifiable. Computed tomographic angiography (CTA) has become a widely accepted method for evaluating carotid artery disease. A meta-analysis of the diagnostic accuracy of CTA for the assessment of carotid stenosis reviewed 28 studies that compared CTA with digital subtraction arteriography (DSA). CTA was found to be highly accurate for diagnosing the degree of carotid stenosis with an overall sensitivity of 97% and specificity of 99%. The goal of this study was to use receiver operating characteristic (ROC) analysis with CTA as the reference standard to establish duplex ultrasound criteria for diagnosing CCA stenosis of ≥ 50% severity. This study was a retrospective one that included 64 patients (42 men, 22 women) with a mean age of 65 ± 12 years who had CCA peak systolic velocity (PSV) > 150 cm/sec and underwent CTA of the cervical and intracerebral vessels within 1 month of the duplex examination. After various exclusions, the final dataset consisted of 62 patients and 115 vessels. Two examiners who were blinded to the duplex velocity parameters reviewed each CTA simultaneously. Percent stenosis was determined by reduction in luminal diameter as well as reduction in the luminal area. Quantitative analysis of CTA images revealed that percent stenosis determined by reduction in luminal area was consistently greater than percent stenosis determined by decrease in vessel diameter: only 14 vessels had stenosis ≥50% by CTA diameter versus 40 as determined by CTA area. Bland-Altman analysis demonstrated poor agreement between percent stenosis determined by vessel diameter compared with percent stenosis determined by reduction in lumen area. The limits of agreement were −13.9% and 41.6%. Thus, percent stenosis by diameter may be nearly 15% below or over 40% above stenosis determined by reduction in luminal area. Therefore, subsequent analysis was performed using percent stenosis by area. Severity of CCA stenosis based on reduction in lumen area was < 50% in 76 vessels: 50% to 59% in 8, 60% to

69% in 8, 70% to 79% in 9, 80% to 89% in 3, 90% to 99% in 5, and occluded in 6 arteries.

T. G. Walker, MD

Duplex ultrasound as the sole long-term surveillance method post-endovascular aneurysm repair: A safe alternative for stable aneurysms

Chaer RA, Gushchin A, Rhee R, et al (Univ of Pittsburgh Med Ctr, PA)

J Vasc Surg 49:845-850, 2009

Objective.—Long-term surveillance with computed tomography (CT) after endovascular aneurysm repair (EVAR) increases both cost and risk. The purpose of this study was to evaluate the safety of an alternative follow-up modality with color flow duplex ultrasound scanning (CDU) as the sole method of imaging.

Methods.—In 2003, we initiated a new follow-up (FU) schedule with yearly CDU as the sole imaging method for selected patients. Indications included a residual sac of less than 4 cm, expanded later to stable sac size for more than 2 years. A stable type II endoleak was not a contraindication. CT scans were obtained selectively-based on suspicious findings of a new endoleak or enlarging sac on CDU. The records of all patients with at least 1 year FU under this schedule were reviewed.

Results.—One hundred eighty-four patients were followed with CDU only for 1 to 4 years for a mean of 24 ± 13 months. The new schedule was initiated at a mean of 34 ± 24 months after EVAR (range 1-112 months). Twenty-three patients had previous endoleaks that had resolved spontaneously or had been treated. During CDU FU, three new endoleaks were detected, one with sac enlargement. All prompted CT evaluation: one type II endoleak with stable sac size could not be identified on CT 3 months later, and two distal type I endoleaks that required limb extension. All three had a prior Ancure endograft. No ruptures or graft occlusions were noted. One abdominal aortic aneurysm (AAA) related death followed graft explantation for infection. There were two additional deaths from malignancy and two from cardiac causes. After the FU switch, freedom from endoleaks was 96%, and from secondary interventions 95% at 48 months by life table method. Mean AAA diameter at baseline was 54 ± 8 mm and decreased to 40 ± 11 mm before the switch to CDU only FU. At last FU mean aneurysm diameter was 39 ± 11 mm. When the current switch criteria were applied to a consecutive series of 200 EVAR patients, 97% would have been eligible for CDU only surveillance by 3 years postoperatively.

Conclusions.—CDU only surveillance post-EVAR is safe and can be initiated early after treatment in patients with shrinking or stable

aneurysms. This policy should result in cost savings advantage and avoid the complications associated with CT.

▶ Endovascular aneurysm repair (EVAR) is currently widely accepted for the treatment of abdominal aortic aneurysms (AAAs). Although this therapy offers immediate advantages over open aneurysm repair with lower perioperative mortality and morbidity, it carries the need for lifelong surveillance for potential complications, including endoleak, change in aneurysm size, graft migration, structural graft failure, and limb outflow impairment caused by limb stenosis or occlusion. The ideal surveillance modality should be noninvasive, cheap, and reproducible, with high sensitivity and specificity for the detection of these various potential adverse events. Computed tomography with intravenous contrast injection (CT) is currently the standard for long-term EVAR surveillance, but is associated with increased cost, radiation exposure, and potential contrast-induced nephropathy with resultant decline in renal function. Color-flow duplex ultrasound scanning (CDU) can also detect endoleaks as well as size changes over time, but is more operator dependent. However CDU has the advantage of being noninvasive, safer, and cheaper than CT scans. Although several studies have evaluated the ability of CDU to detect endoleaks and have established good correlation with CT for the measurement of AAA sac diameter, there currently is no published series of patients followed with CDU only post-EVAR to document the safety of this follow-up protocol. This study was designed to evaluate the safety of a selective policy of EVAR follow-up with CDU as the sole means of imaging surveillance after initial imaging with post-EVAR CT. One hundred eighty-four patients (159 males) were switched to CDU surveillance between 2003 and 2006 and were reviewed. All patients underwent initial post-EVAR surveillance CT scans at 1 and 6 months, with some patients undergoing additional CT imaging. The CDU follow-up schedule was initiated 34 ± 24 months after EVAR (range 1-112 months). The mean AAA diameter at baseline was 54 ± 8 mm and had decreased to 40 ± 11 mm before the decision to implement CDU only surveillance. Forty nine percent of patients had a collapsed aneurysm ($<$ 4 cm) at the time of initiation of the new surveillance protocol. Among the patients on CDU only surveillance, a history of endoleak was present in 36 (19.5%). At the time of the switch, 23 had spontaneously resolved or were treated, and 13 were active persistent type II endoleaks with a stable or shrinking AAA sac. Three new endoleaks were diagnosed during the CDU only surveillance, only 1 presenting with sac enlargement. All prompted CT evaluation: 1 type II endoleak with stable sac size could not be identified on the CT obtained 3 months later, and 2 distal type I endoleaks that required limb extension. No patient had any clinical adverse event during the period of observation No ruptures or graft occlusions were noted. After the switch to CDU surveillance, the freedom from clinically significant endoleaks was 96%, and from secondary interventions 95% at 48 months by life table method. Although CDU has been shown to have an excellent sensitivity and negative predictive value compared with CT for the diagnosis of endoleak, these results have not been uniformly reproduced. Much of this variability may be attributable to differences in proper imaging equipment and experienced technologists, quality control, as well as the implementation of a standard imaging protocol for EVAR. This study was intentionally applied to a selected subset of patients in whom CDU surveillance was felt to be appropriate. The authors caution against indiscriminate CDU surveillance only in all post-EVAR patients. With longer follow-up and accrual of more experience with this regimen, earlier switch to CDU surveillance may be reasonable in certain patients and may have a more significant socioeconomic impact.

T. G. Walker, MD

Venous Interventions

Changes in Portal Systemic Pressure Gradient After Balloon-Occluded Retrograde Transvenous Obliteration of Gastric Varices and Aggravation of Esophageal Varices

Tanihata H, Minamiguchi H, Sato M, et al (Wakayama Med Univ, Wakayamashi, Japan)

Cardiovasc Intervent Radiol 32:1209-1216, 2009

The purpose of this study was to evaluate change in the portal systemic pressure gradient (PSPG) following balloon-occluded retrograde transvenous obliteration (BRTO) and the aggravation of esophageal varices. The PSPG was monitored before and after BRTO in 19 patients. PSPG changes were obtained by subtracting the PSPG before BRTO from that after BRTO. The development of outflow vessels (e.g., left inferior phrenic vein) was classified into two grades: Grade 1, BRTO alone; and Grade 2, coil embolization plus BRTO. After confirming demonstration of the whole gastric varices on angiography and computed tomography, BRTO was conducted using a 5% ethanolamine-iopamidol mixture. Endoscopy was performed to evaluate gastric and esophageal varices before, within 1 month, and 3–6 months after BRTO. Eradication of gastric varices was obtained in all patients and aggravation of esophageal varices was seen in 11 patients. The PSPG was significantly elevated by BRTO ($p = 0.0362$). The PSPG was significantly elevated in patients with Grade 2 compared with those with Grade 1 (7.7 ± 3.7 vs. 3.3 ± 4.3 mmHg, respectively; $p = 0.0314$) and in those with esophageal varices before treatment compared with those without (7.4 ± 4.0 vs. 3.2 ± 3.9 mmHg, respectively; $p = 0.0482$). The cumulative aggravation rate of esophageal varices was significantly higher in 11 patients with a PSPG elevation >5 mmHg than in 8 patients with one of ≤5 mmHg ($p = 0.0105$). In conclusion, BRTO induced a significant elevation in PSPG, with the degree of elevation influencing the aggravation of esophageal varices following BRTO.

▶ Balloon-occluded retrograde transvenous obliteration (BRTO), first reported by Kanazawa et al, is a well-established treatment for gastric varices, in which a sclerosing agent is infused to gastric varices under occlusion of the draining vessel by balloon inflation. Gastric varices most commonly have a spontaneous gastrorenal shunt, with the left adrenal vein targeted as the main draining vessel by balloon inflation. Follow-up examination using gastrointestinal endoscopy after BRTO is necessary, because the incidence of aggravated esophageal varices increases with time. The presence of esophageal varices before BRTO is reported to be a statistically significant prognostic factor for aggravation. The change in hemodynamics caused by BRTO appears to lead to the aggravation of esophageal varices. However, there exist cases in which no aggravation occurs despite the presence of esophageal varices before BRTO and cases in which new esophageal varices appear despite the absence of esophageal varices before BRTO. The mechanism of the aggravation of esophageal varices after BRTO remains unclear. This clinical investigation was designed to evaluate both the development of a portal systemic pressure gradient (PSPG) before and after BRTO and any possible correlation with aggravation of esophageal varices.

This retrospective study evaluated 19 cirrhotic patients with gastric varices, who consecutively underwent prophylactic BRTO and monitoring of hepatic venous wedge pressure and central venous pressure. In each patient, gastric varices were confirmed by endoscopy, and a gastrorenal shunt was demonstrated by contrast-enhanced computed tomography (CT). Of the 19 patients, the technical success rate was 95%, and the median follow-up period was 18 months. Contrast-enhanced CT within 2 days after BRTO demonstrated thrombosis of the whole gastric varices in all patients.

Intraprocedurally, hepatic vein wedge pressure and central venous pressure were measured for approximately 30 seconds during stabilization of the mean pressure value before BRTO and remeasured after demonstration on CT of accumulation of the infused sclerosing agent in the whole gastric varices. The PSPG was obtained by subtracting the central venous pressure from the hepatic vein wedge pressure: both before and after BRTO. The difference was obtained by subtracting the PSPG before BRTO from that after BRTO. At the time of the BRTO procedure, an occluding balloon was inflated in the adrenal vein for blockade of the gastrorenal shunt flow. After confirmation of blockade, balloon-occluded adrenal venography (BOAC) using contrast medium was performed. BOAC usually delineates the left adrenal vein and can delineate other outflow vessels, gastric varices, and inflow vessels. As the criteria of difficulty of BRTO, patients with gastric varices were classified into 2 grades. Grade 1 was considered BRTO alone and grade 2, coil embolization plus BRTO. In grade 2, BOAC did not delineate the whole gastric varices because of the development of outflow vessels, and BRTO was not conducted until the outflow vessels were occluded with microcoils BRTO was performed using a 5% ethanolamine-iopamidol mixture.

The PSPG before BRTO was significantly higher in patients with Child-Pugh B than in those with Child-Pugh A, but no difference existed between patients with grade 1 and those with 2 or between those with and those without esophageal varices before treatment. Enhanced CT was obtained within 2 days following BRTO and demonstrated thrombosis of the gastric varices in all patients. Follow-up endoscopy was performed to evaluate gastric and esophageal varices before, within 1 month, and 3 to 6 months after BRTO. Eradication of gastric varices was obtained in all patients and aggravation of esophageal varices was seen in 11 patients, 8 of whom had esophageal varices before BRTO and 3 who did not. Univariate and multivariate analyses of data before BRTO revealed an elevation of PSPG >5 mm Hg to be a significant risk factor for aggravation of esophageal varices after BRTO, showing a stronger relationship with aggravation than the presence or absence of esophageal varices and grade classification.

A potential disadvantage of BRTO is an increase in portal venous pressure. In this study, the PSPG was elevated in 16 of the 19 cases after treatment, and the elevation in PSPG after BRTO was significantly greater in patients with grade 2 compared with those with grade 1. Namely, the combination of BRTO of the left adrenal vein and coil embolization of the other outflows results in a significant elevation in PSPG, in comparison with BRTO alone. Aggravation of esophageal varices is another disadvantage of BRTO. The aggravation rate has previously been reported to be 16.6% to 66%. The cumulative aggravation rate of esophageal varices in this study was significantly higher in patients with a PSPG elevation >5 mmHg compared to those with an elevation of B5 mmHg. Elevation of PSPG after BRTO is considered to lead to aggravation of esophageal varices.

In conclusion, BRTO induces a significant elevation in PSPG. A significant elevation was observed in patients with development of outflow vessels or the presence of esophageal varices. The degree of elevation was related to aggravation of esophageal varices following BRTO.

T. G. Walker, MD

Endovascular Treatment Options in the Management of Lower Limb Deep Venous Thrombosis

Nazir SA, Ganeshan A, Nazir S, et al (John Radcliffe Hosp, Oxford, UK)
Cardiovasc Intervent Radiol 32:861-876, 2009

Lower limb deep vein thrombosis (DVT) is a common cause of significant morbidity and mortality. Systemic anticoagulation therapy is the mainstay of conventional treatment instituted by most physicians for the management of DVT. This has proven efficacy in the prevention of thrombus extension and reduction in the incidence of pulmonary embolism and rethrombosis. Unfortunately, especially in patients with severe and extensive iliofemoral DVT, standard treatment may not be entirely adequate. This is because a considerable proportion of these patients eventually develops postthrombotic syndrome. This is characterized by chronic extremity pain and trophic skin changes, edema, ulceration, and venous claudication. Recent interest in endovascular technologies has led to the development of an assortment of minimally invasive, catheter-based strategies to deal with venous thrombus. These comprise catheter-directed thrombolysis, percutaneous mechanical thrombectomy devices, adjuvant venous angioplasty and stenting, and inferior vena cava filters. This article

reviews these technologies and discusses their current role as percutaneous treatment strategies for venous thrombotic conditions.

▶ The current standard of care for patients with acutely diagnosed deep venous thrombosis (DVT) is anticoagulation with low molecular weight heparin (LMWH) followed by a 6-month course of oral warfarin. The purpose of anticoagulation is to diminish the severity and duration of symptoms by preventing thrombus propagation and extension, minimize the incidence of recurrent thrombosis, and decrease the risk of pulmonary embolic (PE) complications. Standard medical therapy with anticoagulation is mostly effective at achieving these aims. However, it does not directly promote thrombus dissolution and in patients with severe and extensive DVT, particularly acute iliofemoral disease, standard treatment alone may not be sufficient to resolve the acute emergency. A particularly severe potential sequelae of massive DVT is the ischemic condition known as phlegmasia cerulea dolens, which can be fatal. It is now well documented that conventional systemic thrombolytic therapy to treat phlegmasia is disappointing. In the acute setting, just as worrisome is the propensity toward PE: a potentially catastrophic event that requires prompt and aggressive management. Despite early diagnosis and timely commencement of anticoagulation, PE can occur in as many as 21% of patients with proximal DVT.

Regarding long-term outcomes, conventional solutions appear inadequate for restoration and preservation of venous function in patients with proximal acute DVT. Approximately 95% of patients with iliofemoral DVT treated solely with anticoagulants have valve incompetence after 5 years, and such treatment fails to remove thrombus completely in well over 50% of cases. This stems from the fact that anticoagulation per se does not remove thrombus; this task falls on the endogenous fibrinolytic system to dissolve thrombus at an early stage to preserve valvular function and restore normal unimpeded venous vascular flow. This inability to clear thrombus with standard anticoagulation regimens results in the development of postthrombotic syndrome (PTS), a term used for the constellation of findings such as chronic limb edema, claudication and pain, skin hyperpigmentation, and development of venous stasis. PTS results from chronically elevated venous pressures in the lower limb as a consequence of DVT-related venous outflow obstruction and is exacerbated by valvular incompetence. Despite advances in endovascular techniques, most physicians continue to manage their patients with conventional anticoagulation, irrespective of the site and extent of thrombosis. There is mounting evidence to demonstrate that patients with iliofemoral venous thrombosis suffer more severe postthrombotic morbidity than patients with infrainguinal DVT, but both subsets of patients are almost exclusively managed with anticoagulation alone. The aim of any intervention should be directed at immediate and successful elimination of thrombus. As well as reducing the risk of fatal complications such as PE, this will also increase the likelihood of preservation of valve function. This review highlights the endovascular options in patients with extensive DVT and discusses various strategies for dealing with thrombus. The review is aimed at addressing the distinct lack of awareness of these methods that remain within physician and patient populations. It provides a very extensive examination of the following topics:

(A) Catheter-directed thrombolysis therapy,
(B) Percutaneous mechanical thrombectomy (PMT),
(C) Adjuvant venous angioplasty and stenting, and
(D) Inferior vena cava filters.

For each of these topics, the indications, contraindications, technique, available devices, clinical experience/evidence for use, and complications are summarized. An algorithm for the treatment of DVT is presented in Fig 1 in the original article.

In summary this review highlights the main interventional procedures that can be used in the treatment of DVT. While procedural details need to be customized to specific situations, the article presents a possible technical scheme for the management of those patients most likely to derive benefit from their implementation. The authors maintain that although endovascular strategies are increasingly being used in the clinical setting, they are still underutilized and feel that this will likely continue to be the case until there is compelling evidence from well-designed prospective randomized clinical trials validating their safety and efficacy.

T. G. Walker, MD

Institutional protocol improves retrievable inferior vena cava filter recovery rate

Ko SH, Reynolds BR, Nicholas DH, et al (Univ of Pittsburgh School of Medicine, PA)
Surgery 146:809-816, 2009

Background.—In the trauma population, the use of retrievable inferior vena cava filters (RIVCF) is rapidly gaining acceptance in patients at high risk for venous thromboembolism. This study reports the impact of an institutional protocol on retrieval rates of RIVCF at a level I trauma center.

Methods.—A review of an institutional Trauma Registry identified 94 consecutive patients who received RIVCF between January 2004 and February 2007 (group I) before the protocol was instituted. Under the protocol, 61 consecutive trauma patients received RIVCF between August 2007 and July 2008 (group II) and were prospectively followed.

Results.—Filter retrieval eligibility criteria were met in 81% (76/94) of patients in group I and in 61% (37/61) of patients in group II. Of those eligible, retrieval-attempt rates were 42% (32/76) in group I versus 95% (35/37) in group II ($P < .001$). Clinician oversight of the filter accounted for 89% (39/44) of failure of retrieval attempts; patient noncompliance accounted for the rest in group I. In group II, the latter accounted for all such failures. Retrieval was successful in 37% (28/76) and in 84% (31/37)

TABLE 4.—Causes of Filter Nonretrieval

n	Group I 66	Group II 30	*P* value*
Not eligible for filter removal			
In-hospital death	4 (6.1)	5 (16.7)	.13
Physician preference	0	11 (36.7)	—
Medical reasons	10 (15.2)	4 (13.3)	.99
Patient refusal of retrieval	4 (6.1)	4 (13.3)	.25
Patient noncompliance	5 (7.6)	2 (6.7)	.99
Clinician oversight	39 (59.1)	0	—
Intra-operative variables			
Technical failure	2 (3.0)	2 (6.7)	.37
IVC thrombosis	1 (1.5)	2 (6.7)	.23
Thrombosis on filter	1 (1.5)	0	—

IVC, Inferior vena cava.
*Two-sided Fisher exact test.

of the eligible patients in groups I and II, respectively ($P < .001$). No retrieval procedure-related complications occurred.

Conclusion.—An institutional protocol for prospective monitoring of RIVCF significantly increases filter retrieval rate.

▶ Placement of inferior vena cava (IVC) filters has been used effectively to decrease the risk of pulmonary embolism (PE), but has associated complications. Permanent IVC filter placement is associated with long-term complications, such as deep vein thrombosis (DVT), IVC thrombosis, penetration into surrounding structures, such as the IVC wall, duodenum, and aorta, and filter fracture and migration. Temporary, retrievable IVC filters have become available in recent years, and are increasingly being used for inferior vena caval interruption for the prevention of venous thromboembolism (VTE), while allowing for eventual IVC filter removal, thereby eliminating the potential for long-term complications.

Trauma victims have the highest rate of VTE of all hospitalized patients, with a reported incidence as high as 58%. Because the risk period for VTE is relatively limited in this patient cohort, temporary IVC filter utilization is an attractive alternative. Indeed, since the availability of retrievable IVC filters (RIVCF), a marked increase in IVC filter placement has been observed; however, multiple studies have shown that most of these filters are, in fact, not removed. Although the definition of retrieval rate used in the literature varies among different studies, the published recovery rates range from 15% to 59%, while in a multi-center trauma study, the retrieval rate was only 21%. This observation raises concerns because patients will be exposed to the unnecessary risk of the long-term sequelae of permanent filters, particularly in those who received filters for prophylaxis. Although there is the assumption that temporary IVC filters that became permanent are benign compared with permanent IVC filters, this has never been proven. The permanent filters, on the other hand, have been shown to be associated with an increased risk of recurrent DVT in a long-term, prospective study.

Multiple reasons account for failure to remove temporary filters. It is well known that trauma patients are very difficult to follow. Many patients move, die, or are lost to follow-up. The relatively large geographic area the trauma centers serve also contributes to the difficulties. Patients are often transferred to community hospitals or referred to local hospitals for continued care. Even when patients return for follow-up, the poorly defined safe "indwell time" may result in declaring a filter non-retrievable.

The authors were able to document in this study that clinician oversight was the primary factor that contributed to the failure to remove RIVCF before institution of their protocol. In the present study, before institution of the IVC filter retrieval protocol, the retrieval rate was only 37%, with fewer than half of patients eligible for filter removal actually undergoing a retrieval attempt. This oversight was felt to occur because as trauma patients pass through several teams of clinical services from intensive care unit, to step-down unit, to the floor, some of the information gets lost, including the status of the filter. Lack of a designated person or a protocol in place to track the filter exacerbates loss of follow-up of the filter.

After this analysis, a protocol was instituted to improve the retrieval rates, by compiling a prospective database of patients who received a RICVF and coordinating the timely removal of the RICVF when indicated either during the inpatient stay or as an outpatient. If the treating clinicians decided that the filter was to be left in permanently owing to a very high risk for VTE or ongoing contraindication to low-molecular-weight heparin(LMWH), this decision was documented in the medical record and in the database. Inpatients that had an RIVCF placed were highlighted on the daily trauma-rounding list. All trauma providers were educated on an ongoing basis on the indications for RIVCF insertion and retrieval. These efforts were supplemented by electronic tracking of patients with RIVCF; an automated e-mail was sent to the admitting attending of any patient who had placement of a RIVCF to serve as a reminder to plan its eventual retrieval when clinically indicated. This project also required coordination between the divisions of trauma surgery, vascular surgery, and the interventional radiology service for both inpatients and outpatients. RIVCF retrieval rates were compared between the trauma patients who underwent temporary IVC filtration before (group I) and after (group II) implementation of this institutional protocol at University of Pittsburgh Medical Center—Presbyterian University Hospital (UPMC-PUH).

As a result, the overall retrieval attempt rate increased from 42% to 95% and overall retrieval rate from 37% to 84% in appropriate patients, a marked improvement. A prospective protocol can resolve many of these problems and allows clinicians to follow the status of IVC filters even after discharge. Such a protocol mediates a systematic effort to remove filters by establishing an active line of communication between the clinical services and alerts the status of the filter during hospital course and after patient's discharge. With the increasing "indwell time" of these devices and the number of patients being discharged from the hospital or transferred to another facility with filters, the role of a prospective protocol is critical in ensuring filter retrieval when appropriate.

Despite the improved rate of filter recovery with the protocol, it should be noted that only half of all the patients who received RIVCF had it removed in the present series, as detailed in Table 4. A variety of reasons, from patient factors to physician preferences, preclude patients from filter removal. Placing these filters with the assumption that most of these will be retrieved is obviously wrong, especially in the absence of high-risk criteria.

The authors conclude that an institutional protocol for prospective monitoring of RIVCF significantly improved the filter removal rate. However, a significant portion, nearly 50% in this series, is still left permanently. IVC filtration should be used judiciously, especially when placed for prophylactic indications, because the long-term complications of non-retrieved devices remain to be determined. Indications for temporary IVC interruption in trauma patients need to be reevaluated.

T. G. Walker, MD

Peripheral Stent Placement in Hemodialysis Grafts

Kariya S, Tanigawa N, Kojima H, et al (Kansai Med Univ, Osaka, Japan; et al)
Cardiovasc Intervent Radiol 32:960-966, 2009

The purpose of the present study was to evaluate the clinical outcome of peripheral stent placement after failed balloon angioplasty in patients with grafts who are on hemodialysis. We examined 30 Wallstents that were placed in 26 patients because balloon angioplasty failed or early restenosis (<3 months) occurred within 3 months. We retrospectively reviewed 267 consecutive balloon angioplasties performed in 71 patients with graft access between August 2000 and March 2007. Stent placements accounted for 30 (11.2%) of the 267 balloon angioplasties. The clinical success rate of stent placement was 93.3% (28 of 30 stent placements). The 3-, 6-, and 12-month primary patency rates were 73.3%, 39.3%, and 17.7%,

TABLE 3.—Primary Patency of Stent Placements Compared with Balloon Angioplasty Alone

Category	No. of Patients	Primary Patency Period (Months)[a]	Primary Patency Rate (%)[b] 3 Months	6 Months	12 Months	*P*-value
Stent placement	25	7.0 ± 9.8	73.3 ± 9.4 (16)	39.3 ± 10.7 (8)	17.7 ± 8.8 (3)	0.0279[c]
Subgroup						
Indication A[d]	10	6.5 ± 3.8	80.0 ± 12.6 (8)	57.1 ± 16.4 (5)	17.1 ± 14.5 (1)	
Indication B[e]	15	7.3 ± 12.5	68.4 ± 13.2 (8)	25.7 ± 12.7 (3)	17.1 ± 11.0 (2)	
Nonthromboticaccess	18	7.2 ± 11.0	81.4 ± 9.8 (12)	44.8 ± 13.3 (6)	18.6 ± 11.4 (2)	
Thrombotic access	7	6.6 ± 6.7	57.1 ± 18.7 (4)	28.6 ± 17.1 (2)	–	
Balloon angioplasty alone	44	12.7 ± 12.9	80.9 ± 6.1 (32)	72.8 ± 7.0 (27)	44.1 ± 8.4 (13)	

[a]Data are mean ± standard deviation.
[b]Data are primary patency rate ± standard error. Numbers in parentheses are numbers at risk at start interval.
[c]Determined by Breslow–Gehan–Wilcoxon test. $P = 0.0279$ vs. balloon angioplasty alone.
[d]Technical failure of balloon angioplasty based on insufficient access flow and insufficient conversion of continuous palpable thrill.
[e]Restenosis within 3 months after previous balloon angioplasty for the same lesion.

TABLE 4.—Comparison of Primary Patency Between Previous Balloon Angioplasty (Ba) Alone and Stent Placement in Indication B Group: Restenosis within 3 Months After Previous Ba for the Same Lesion ($N = 15$)

Intervention	Primary Patency Period (Month)[a]	P-value[b]	Primary Patency Rate (%)[c] 1 Month	2 Months	3 Months	6 Months	12 Months	P-value[d]
Previous BA	1.9 ± 0.7	0.0468	86.7 ± 8.8 (13)	53.3 ± 12.9 (8)	0 (0)	–	–	0.0059
Stent placed	7.3 ± 12.5		93.3 ± 6.4 (13)	85.6 ± 9.5 (10)	68.4 ± 13.2 (3)	25.7 ± 12.7 (3)	17.1 ± 11.0 (2)	

[a]Data are mean ± standard deviation.
[b]Determined by Wilcoxon signed-rank test. Primary patency periods after stent placement were compared with those after previous BA alone.
[c]Primary patency rate ± standard error. Numbers in parentheses are those at risk at start interval.
[d]Determined by Breslow–Gehan–Wilcoxon test. Comparison of primary patency rates after stent placement with those after previous BA alone.

respectively. The 1-, 2-, and 3-year secondary patency rates were 90.2%, 83.8%, and 83.8%, respectively. Primary patency was significantly prolonged by stent placement after early restenosis compared with previous balloon angioplasty alone ($P = 0.0059$). Primary patency after stent placement was significantly lower than after successful balloon angioplasty without indications for stent placement ($P = 0.0279$). Secondary patency rates did not significantly differ between stent placement and balloon angioplasty alone. The mean number of reinterventions required to maintain secondary patency after stent placement was significantly larger than that after balloon angioplasty alone (Mann-Whitney U test, $P = 0.0419$). We concluded that peripheral stent placement for graft access is effective for salvaging vascular access after failed balloon angioplasty and for prolonging patency in early restenosis after balloon angioplasty. However, reinterventions are required to maintain secondary patency after stent placement. Furthermore, peripheral stent placement for graft access cannot achieve the same primary patency as balloon angioplasty alone.

▶ Percutaneous transluminal angioplasty has become the first choice of treatment for failure of vascular access. The introduction of cutting and ultrahigh pressure balloons appears to have increased the success rate of balloon angioplasty. However, stents are still occasionally placed to treat stenosis after failed balloon angioplasty. The Society of Interventional Radiology (SIR) guidelines do not recommend the routine use of stents to prevent restenosis and state that the role of stents has yet to be fully defined. Three prospective randomized studies have found that stent placement does not confer an advantage over successful angioplasty. Thereafter, some studies have described the placement of metallic stents for peripheral lesions after failed balloon angioplasty, which seems to have been effective.

Stent placement for failed balloon angioplasty has been studied in detail, and therapeutic outcomes have been documented. The authors of these publications seem to promote stent use for failed balloon angioplasty. However, repeated reinterventions seem to be required after placement of stents, which limits their effectiveness.

This retrospective study evaluates the clinical results of peripheral stent placement after failed balloon angioplasty and examines whether or not the primary patency of stent placement in this circumstance can equal that of balloon angioplasty alone. The authors investigated 30 stent placements for graft access in 26 patients with 30 peripheral stenoses, including 2 at the stented segment. They also retrospectively reviewed 267 consecutive balloon angioplasties performed due to access failures in 71 patients between August 2000 and March 2007; the clinical success rates, as well as complications, primary patency, and secondary patency were documented in these patients. They then compared the patency before and after stent placement in early restenosis and compared stent placement with balloon angioplasty for the primary patency, secondary patency, and number of reinterventions required to maintain secondary patency.

According to DOQI guidelines, all angioplasties were performed on patients exhibiting >50% stenosis and clinical/physiologic abnormalities. Stents were placed after failed balloon angioplasty defined as follows: Indication A was failed balloon angioplasty based on insufficient restoration of access flow assessed by fistulography and insufficient conversion of a continuous palpable thrill at the distal (venous) end of grafts. Indication B was restenosis occurring within 3 months after the last balloon angioplasty for the same lesion. However, when the vascular diameter and access flow after the present balloon angioplasty were more favorable than those after the last balloon angioplasty, stent placement was excluded.

For angioplasty, the balloon catheters ranged in size from that of the normal vessel diameter to 20% larger (6 mm ± 10 mm in diameter); PTA balloons used included used Ultra-thin Diamond (Boston Scientific, Natick, MA, USA; rated burst pressure, 15 atm), Blue Max (Boston Scientific; rated burst pressure, 20 atm), or Peripheral Cutting (Boston Scientific; rated burst pressure, 10 atm) balloons. All balloons were dilated using a pressure inflation device (Everest; Medtronic, Minneapolis, MN, USA) to below the rated burst pressure until the balloon waist disappeared and then were inflated for 1 min (or for 5 min in the event of recoil).

In the stenting groups, the authors used self-expanding stents and Wallstents (Boston Scientific, Natick, MA) of 6- to 10-mm nominal diameter that were adapted to the largest diameter of the segment covered with the stent so that the entire stent could touch the luminal wall. The stent was kept as short as possible but long enough to bridge the entire lesion with a slight overlap at its proximal and distal ends. Stents were dilated after deployment using the original balloon to ensure close contact with the vessel wall.

Clinical success was defined as the resumption of normal dialysis for at least 1 session, in accordance with published Society of Interventional Radiology (SIR) guidelines. Clinical follow-up of all patients included a physical examination, venous dialysis pressure measurements at each hemodialysis session, and monthly evaluations of dialysis dose and urea recirculation. Ultrasonpgraphy was performed when results were abnormal during clinical follow up. Follow-up continued until the patient died, surgical revision excluded the stent-implanted segment, or the graft was abandoned. "Post-intervention primary patency" and "post-intervention secondary patency" were considered as primary and secondary patency, respectively. Loss of patency was defined as treatment of a lesion anywhere within the access circuit, from the arterial inflow to the superior vena cava-right atrial junction, according to published SIR guidelines.

The clinical success rate of stent placement was 93.3% (28 of 30 stent placements). The 3-, 6-, and 12-month primary patency rates were 73.3%, 39.3%, and 17.7%, respectively. The 1-, 2-, and 3-year secondary patency rates were 90.2%, 83.8%, and 83.8%, respectively. Primary patency was significantly prolonged by stent placement after early restenosis compared with previous balloon angioplasty alone. Table 3 reports the mean primary patency periods, the primary patency rates for successful stent placement, and the subgroups. Table 4 reports a comparison of the access patency in indication B after stent placement and previous balloon angioplasty for the same lesions.

If early restenosis occurs, stent placement can be a useful option because the present study found that primary patency was prolonged significantly by stent placement for peripheral stenoses recurring within 3 months. Primary patency persisted significantly longer after stent placement than after previous balloon angioplasty alone. However, the duration of primary patency after stent placement following failed balloon angioplasty was significantly shorter than that for successful balloon angioplasty alone for stenosis without stent placement. Even when stents were placed for stenosis after failed balloon angioplasty, the primary patency was never the same as that for successful balloon angioplasty alone.

The published secondary patency rates in peripheral access grafts after stent placement at 1-year range from 60% to 90%. The results for secondary patency rates in this study are comparable with these. Although peripheral stents allow long-term secondary patency, multiple reinterventions are required to maintain it. Furthermore, more reinterventions may be needed to maintain secondary patency compared with balloon angioplasty alone. The data in this study indicate that although stent patency was sometimes quite long, stenosis at the stented segment is predictable (78.6%).

The authors conclude that peripheral stent placement is effective for salvaging hemodialysis access after failed balloon angioplasty. Peripheral stent placement for early (< 3 months) recurring stenosis significantly improves primary patency compared with previous angioplasty alone. However, reinterventions were required to maintain secondary patency after stent placement. They emphasize that peripheral stent placement cannot achieve the same primary patency as successful balloon angioplasty without stent placement. They recommend that although peripheral stent placement is effective for failed balloon angioplasty, the success rate of balloon angioplasty needs to be improved to reduce the use of stents.

T. G. Walker, MD

Radiofrequency Endovenous ClosureFAST versus Laser Ablation for the Treatment of Great Saphenous Reflux: A Multicenter, Single-blinded, Randomized Study (RECOVERY Study)

Almeida JI, Kaufman J, Göckeritz O, et al (Miami Vein Ctr, FL; Oregon Health and Science Univ, Portland; Venenzentrum Elsterpark, Leipzig, Germany; et al)
J Vasc Interv Radiol 20:752-759, 2009

Purpose.—The present study was designed to address the hypothesis that radiofrequency (RF) thermal ablation, as represented by the ClosureFAST system, is associated with improved recovery and quality-of-life (QOL) parameters compared with 980-nm endovenous laser (EVL) thermal ablation of the great saphenous vein (GSV).

Materials and Methods.—Eighty-seven veins in 69 patients were randomized to ClosureFAST or 980-nm EVL treatment of the GSV. The study was prospective, randomized, single-blinded, and carried out at five American sites and one European site. Primary endpoints (postoperative pain, ecchymosis, tenderness, and adverse procedural sequelae) and secondary endpoints (venous clinical severity scores and QOL issues) were measured at 48 hours, 1 week, 2 weeks, and 1 month after treatment.

Results.—All scores referable to pain, ecchymosis, and tenderness were statistically lower in the ClosureFAST group at 48 hours, 1 week, and 2 weeks. Minor complications were more prevalent in the EVL group ($P = .0210$); there were no major complications. Venous clinical severity scores and QOL measures were statistically lower in the ClosureFAST group at 48 hours, 1 week, and 2 weeks.

Conclusions.—RF thermal ablation was significantly superior to EVL as measured by a comprehensive array of postprocedural recovery and QOL parameters in a randomized prospective comparison between these two thermal ablation modalities for closure of the GSV.

▶ Catheter-based thermal ablation with either radiofrequency [RF] or laser as the energy source has arguably become the standard treatment for symptomatic great saphenous vein (GSV) incompetence. Historically, RF catheters were the first devices to become available to interventionalists for GSV ablation, yet relatively early follow-up after RF ablation demonstrated only 87% occlusion of the

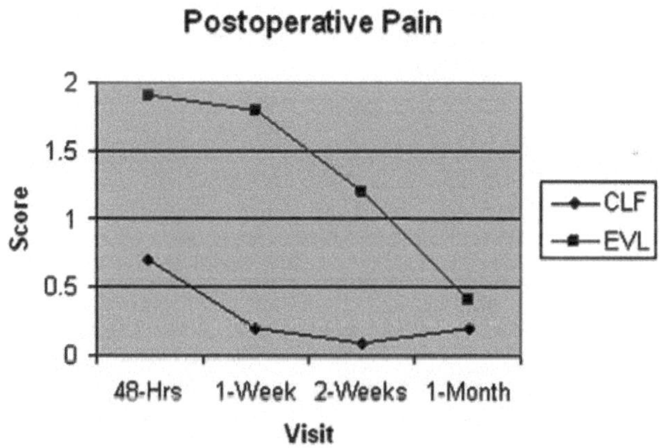

Visit	Postoperative Pain CLF mean ± SD (m)	EVL mean ± SD (m)	*t* test *P* value
48 Hour	0.7 ± 0.9 (46)	1.9 ± 1.6 (41)	<.0001
1 Week	0.2 ± 0.6 (40)	1.8 ± 1.8 (36)	<.0001
2 Weeks	0.1 ± 0.4 (43)	1.2 ± 1.7 (39)	<.0001
1 Month	0.2 ± 0.8 (46)	0.4 ± 1.2 (41)	.2962

FIGURE 1.—Postoperative pain measured at each follow-up visit. (Courtesy of Almeida JI, Kaufman J, Göckeritz O, et al. Radiofrequency endovenous closureFAST versus laser ablation for the treatment of great saphenous reflux: a multicenter, single-blinded, randomized Study (RECOVERY study). *J Vasc Interv Radiol.* 2009;20:752-759.)

TABLE 4.—Complications Identified at Follow-up Visits

Visit	Variable	ClosureFAST	EVL	*P* Value*
48 Hours	Hyperpigmentation	0	0	
	Phlebitis	0	5 (12.2)	.020
	Paresthesia	0	1 (2.4)	.471
	Erythema	0	3 (7.3)	.101
	Infection	0	0	
	TE/DVT	0	1 (2.4)	.471
1 Week	Hyperpigmentation	0	0	
	Phlebitis	0	1 (2.4)	.471
	Paresthesia	1 (2.2)	0	1.000
	Erythema	0	0	
	Infection	0	0	
	TE/DVT	0	1 (2.4)	.471
2 Weeks	Hyperpigmentation	0	0	
	Phlebitis	0	2 (4.9)	.219
	Paresthesia	1 (2.2)	2 (4.9)	.600
	Erythema	0	3 (7.3)	.101
	Infection	0	0	
	TE/DVT	0	1 (2.4)	.471
1 Month	Hyperpigmentation	1 (2.2)	0	1.000
	Phlebitis	0	0	
	Paresthesia	0	0	
	Erythema	0	0	
	Infection	0	0	
	TE/DVT	0	0	
Sequelae at any follow-up	Hyperpigmentation	1 (2.2)	0	>.999
	Phlebitis	0	6 (14.6)	.009
	Paresthesia	1 (2.2)	2 (4.9)	.0600
	Erythema	0	4 (9.8)	.045
	Infection	0	0	
	TE/DVT	0	1 (2.4)	.471
	Total limbs	2 (4.4)	9 (22.0)	.021

Note.—Values in parentheses are percentages. DVT = deep vein thrombosis; TE = thromboembolism.
*Fisher exact test.

GSV at 5 years, with a 21% varicose vein recurrence rate in a multicenter registry completed by Merchant and Pichot.[1] Four randomized controlled trials subsequently compared endovenous RF ablation versus surgical vein stripping, all of which reported superior results with RF ablation. Findings included faster recovery, less postoperative pain, fewer adverse events, and superior quality-of-life (QOL) scores. The earlier-generation RF ablation system, however, had 2 disadvantages compared with using endovenous laser ablation (EVLA) for treating GSV incompetence: these were slow pullback speed and the occasional need to remove the catheter to clear coagulum from the bipolar electrodes. Subsequently, the ClosureFAST RF catheter (VNUS Medical Technologies, San Jose, California) has been introduced and has led to dramatic improvement in the procedure, including elimination of the laborious pullback, short energy cycle, and rapid treatment as a result of a constant temperature level of 120°C.

When EVLA was adopted for treating GSV incompetence, 3-year follow-up on 499 limbs treated in a single center demonstrated successful ablation in 93% of 121 limbs seen at 2 years, with no recurrences in 40 limbs followed to 3 years.[2] Reasons for the more effective ablation with EVLA were unclear until articles on the dose-response relationship between laser energy and durability of vein occlusion were subsequently published. Bruising, transient pain, and thigh induration are common adverse events after EVLA, which may be caused by laser-induced perforation of the vein wall and extravasation of blood into surrounding tissue. After EVLA, one can generally expect 70% of limbs to experience some degree of pain, and 50% to require analgesics for pain management.

This multicenter, prospective, randomized study compared recovery and QOL factors between RF and EVL ablation. The ClosureFAST device was compared with a 980-nm EVLA at comparable energy delivery for closure of incompetent GSVs. The presence and intensity of postoperative pain, ecchymosis, and tenderness; adverse procedural sequelae such as deep vein thrombosis, paresthesia, phlebitis, hyperpigmentation, and infection; and periprocedural analgesic agent use and QOL were measured at 48 hours, 1 week, 2 weeks, and 1 month after treatment. There were no significant demographic differences between neither the groups nor in Clinical/Etiology/Anatomy/Pathophysiology (CEAP) classes, mean length of treated segment, or target vein diameter.

As Fig 1 demonstrates, the ClosureFAST group reported significantly lower pain levels than the EVLA group during follow-up. This was also true for postoperative tenderness and ecchymosis. Complications, as demonstrated in Table 4, were statistically more prevalent in the EVLA group than the ClosureFAST group (22.0% vs. 4.4%; $P = .0210$). Changes in global quality of life scores were better with ClosureFAST treatment at 7 and 14 days after treatment. Vein occlusion and elimination of reflux were achieved in 100% of limbs irrespective of treatment modality.

All indices measuring postinflammatory sequelae demonstrated a statistically significant difference favoring RF ablation versus EVLA, and the differences did not result from differences in energy delivery. RF treatment produced significantly less pain, bruising, and tenderness, as well as fewer adverse sequelae than laser treatment.

T. G. Walker, MD

References

1. Merchant RF, Pichot O. Long-term outcomes of endovenous radiofrequency obliteration of saphenous reflux as a treatment for superficial venous insufficiency. *J Vasc Surg*. 2005;42:502-509.
2. Min RJ, Khilnani N, Zimmet S. Endovenous laser treatment of saphenous vein reflux: long-term results. *J Vasc Interv Radiol*. 2003;14:991-996.

Short- and Long-term Retrievability of the Celect Vena Cava Filter: Results from a Multi-institutional Registry

Lyon SM, Riojas GE, Uberoi R, et al (Monash Univ, Melbourne, Australia; Hosp Universitario Universidad Automa de Nuevo Leon, Monterrey; John Radcliffe Hosp, Oxford)

J Vasc Interv Radiol 20:1441-1448, 2009

Purpose.—To evaluate retrievability of the Celect vena cava filter over time and to assess the safety of the retrieval procedure in a prospective multicenter registry.

Materials and Methods.—Between October 2005 and March 2008, Celect filters were placed in 95 patients (61 men; mean age, 51 years ± 18.5) with a temporary need for an inferior vena cava (IVC) filter. All patients satisfied requirements for filter placement; the primary indications for placement were pulmonary embolism (PE) with a contraindication to or failure of anticoagulation ($n = 40$), high risk for further PE ($n = 29$), trauma ($n = 23$), or massive PE with residual deep vein thrombosis and risk for further PE ($n = 3$). Filter orientation, vena cava injury, and other device-related incidents were evaluated at implantation and

TABLE 2.—Participation Criteria

Inclusion Criteria

1. Any patient considered at high risk for PE and is under consideration for placement of a permanent or short-term IVC filter was required to have at least one of the following seven indications.
   a. Evidence of PE or IVC, iliac, or femoropopliteal DVT and a contraindication to, or complication or failure of, anticoagulation.
   b. Massive PE with residual DVT in a patient at risk for further PE
   c. Free-floating iliofemoral or IVC thrombus
   d. Severe cardiopulmonary disease and deep venous thrombus (eg, cor pulmonale with pulmonary hypertension)
   e. Poor compliance with anticoagulant medications
   f. Severe trauma without documented PE or DVT in patient with a closed head injury, spinal cord injury, or multiple long bone or pelvic fractures
   g. High risk (eg, immobilized, intensive care, prophylactic preoperative placement in patients with multiple risk factors for VTE)
2. Patient must have patent internal jugular vein
3. Patient or legal guardian must have signed the informed consent document
4. Patient or legal guardian agrees that the patient will have a clinical and imaging examination performed before filter retrieval
5. Patient or legal guardian agrees that the patient will return for clinical and imaging follow-up at 30 days and 3 months after filter retrieval. If the filter is not retrieved, the clinical and imaging follow-up would be at 30 days, 3 months, 6 months, and 12 months

Exclusion Criteria

1. Age <18 y
2. Pregnancy
3. Uncontrollable coagulopathy
4. Life expectancy <6 months
5. Vena cava diameter >30 mm or <15 mm on at least an AP view measured by vena cava sizing catheters
6. Contrast agent allergy that cannot be adequately premedicated
7. Simultaneous participation in another investigative drug or device trial or previous placement of an IVC filter

Note.—AP = anteroposterior; DVT = deep vein thrombosis; VTE = venous thromboembolism.

TABLE 3.—Compliance with Requirements for IVC Filter Placement

Indication	No. of Pts.
High risk (eg, immobilized, intensive care, prophylactic preoperative placement in patients with multiple risk factors for VTE)	51
Evidence of PE or IVC, iliac, or femoropopliteal DVT and a contraindication to, or complication or failure of, anticoagulation	49
Severe trauma without documented PE or DVT in patient with a closed head injury, spinal cord injury, or multiple long bone or pelvic fractures	30
Massive PE with residual DVT in a patient at risk for further PE	7
Free-floating iliofemoral or IVC thrombus	5
Severe cardiopulmonary disease and DVT (eg, cor pulmonale with pulmonary hypertension)	0
Poor compliance with anticoagulant medications	0

Note.—DVT = deep vein thrombosis; VTE = venous thromboembolism.

TABLE 4.—Primary Indications for Filter Placement ($N = 95$)

Indication for Placement	Incidence	Retrieval Attempts	Successful Retrieval	Mean Time to Retrieval (d)
PE or IVC/iliac/femoropopliteal DVT and anticoagulation contraindication/complication/failure	40 (42.1)	19 (47.5)	18 (94.7)	181.4
High risk*	29 (30.5)	19 (65.5)	18 (94.7)	144.6
Trauma†	23 (24.2)	18 (78.3)	18 (100)	238.8
Massive PE with residual DVT and risk of further PE	3 (3.2)	2 (66.7)	2 (100)	136.5

Note.—Values in parentheses are percentages. VTE = venous thromboembolism.

*Includes immobilized and intensive care patients and cases of prophylactic preoperative placement in patients with multiple risk factors for VTE.

†All trauma patients were without documented PE or DVT, but with multiple long bone or pelvic fractures.

retrieval. The degree of difficulty associated with retrieval was also assessed.

Results.—Filter retrieval was attempted in 58 patients (mean indwell time of 179 days; median, 168.5 d; range, 5–466 d). Fifty-six filters (96.6%) were successfully retrieved. Unsuccessful retrieval attempts were attributed to filter tilt ($n = 1$) or excessive tissue growth with the hook embedded in the endothelium ($n = 1$). No adverse events were associated with the inability to retrieve these filters. A Kaplan-Meier product-limit survival estimate revealed that the probability of successful filter

TABLE 6.—Incidences of Filter Tilt at Implantation and Retrieval

Tilt (°)	All Patients ($N = 95$)	Attempted Retrieval Cohort ($n = 58$) At Implantation	At Retrieval
0	36 (37.9)	18 (31.0)	22 (37.9)
≤5	18 (18.9)	13 (22.4)	17 (29.3)
6–10	27 (28.4)	16 (27.6)	15 (25.9)
11–15	8 (8.4)	6 (10.3)	2 (3.4)
≥16	6 (6.3)	5 (8.6)	2 (3.4)

TABLE 7.—Details of Filter Retrieval ($n = 58$)

Retrieval Detail	Incidence
Route of retrieval	
Right internal jugular	57 (98.3)
Left internal jugular	1 (1.7)
Technical difficulty with retrieval	
None	36 (62.1)
Minimal	12 (20.7)
Moderate	4 (6.9)
Major	4 (6.9)
Not assessed*	2 (3.4)

*Technical difficulty associated with retrieval was not reported for unsuccessful retrieval attempts.

retrieval remained at 100% at 50 weeks and at more than 74% at 55 weeks after implantation. No adverse events were related to the retrieval procedures.

Conclusions.—Retrieval of the Celect filter was performed safely as long as 466 days after implantation.

▶ Although inferior vena cava (IVC) filters were originally developed to be permanently implanted, retrievable IVC filters have been used clinically for more than 15 years. Currently, 5 filters are cleared for marketing by the US Food and Drug Administration with retrieval indications: the Recovery G2 (Bard Peripheral Vascular, Tempe, Arizona), OptEase (Cordis, Miami Lakes, Florida), Option (Angiotech, Vancouver, BC, Canada), Günther Tulip (William Cook Europe, Bjaeverskov, Denmark), and Celect (William Cook Europe) filters.

The Günther Tulip filter was the first retrievable filter; it was first evaluated in 1993. Since then, clinical data have continued to support the retrievability of the Günther Tulip filter. The largest and most recent study of the filter indicates retrieval success rates of 99% at 4 weeks after implantation, 94% at 12 weeks after implantation, and 37% at 52 weeks after implantation. Even with these successes, design improvements have been sought to further extend the time period and success rate for safe retrieval. The Celect filter, which is equivalent to its predecessor (the Günther Tulip filter) with regard to intended use, materials of construction (cobalt chromium alloy), and placement and retrieval procedures, is intended to have an extended safe retrieval period as a result of design changes in the secondary legs. Available data on the Celect filter have demonstrated an in vitro embolus capturing efficiency equivalent to that of the Günther Tulip filter and safety and long-term retrievability in an ovine model. However, until this study, there are no prospective multicenter registry studies to evaluate the retrievability of the Celect IVC filter in a clinical setting.

Ninety-five patients at 6 institutions underwent placement of a Celect IVC filter, with the intent to subsequently retrieve the filter. The study group included 61 men (64%) and 34 women (36%) with a mean age of 51 years ± 18.5 years. Patients were prospectively enrolled in the study after satisfying all the study inclusion criteria and none of the study exclusion criteria (Table 2); each patient satisfied 1 or more of the indications for filter placement (Tables 3 and 4). Each patient's involvement in the study ended at 3 months after retrieval or when a decision was made that the device would remain as a permanent implant. Celect filters were successfully placed in all patients (N = 95) according to the manufacturer's instructions for use. In 44 patients (46.3%), filters were introduced by the internal jugular approach, and in 51 patients (53.7%), filters were introduced by the femoral approach. Most filters (98.9%) were implanted in the IVC in an infrarenal location; 1 filter (1.1%) was implanted in the IVC in a suprarenal location.

Patients were followed up at 30 days, 3 months, 6 months, 12 months, and per the standard of care at each institution thereafter, until a decision to retrieve the filter was made. The follow-up visits included a detailed clinical examination, plain abdominal radiography, and duplex ultrasound imaging of the vena cava. Patients underwent a filter retrieval attempt when deemed clinically appropriate by their physician. Cavograms were obtained before retrieval to assess for possible filter migration, tilt, thrombus, injury to the vena cava, or other events related to the filter. Filter retrieval was not recommended in patients with trapped thrombi occupying greater than 25% of the filter's cone volume. Filter retrieval occurred via the jugular approach and was performed with the Günther Tulip filter retrieval set or another appropriate snare device. For each retrieval attempt, the technical difficulty associated with retrieval was indicated by the physician as none, minimal, moderate, major, not possible, or not assessed. A cavogram was obtained after retrieval to assess for procedural trauma or other complications related to retrieval. After retrieval, filters were examined for trapped thrombus, endothelium formation, and mechanical damage. Postprocedural care was provided according to the standard of care at each institution. In the event in which a filter could not be retrieved, a reason for retrieval failure was provided, and the filter was left as a permanent implant. Patients were seen for follow-up at 30 days and 3 months after retrieval. At the 30-day follow-up, patients underwent a detailed clinical examination and a duplex ultrasonography (US) examination of the vena cava to assess for thrombus formation. At the 3-month follow-up, patients underwent a detailed clinical examination.

The primary objective of the study was to verify the safety and performance of the Celect filter by assessing the incidence and severity of adverse events associated with retrieval of the device. The composite major complication rate (ie, hemorrhage, pulmonary embolism [PE], vena cava perforation, procedure-related death, occlusion, significant migration [> 20 mm], and filter fracture) was evaluated in relation to the retrieval procedure (ie, before vs after retrieval). Additional measures included technical success of the implantation and retrieval procedures, incidence, and degree of filter tilting relative to the longitudinal axis of the IVC, trauma to the IVC during the implantation and retrieval procedures, acute trauma to the internal jugular vein (ie, vascular access site for retrieval), thrombosis or stenosis at the implantation site or IVC, and complications at the vascular access site (eg, thrombosis, infection, and bleeding). The incidence and degree of filter tilting at implantation was compared with that at retrieval, and filters were categorized based on their change in tilt angle (ie, increased, decreased, or unchanged). These data are summarized in Table 6. Analysis of successful retrieval by time was performed with the use of the Kaplan-Meier product-limit method.

Filter retrieval was attempted in 58 patients (61%). At retrieval, 22 patients (38%) were taking anticoagulation medication. Trapped thrombus was observed in 4 filters (7.0%), although not to a degree that prevented filter retrieval. Strut penetration (ie, transmural incorporation or filter incorporation into the IVC wall) was noted in 21 patients based on preretrieval imaging. Importantly, there was no evidence of IVC perforation (ie, strut protruding out of the IVC wall) or extravasation or hemorrhage in any patient, and there were no patient-reported complications typically associated with perforation (eg, pain). Of these patients with penetration, 20 (95%) had successful filter retrievals. One patient had an unsuccessful filter retrieval procedure, which was associated with excessive tissue growth. Details of the retrieval procedures are summarized in Table 7.

Filter tilt often contributes to failed IVC filter retrieval attempts. Therefore, filter tilt was assessed at implantation and retrieval in this study, and the change in filter tilt from implantation to retrieval was reported, providing an indirect assessment of the self-centering ability of the Celect filter. Both failed retrieval attempts in this study were attributed at least in part to filter tilt. Data on the change in filter tilt from the time of implantation to the time of retrieval (Table 6) indicate that the Celect filter self-centered to some extent in 31% of patients and remained at the same degree of tilt in 51.7% of patients.

Complications that occurred before and after retrieval procedures were assessed. There were no reports of significant filter migration, filter fracture, IVC occlusion, or IVC perforation in this study. Minor events were reported (eg, fever for 1 day, elective surgical procedure, pneumonia); however, none of these events were related to the device or the retrieval procedure. Eighteen deaths were reported before a decision to retrieve the filter could be made; only 2 deaths were adjudicated as potentially device- or procedure related. In the 90 days after retrieval, there were no procedure-related events, providing evidence of retrieval safety. Patients who commonly receive filters are those with advanced malignancies, thrombotic tendencies, trauma, and critically illnesses; therefore, adverse events are not uncommon. Moreover, the events reported in this study are consistent with the patient population, and the number of events reported in this study is not considered to be high.

The retrieval data presented here suggest that, even after extended indwell times, filter retrieval is successful and safe. These data establish the Celect filter as a safe retrievable filter and are clinically valuable in weighing the probability of successful filter retrieval (estimated from implantation duration) versus the overall medical condition of each patient. For instance, comparison of the retrievability of the Celect filter with the retrievability of the Günther Tulip filter, which has a long history of clinical use with few complications, suggests that the Celect filter is associated with higher successful retrieval rates at longer indwell times. As new filters come to market and the trend continues toward increased filter placement with the intention of retrieval, studies must continue to provide information regarding retrieval success rate by indwell time and the complication rates before and after retrieval so that physicians have sufficient information to select the best filter for each patient.

T. G. Walker, MD

The Case for Primary Placement of Tunneled Hemodialysis Catheters in Acute Kidney Injury

Coryell L, Lott JP, Stavropoulos SW, et al (Univ of Pennsylvania Med Ctr, Philadelphia)

J Vasc Interv Radiol 20:1578-1581, 2009

Purpose.—Nontunneled hemodialysis catheters (NTDCs) are widely used for initial hemodialysis access in new-onset renal failure. The National Kidney Foundation recommends NTDC use for hemodialysis duration of less than 1 week in acute kidney injury because of the increased infection risk compared with tunneled hemodialysis catheters (TDCs) with longer use. The present study was performed to determine whether primary placement of TDCs in this setting is more appropriate, and whether there are predictors of recovery of renal function in less than 1 week.

Materials and Methods.—In the authors' practice, patients referred to the interventional radiology unit in whom no contraindications exist receive a TDC; 76 patients who received a primary TDC for acute kidney injury and who eventually recovered renal function were retrospectively reviewed herein. Causes of renal failure, various renal function parameters, and demographics were collected, as were TDC dwell times, in an effort to determine predictors of recovery and/or extended duration of use.

Results.—Mean TDC dwell time in patients who eventually recovered from acute kidney injury was 34 days; only 15 of 76 (20%) recovered within 1 week. At TDC placement, there were no significant differences between patients who recovered in less than (vs greater than) 1 week.

Conclusions.—The present results support primary placement of TDCs in patients with acute kidney injury who require hemodialysis and in whom no contraindications exist, as no predictors of recovery of renal function in less than 1 week were identified.

▶ Nontunneled hemodialysis catheters (NTDCs) are widely used as the initial hemodialysis access in new-onset renal failure. The most recent National

TABLE 1.—Characteristics of Patients by Duration of Renal Recovery

Characteristic	Recovery <1 Week ($n = 15$)	Recovery ≥1 Week ($n = 61$)	*P* Value
Age (y)	56 ± 4.9	56 ± 1.9	.99
Female sex (%)	40	39	.96
Creatinine (mg/dL)	6 ± 0.54	6 ± 0.4	.7
24-hour urine volume (mL)	751 ± 201	791 ± 118	.87
Oliguria duration (d)	2.4 ± 1.4*	1.9 ± 1.3†	.37
Chronic renal insufficiency	8/15 (53)	22/52 (42)	.45
Location in hospital (%)			.14
Medicine floor	10 (67)	24 (39)	
Cardiac care unit	3 (20)	5 (8)	
Medical intensive care unit	1 (7)	12 (20)	
Intermediate medical care unit	1 (7)	2 (3)	
Cardiac intensive care unit	0	4 (7)	
Surgical intensive care unit	0	4 (7)	
Unknown	0	10 (16)	

Note.—Values presented as means ± SD where applicable. Values in parentheses are percentages.
*Among seven of 15 patients (47%) with oliguria.
†Among 15 of 61 patients (25%) with oliguria.

Kidney Foundation Kidney Disease Outcomes Quality Initiative recommendations for hemodialysis access in acute kidney injury (the preferred term rather than "acute renal failure") state that tunneled hemodialysis catheters (TDCs) should be used for hemodialysis durations greater than 1 week and that conversion of NTDCs to TDCs should be performed if the duration of hemodialysis exceeds 1 week. The recommendation to limit NTDC use to short duration stems from an increased risk of infection compared with TDCs and the greater blood flow rates that are achievable with TDCs. The risk of NTDC infection also increases substantially beyond 1 week.

Accordingly, the decision for primary placement of an NTDC or a TDC in acute kidney injury hinges on whether renal function is expected to return in less than 1 week or greater than 1 week. However, few studies have examined potential predictors of renal recovery in patients with acute kidney injury and, consequently, estimating the temporal course of renal recovery remains difficult. Given the uncertainty concerning the duration of renal recovery in this population, this study sought to determine whether primary placement of TDCs in this setting is more appropriate than NTDC placement and whether any predictors of recovery of renal function in less than 1 week could be identified.

A retrospective study was conducted to assess the relationship between potential demographic and clinical predictors of renal recovery and duration of renal recovery in patients with acute kidney injury requiring hemodialysis. In the authors' institution, by convention all patients referred to the interventional radiology service for acute dialysis access receive a TDC in the absence of any contraindications such as uncorrectable coagulopathy. The authors identified 111 patients who received a primary TDC for acute kidney injury and who eventually recovered renal function. Thirty-five of these 111 patients were excluded from the study group for reasons, such as prior placement of TDC, subsequent necessary replacement of a TDC, or incomplete data available for

TABLE 2.—Etiology of Renal Failure by Duration of Renal Recovery*

Etiology	<1 Week ($n = 15$)	≥1 Week ($n = 61$)
Hypovolemia/hypotension	3	6
Drug toxicity	2	2
Heart failure	2	2
Hepatorenal syndrome	1	2
Hypovolemia/hypotension, drug toxicity	1	2
Contrast nephropathy	1	1
Contrast nephropathy, infection, hypovolemia/hypotension	1	0
Infection, hypovolemia/hypotension	1	0
Sepsis, heart failure	1	0
Shock, heart failure, drug toxicity	1	0
Tumor lysis, shock	1	0
Rhabdomyolysis	0	5
Acute kidney injury on chronic renal insufficiency	0	3
Acute kidney injury on chronic renal insufficiency, hypovolemia/hypotension	0	2
Heart failure, hypovolemia/hypotension	0	2
Membranoproliferative glomerulonephritis	0	2
Multiple myeloma	0	2
Rapidly progressive glomerulonephritis	0	2
Thrombotic thrombocytopenic purpura	0	2
Contrast nephropathy, heart failure	0	2
Heart failure, acute kidney injury on chronic renal insufficiency	0	1
Heart failure, hypovolemia/hypotension, contrast nephropathy	0	1
Heart failure, rapidly progressive glomerulonephritis	0	1
Hepatorenal syndrome, drug toxicity	0	1
Hepatorenal syndrome, hypotension/hypovolemia	0	1
HIV nephropathy	0	1
Multiple myeloma, membranoproliferative glomerulonephritis, hepatorenal syndrome	0	1
After renal obstruction	0	1
Sepsis	0	1
Sepsis, drug toxicity	0	1
Sepsis, hypovolemia/hypotension	0	1
Systemic lupus erythematosus nephritis	0	1
Tumor lysis	0	1
Unknown	0	10

*Analysis of variance yields a global *P* value of .48, which is not significant, so there are no associations for any etiologies. Although no association between etiology and recovery time was found, this should be interpreted with caution because the study is not adequately powered.

review. The final study population therefore consisted of 76 patients. Charts were reviewed to determine when the decision was made for TDC placement. Data for potential predictors of renal recovery were then abstracted from patient charts, including serum creatinine level, 24-hour urine output, oliguria duration if applicable(defined as <400 mL in 24 hours), etiology of renal failure, active comorbidities, age, sex, and patient location within the hospital. The dependent variable of interest, TDC dwell time, was calculated according to the date the TDC was placed and the date the TDC was removed in the interventional radiology unit. A renal attending physician made the decision when renal function had recovered and therefore when the TDC for hemodialysis should be removed in all cases.

There were no insertion complications in the study. Mean TDC dwell time in patients who eventually recovered from acute kidney injury was 34 days; only 15 of 76 (20%) recovered within 1 week. Thirty-three of 76 patients (43%) recovered within 2 weeks, 45 (59%) within 3 weeks, and 56 (74%) within 4 weeks. The remaining 20 patients (26%) recovered in longer than 4 weeks. The characteristic of the patients by the duration of renal recovery is detailed in Table 1. The distribution of etiologies of renal failure in patients who recovered from acute renal injury in less than 1 week was not significantly different from that in patients whose recovery took longer than 1 week (Table 2). Of note, 8 of 15 patients (53.3%) who recovered in less than 1 week had underlying chronic kidney disease per their patient chart, similar to the 22 of 52 patients (42.3%) who recovered in greater than 1 week ($P = .45$).

One prospective study[1] has shown that neither patient characteristics (eg, age, sex, comorbid conditions), severity of illness (eg, Acute Physiology and Chronic Health Evaluation III status, number of failed organs), nor mode and duration of renal replacement therapy were related to recovery of renal function. Similarly, the authors were unable to identify any predictors of renal recovery in thist study, possibly as a result of insufficient power to detect statistically significant differences between patients whose renal function recovered in less than 1 week versus longer than 1 week. They state that they believe the benefit to the 80% of patients who needed TDCs in this study (ie, avoiding multiple procedures and catheter changes and thereby increased risk of infection and thrombosis) outweighs the increased cost to the 20% who could have received NTDCs in retrospect. The authors feel that these results argue for primary placement of TDCs in all patients with acute kidney injury who require hemodialysis and in whom no contraindications exist, as we were unable to identify any predictors of recovery of renal function in less than 1 week. This suggests that revision of current Kidney Disease Outcomes

Quality Initiative recommendations may be warranted.

T. G. Walker, MD

Reference

1. Schiffl H. Renal recovery from acute tubular necrosis requiring renal replacement therapy: a prospective study in critically ill patients. *Nephrol Dial Transplant.* 2006;21:1248-1252.

7 Neuroradiology

Introduction

It's hard to believe that another year has passed and that we have completed another volume of the Year Book of Diagnostic Radiology. Just for fun, I had a look at what I selected for last year's (2009) volume to see what had changed. For one thing, I selected a lot more head and neck articles. Why? There were some significant developments in our understanding of imaging disorders in this small but critical anatomic area. For example, one documents the revelation that maxillary mucosal cysts are not a manifestation of rhinosinusitis. Yep! That was a new one on me ... and I just heard one of our head and neck surgeons give a lecture in which he pleaded with our neuroradiology colleagues, "Please don't say in your report 'maxillary mucosal cyst versus polyp' as the former isn't important. The latter is!" 'Nuff said. Another key article was on the sphenoid sinus septations and their relationship with internal carotid arteries, something we should be paying special attention to because they have real implications for surgery.

Stroke, as always, was a big topic this year. We've commented before on the fluid attenuated inversion recovery (FLAIR) hyperintense vessels as a sign of stroke. A very nice article just published showed that these "dot signs" in the distal vessels actually correlate with collateral circulation, not clot. Other insights include how to predict the uncommon but potentially life-threatening complication of cerebral hyperperfusion syndrome after carotid endarterectomy. Finally, we learned this year that reperfusion is a more accurate predictor of follow-up infarct volume than recanalization in acute stroke.

I didn't select a lot of articles on brain neoplasms this year. But a significant article on pineal cysts was published and worth reviewing. These cysts, often identified as incidental findings on MR scans, pose a dilemma for both radiologists and clinicians alike. With the identification of a new pineal parenchymal tumor (intermediate in differentiation between pineocytoma and pineoblastoma), the findings of a cystic mass in the pineal gland become more important. We are publishing an article on these PPTIDs (pineal parenchymal tumors of intermediate differentiation), and I'll give you a preview: with one exception, they all appeared larger and more aggressive (commonly had local invasion) than pineocytomas or nonneoplastic pineal cysts. So what to do when you find a truly benign, cystic-appearing pineal mass? In children and young adults, they rarely change over time—and the same is probably true for middle-aged and older adults.

There are a number of very nice articles covering a spectrum of disease from treatment of lumbar synovial cysts to distinguishing between abscess and necrotic brain tumor on 3T imaging. A little bit of this, a little bit of that—all interesting, all useful. Sit back, read, and enjoy!

Anne G. Osborn, MD

Brain: Atherosclerosis, Carotid Stenosis

Prediction of Cerebral Hyperperfusion after Carotid Endarterectomy Using Middle Cerebral Artery Signal Intensity in Preoperative Single-Slab 3-Dimensional Time-of-Flight Magnetic Resonance Angiography

Kuroda H, Ogasawara K, Hirooka R, et al (Iwate Med Univ, Morioka, Japan)
Neurosurgery 64:1065-1071, 2009

Objective.—Cerebral hyperperfusion after carotid endarterectomy (CEA) occurs in patients with preoperative impairments in cerebral hemodynamics. Signal intensity of the middle cerebral artery (MCA) on single-slab 3-dimensional time-of-flight magnetic resonance angiography (MRA) can assess hemodynamic impairment in the cerebral hemisphere. The purpose of the present study was to determine whether the signal intensity of the MCA on preoperative MRA could identify patients at risk for cerebral hyperperfusion after CEA.

Methods.—The signal intensity of the MCA ipsilateral to CEA on preoperative MRA was graded according to the ability to visualize the MCA in 81 patients with ipsilateral internal carotid artery stenosis (≥70%). Cerebral blood flow was also quantified using single-photon emission computed tomography before and immediately after CEA and on the third postoperative day.

Results.—Cerebral hyperperfusion immediately after CEA (cerebral blood flow increase ≥100% compared with preoperative values) was observed in 10 patients. Multivariate analysis revealed that only reduced signal intensity of the MCA was significantly associated with the development of postoperative cerebral hyperperfusion (95% confidence interval, 1.015–1.401; $P = 0.0319$). When the reduced signal intensity of the MCA on MRA was defined as an impairment in cerebral hemodynamics, MRA grading resulted in 100% sensitivity and 63% specificity, with a 28% positive predictive value and a 100% negative predictive value for the development of post-CEA hyperperfusion. Hyperperfusion syndrome developed on the fourth and sixth postoperative days in 2 of the 10 patients who exhibited hyperperfusion immediately after CEA.

Conclusion.—Signal intensity of the MCA, as assessed by this simple MRA method, may identify patients at risk for post-CEA cerebral hyperperfusion.

▶ Cerebral hyperperfusion syndrome (CHS) is an uncommon but dreaded complication of carotid reperfusion procedures (it does occur in other settings, such as status epilepticus, for example). It is defined as a major increase in ipsilateral cerebral blood flow (CBF) well above normal metabolic demands. Carotid endarterectomy (CEA), angioplasty, stenting, and occasionally even thrombolysis can cause CHS. Estimates are that between 5% and 10% of CEA patients may develop CHS.

As CEA is a common procedure, it would be very helpful if there were a way to predict which patients are at risk for CHS. The authors took a group of 81 patients with hemodynamically significant carotid stenosis (defined in the NASCET trials as ≥70%) and evaluated both signal intensity in the middle cerebral artery (MCA) and CBF. Interestingly, it was reduced MCA signal intensity that was most strongly associated with CHS (Fig 1 in the original article). This is an important finding, as it is an easily obtainable clue that a patient is "at risk" for developing CHS. Most patients have preoperative magnetic resonance imaging (MRIs) and adding magnetic resonance angiography (MRA), and then evaluating MCA signal intensity is much easier and simpler (as the authors note) than doing perfusion MR (which some institutions do not do).

Another option—something that we use—is CTA with computed tomography (CT) perfusion. Works very nicely. You can see the increased intravascular contrast in the MCA easily and equally nicely identify increased CBF. It is particularly helpful if you have a preoperative CT/CTA with which the post-endarterectomy study can be compared.

In a related study published about the same time, another group of authors noted that delayed/prolonged mean transit time (MTT) of more than 3 seconds was associated with occurrence of CHS.[1]

A. G. Osborn, MD

Reference

1. Tsent Y-C, Hsu HL, Lee TH, Hsieh IC, Chen CJ. Prediction of cerebral hyperperfusion syndrome after carotid stenting: a cerebral perfusion computed tomography study. *J Comput Assist Tomogr.* 2009;33:540-545.

Distinction between pyogenic brain abscess and necrotic brain tumour using 3-tesla MR spectroscopy, diffusion and perfusion imaging

Chiang I-C, Hsieh T-J, Chiu M-L, et al (Kaohsiung Municipal Hsiao-Kang Hosp, Taiwan; Kaohsiung Med Univ, Taiwan)

Br J Radiol 82:813-820, 2009

The purpose of this study is to compare the effectiveness of relative cerebral blood volume, apparent diffusion coefficient and spectroscopic imaging in differentiating between cerebral abscesses and necrotic tumours. In the prospective study, a 3-tesla MR unit was used to perform proton MR spectroscopy, diffusion and perfusion imaging in 20 patients with cerebral abscesses and 26 patients who had solitary brain tumours (14 high-grade gliomas and 12 metastases). We found the mean apparent diffusion coefficient value at the central cavities of the cerebral abscesses to be significantly lower than in necrotic tumours. The mean relative cerebral blood volume values of the necrotic tumour wall were statistically significantly higher than the mean relative cerebral blood volume values of the cerebral abscess wall by the Student's *t*-test. The proton spectra obtained revealed amino acids only in the cerebral abscesses. Although the conventional MRI characteristics of cerebral abscesses and necrotic tumours may sometimes be similar, diffusion, perfusion-weighted and spectroscopic MRI enables distinction between the two.

▶ The differential diagnosis of ring-enhancing parenchymal lesions on CT and MR is extensive. While trainees memorize the mnemonic "MAGICAL DOCTOR" to remind themselves of some of the many entities (metastasis, abscess, glioma...demyelinating disease, resolving hematoma, etc) that can present with this finding, the fact is that the major differential diagnosis is infection (abscess) versus necrotic neoplasm (glioblastoma or metastasis) (Fig 1a,c in the original article).

Conventional contrast-enhanced MR—whether obtained at 1.5 or 3 T—is not only highly sensitive in detecting brain lesions but also quite nonspecific. Adding diffusion scans is helpful (abscesses generally restrict while neoplasms generally do not), but in our experience metastases sometimes restrict too (as do highly cellular neoplasms). Looking at apparent diffusion coefficient (ADC) values as well is helpful; the ADCs within abscesses are generally lower than necrotic tumors.

Adding spectroscopy to our diagnostic armementarium is very helpful, as the authors demonstrate. Detecting amino acid spectra is very useful and indicates abscess (Fig 3 in the original article). Perfusion scans show that mean relative cerebral blood volumes (rCBVs) of tumor walls are significantly higher than the rim surrounding abscess cavities (Fig 1b,d in the original article). Cool.

A. G. Osborn, MD

Incidental MRI anomalies suggestive of multiple sclerosis: The radiologically isolated syndrome

Okuda DT, Mowry EM, Beheshtian A, et al (Univ of California, San Francisco)
Neurology 72:800-805, 2009

Background.—The discovery and broad application of MRI in medicine has led to an increased awareness in the number of patients with incidental white matter pathology in the CNS. Routinely encountered in clinical practice, the natural history or evolution of such individuals with respect to their risk of developing multiple sclerosis (MS) is unclear.

Objective.—To investigate the natural history of patients who exhibit incidental imaging findings highly suggestive of MS pathology.

Methods.—Detailed clinical and radiologic data were obtained from asymptomatic patients with MRI anomalies suggestive of MS.

Results.—The cohort consisted of 41 female and 3 male subjects (median age = 38.5, range: 16.2–67.1). Clinical evaluations were performed in 44 patients at the time of initial imaging; longitudinal clinical follow-up occurred for 30 patients, and longitudinal MRI data were acquired for 41 patients. Neurologic examination at the time of the initial MRI scans was normal in nearly all cases. While radiologic progression was identified in 59% of cases, only 10 patients converted to either clinically isolated syndrome or definite MS. The presence of contrast-enhancing lesions on the initial MRI was predictive of dissemination in time on repeat imaging of the brain (hazard ratio [HR] = 3.4, 95% confidence interval [1.3, 8.7], $p = 0.01$).

Conclusion.—Individuals with MRI anomalies highly suggestive of demyelinating pathology, not better accounted for by another disease process, are very likely to experience subsequent radiologic or clinical events related to multiple sclerosis. Additional studies will be necessary to fully define this risk.

► This was one of my favorite articles of the whole year. Why? We all struggle with the imaging findings of multifocal T2/fluid-attenuated inversion recovery (FLAIR) hyperintensities in the brain (sometimes called "unknown bright objects" or "UBOs"). There are literally dozens of entities that can cause these findings, including normal aging (the conventional wisdom has been one spot per decade, which I do not personally use).

Multiple sclerosis (MS), one of the causes of "UBOs" in the brain, is usually imaged in patients who are suspected of harboring the disease. But what about "MS-like" lesions in patients who have no clinical evidence for demyelinating disease at the time of imaging?

TABLE 1.—Proposed Diagnostic Criteria for the Radiologically Isolated Syndrome

A. The presence of incidentally identified CNS white matter anomalies meeting the following MRI criteria:
1. Ovoid, well-circumscribed, and homogeneous foci with or without involvement of the corpus callosum
2. T2 hyperintensities measuring >3 mm and fulfilling Barkhof[7] criteria (at least 3 out of 4) for dissemination in space
3. CNS white matter anomalies not consistent with a vascular pattern
B. No historical accounts of remitting clinical symptoms consistent with neurologic dysfunction
C. The MRI anomalies do not account for clinically apparent impairments in social, occupational, or generalized areas of functioning
D. The MRI anomalies are not due to the direct physiologic effects of substances (recreational drug abuse, toxic exposure) or a medical condition
E. Exclusion of individuals with MRI phenotypes suggestive of leukoaraiosis or extensive white matter pathology lacking involvement of the corpus callosum
F. The CNS MRI anomalies are not better accounted for by another disease process

The authors present 44 patients (90% female) who had MR anomalies highly suggestive of demyelinating pathology and who had normal neurologic examinations at the time of initial imaging. Remember the old saying, "If it quacks and walks like a duck, it probably is a duck." Turns out it is the same here. If it looks like an MS lesion (Table 1), it probably will turn out to be MS. This new form of MS, the radiologically isolated syndrome, is important for radiologists to recognize and for clinicians to follow-up.

The presence of contrast-enhancing lesions on the initial MR was especially ominous (hazard ratio of 3.4) (Fig 1 in the original article).

A. G. Osborn, MD

MR imaging of hypoglycemic encephalopathy: lesion distribution and prognosis prediction by diffusion-weighted imaging

Ma J-H, Kim Y-J, Yoo W-J, et al (Catholic Univ of Korea, South Korea; et al)

Neuroradiology 51:641-649, 2009

Introduction.—The aim of this study was to evaluate the patterns of hypoglycemic encephalopathy on diffusion-weighted imaging (DWI) and the relationship between the imaging patterns and clinical outcomes.

Methods.—This retrospective study included 17 consecutive patients that had hypoglycemic encephalopathy with DWI abnormalities. The topographic distributions of the DWI abnormalities of the cortex, deep gray matter, and white matter structures were assessed. In addition, possible correlation between the patterns of brain injury on DWI and clinical outcomes was investigated.

Results.—There were three patterns of DWI abnormalities: involvement of both gray and white matter ($n = 8$), selective involvement of gray matter ($n = 4$), and selective involvement of white matter ($n = 5$). There was no significant difference in the initial blood glucose levels among patients for each of the imaging patterns. Most patients (16/17) had bilateral symmetrical abnormalities. Among patients with bilateral symmetrical gray and/or white matter injuries, one had moderate to severe disability and 14 remained in a persistent vegetative state. The two patients with a focal unilateral white matter abnormality and a localized splenial abnormality recovered without neurological deficits.

Conclusion.—The results of this study showed that white matter was more sensitive to hypoglycemia than previously thought and there was no specific association between the patterns of injury and clinical outcomes whether the cerebral cortex, deep gray matter, and/or white matter were affected. Diffuse and extensive injury observed on the DWI predicts a poor neurologic outcome in patients with hypoglycemic injuries.

▶ Hypoglycemia is an uncommon but potentially serious metabolic encephalopathy. It basically results from an imbalance between glucose supply and utilization that results in brain injury. The release of excitatory neurotransmitters increases glucose utilization even more, often resulting in patchy or diffuse laminar cortical necrosis.

The classic teaching in adult hypoglycemia is that the parieto-occipital cortex is most commonly involved, with the basal ganglia (especially the putamina) also frequently affected; the thalami and white matter are classically spared. When standard magnetic resonance (MR) sequences are used, that seems to hold true in most cases. However, if diffusion-weighted imaging (DWI) is added, that changes.

The authors identified 3 patterns of DWI abnormalities, the most common of which were involvement of both gray and white matter (Fig 1 in the original article) or gray matter alone. However, 5 of their 17 patients demonstrated selective white matter involvement. They showed 1 case (Fig 3 in the original article) where there was striking involvement of the deep white matter with complete sparing of the cortex and deep gray nuclei.

Bottom line: when a patient comes in with coma and seizures of unknown etiology and MR scan shows selective white matter involvement, we should suspect not just toxic disease but hypoglycemia as well. With the near-epidemic rise of adult type 2 diabetes throughout the world, we are likely to see more cases of hypoglycemic encephalopathy, including the "white matter only" variant that the authors describe.

A. G. Osborn, MD

MRI lesion profiles in sporadic Creutzfeldt–Jakob disease

Meissner B, Kallenberg K, Sanchez-Juan P, et al (Georg-August Univ Goettingen, Germany; Fundación "Marqués de Valdecilla" IFIMAV and Centro de Investigación Biomédica en Red sobre Enfermedades Neurodegenerativas, Santander, Spain; et al)
Neurology 72:1994-2001, 2009

Background.—With respect to sporadic Creutzfeldt–Jakob disease (sCJD), six molecular subtypes (MM1, MM2, MV1, MV2, VV1, and VV2) have been described, which vary with respect to age at disease onset, disease duration, early symptoms, and neuropathology. MRI signal alterations were reported to correlate with distinct Creutzfeldt–Jakob disease (CJD) subtypes. This multicenter, international study aimed to describe the brain MRI findings associated with each of the sCJD molecular subtypes.

Methods.—Pathologically confirmed sCJD cases with codon 129 genotype (MM, MV, and VV), PrP^{Sc} type, and fluid-attenuated inversion recovery (FLAIR) or diffusion-weighted imaging (DWI) were collected in seven countries. All MRI scans were assessed for signal changes according to a standard protocol encompassing seven cortical regions, basal ganglia, thalamus, and cerebellum.

Results.—MRI scans were evaluated in 211 CJD patients (98 MM1, 23 MM2, 19 MV1, 30 MV2, 9 VV1, and 32 VV2). Basal ganglia hyperintensities occurred most frequently in MV2, VV2, and MM1 subtypes (79, 77, and 70%). Wide cerebral cortical signal increase was most common in VV1, MM2, and MV1 subtypes (86, 77, and 77%). Thalamic hyperintensities occurred most often in VV2 (45%) and MV2 (43%). The most consistent finding across most subtypes was high signal in basal ganglia, with these abnormalities found in 63% (FLAIR) and 71% (DWI).

Conclusion.—Cortical signal increase and hyperintensities in the basal ganglia and thalamus are detected by MRI across all molecular sporadic Creutzfeldt–Jakob disease subtypes. Our findings argue that characteristic MRI lesion patterns may occur for each molecular subtype.

▶ Sporadic Creutzfeldt-Jakob disease (sCJD) is a rare but feared fatal spongiform encephalopathy. Several molecular subtypes of sCJD exist. This interesting article was designed to see if imaging findings (MRI "lesion profiles") were correlated with disease subtype.

By now, most radiologists are familiar with imaging manifestations of sCJD: bilateral hyperintensity (especially on fluid-attenuated inversion recovery [FLAIR]) in the anterior basal ganglia. Less commonly recognized is the cortical hyperintensity that may be comparatively subtle. Thalamic hyperintensities are also less common. Turns out that basal ganglia hyperintensity is pretty consistent and found across all but one molecular subtype of CJD (Fig 1 in the original article). Of particular interest was the authors' finding that the one molecular subtype of CJD that generally demonstrated an absence of basal ganglia signal abnormalities (type VV1) also showed widespread cortical lesions (found in more than three cortical areas). Thalamic lesions ("pulvinar sign") were identified only in the MV2 subtype.

Bottom line: Basal ganglia and cortical hyperintensities (limited OR widespread) are the most common MR findings in sCJD and are typically found in MM1 subtype patients. These seem to have the most rapid disease course. VV1 and MM2 subtypes were most often associated with cortical disease.

A. G. Osborn, MD

Brain: Miscellaneous

Rathke's cleft cyst: clinicopathological and MRI findings in 22 patients

Wen L, Hu L-b, Feng X-y, et al (Third Military Med Univ, ChongQing, P.R. China; The Second People's Hosp of ChongQing, China; Med Ctr of FuDan Univ, ShangHai, P.R. China; et al)

Clin Radiol 65:47-55, 2010

Aim.—To evaluate clinical findings and magnetic resonance imaging (MRI) characteristics of Rathke's cleft cyst (RCC) in 22 patients.

Materials and Methods.—Twenty-two patients were imaged using non-enhanced MRI and 17 underwent an additional contrast-enhanced MRI examination. Fifteen patients received an additional non-enhanced computed tomography (CT) examination, and amongst these, two underwent contrast-enhanced CT. Two radiologists read the images retrospectively. The imaging data were studied with regards to location, size, margin, signal intensity, enhancement characteristics, haemorrhage, and presence of calcifications. Clinical data, such as presenting signs and symptoms, physical findings, and medical histories, were collected. Histopathological studies were performed and analysed by two pathologists.

Results.—Nine lesions were located in the intrasellar region, 12 in both the intra- and suprasellar regions and one in the suprasellar region. The maximum diameter of the RCCs varied from 0.7 to 4 cm, with an average size of 1.7 ± 0.7 cm. MRI features of RCC were divided into three groups based on T1-weighted imaging (T1WI): hypo- ($n = 6$), iso- ($n = 9$), and hyperintensity group ($n = 7$). Patients in the latter two groups were statistically younger than that in the former group. The lesion size in the iso- and hyperintensity groups was significantly less than that in the hypointensity group ($F = 6.421$, $p = 0.007$). Only two cases showed enhancement after contrast injection in the cohort. One lesion with haemorrhage was found as were two cases with intracystic nodules.

Conclusion.—Although MRI features of RCCs are variable, RCCs should be suspected when the following conditions occur: lesions located in the intrasellar region or involving both intra and suprasellar regions, less than 1.5 cm in diameter, iso- or hyperdense on T1WI and no signal

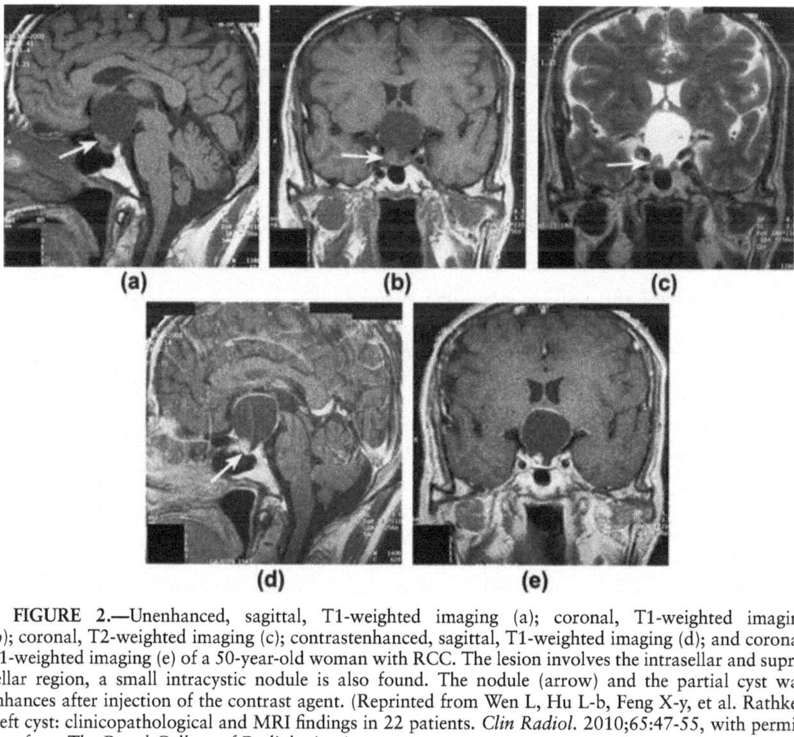

FIGURE 2.—Unenhanced, sagittal, T1-weighted imaging (a); coronal, T1-weighted imaging (b); coronal, T2-weighted imaging (c); contrastenhanced, sagittal, T1-weighted imaging (d); and coronal, T1-weighted imaging (e) of a 50-year-old woman with RCC. The lesion involves the intrasellar and suprasellar region, a small intracystic nodule is also found. The nodule (arrow) and the partial cyst wall enhances after injection of the contrast agent. (Reprinted from Wen L, Hu L-b, Feng X-y, et al. Rathke's cleft cyst: clinicopathological and MRI findings in 22 patients. *Clin Radiol.* 2010;65:47-55, with permission from The Royal College of Radiologists.)

enhancement after contrast injection. In addition, the first case of a RCC with a markedly enhanced intracystic nodule is reported.

▶ We know about Rathke's cleft cysts (RCCs). Most folks who deal with brain imaging have certainly seen at least a few cases. But have we diagnosed them preoperatively? If the experience of this group is any indication, the answer probably is: Not!! The authors report 22 proven cases of RCCs, only 4 of which (18%) were correctly diagnosed before surgery. The rest were called pituitary macro- or microadenomas, arachnoid cysts, or craniopharyngiomas. Most presented with headache followed by visual impairment.

This should give us all pause. We know that all cysts (not just RCCs) can have widely varying signal intensity on magnetic resonance (MR) depending on their contents. If an intracystic nodule is present, that helps establish the diagnosis of RCC. So does a "claw sign" (thin rim of enhanced, compressed pituitary tissue surrounding the cyst).

In my experience, we have become better at diagnosing RCCs over the years. However, we have found odd things: RCCs can undergo "cyst apoplexy" and hemorrhage, mimicking pituitary apoplexy. The authors noted this too. Calcification is rare (helping distinguish RCC from craniopharyngioma). Even rarer is something they saw that I have not: An intracystic nodule that enhanced on MR (Fig 2).

What can help improve our diagnostic accuracy? The authors suggest dynamic MR, which I have found helpful as RCCs themselves do not enhance, even slowly. But preoperative diagnosis of RCC versus its "mimics" is still a challenge.

A. G. Osborn, MD

Serial Follow-Up MRI of Indeterminate Cystic Lesions of the Pineal Region: Experience at a Rural Tertiary Care Referral Center

Cauley KA, Linnell GJ, Braff SP, et al (Univ of Massachusetts Med Ctr, Worcester; Med Ctr Campus, Burlington, VT)

AJR Am J Roentgenol 193:533-537, 2009

Objective.—The purpose of this study was to retrospectively assess the outcomes of temporal MRI follow-up of indeterminate cystic lesions of the pineal region.

Materials and Methods.—Cases of indeterminate pineal lesions were identified by a computerized search of radiology reports at our institution from 1998 to 2007. Twenty-six indeterminate pineal lesions, one each in 26 patients (six males and 20 females), were followed in the current study. Cases were included in the study if the radiology report detailed an incidentally noted but indeterminate or worrisome cystic lesion of the pineal region and if temporal follow-up MRI was recommended by the interpreting neuroradiologist. For a case to be included in our data set, a follow-up MRI study must have been obtained at least 6 months after the initial study.

Results.—Follow-up imaging ranged from 7 months to 8 years. Lesions ranged in description and appearance from "probable benign cyst" with typical cystic imaging features to "pineal mass" with features of solid enhancement. We found that, within the limitations of the technique, there were no significant changes in the size or character of any indeterminate pineal region lesion followed at our institution over the time interval of 1998–2007.

Conclusion.—In our study, all pineal lesions found incidentally and for which follow-up imaging had been recommended were stable over time. These data suggest that neoplastic lesions of the pineal region either are too rare or grow too slowly to be identified by traditional temporal imaging follow-up. With increasing interest in optimal allocation of health care resources, our findings suggest that incidentally identified pineal region cystic lesions, both typical and atypical, can be followed clinically.

▶ Oh, those pesky pineals! With good-quality magnetic resonance (MR) imaging, pineal cysts (aka, cystic-appearing pineal glands) are seen in up to 20% to 25% of normal MR scans. As the authors note, "typical" pineal cysts generally do not pose a significant diagnostic dilemma, but more complex-appearing lesions can appear similar to pineocytoma (I personally do not think germinoma is nearly the diagnostic problem that pineocytoma is). So, to "play it safe," many radiologists recommend that pineal lesions—especially those with more complex imaging features like hemorrhage or nodular enhancement (Fig 4 in the original article)—be followed up.

In this study, the authors followed up what they called "indeterminate" cystic lesions (the potentially more worrisome part of the pineal "incidentaloma" imaging spectrum) and found that they all were stable over time, concluding that neoplastic lesions of the pineal region are either too rare or grow too slowly to be differentiated by traditional imaging follow-up. They concluded that both "typical" and "atypical" cystic pineal lesions can be followed clinically, being reimaged only if the patient developed symptoms referable to the lesion.

They did discuss the pediatric subpopulation and their small numbers(only 5 patients under 16 years), stating that the frequency of pineal tumors is thought to be higher in children. In another article I reviewed, the pediatric and young adult population is addressed—with the same conclusion: these lesions are generally stable over time and clinical (not imaging) follow-up is sufficient.

A. G. Osborn, MD

The natural history of pineal cysts in children and young adults

Al-Holou WN, Maher CO, Muraszko KM, et al (Univ of Michigan, Ann Arbor)

J Neurosurg Pediatr 5:162-166, 2010

Object.—The authors reviewed their experience with pediatric pineal cysts to define the natural history and clinical relevance of this common intracranial finding.

Methods.—The authors identified all patients with pineal cysts who had been clinically evaluated at their institution over an 11.5-year interval and were < 25 years of age at the time of diagnosis. All inclusion criteria were met in 106 patients, and included repeated MR imaging as well as repeated clinical evaluation over at least a 6-month interval.

Results.—The mean age at diagnosis was 11.7 ± 7.2 years. Forty-two patients were male and 64 were female. On follow-up MR imaging evaluation at a mean interval of 3.0 years from the initial study, 98 pineal cysts had no increase in size and no change in imaging appearance. Six pineal cysts increased in size and 2 others had a change in imaging characteristics without associated growth. Younger age was associated with cyst change or growth on follow-up imaging ($p = 0.02$). The mean age of patients with cysts that changed or grew was 5.5 years, and the mean age of patients with stable pineal cysts was 12.2 years. Initial cyst size and appearance on MR imaging were not significant predictors of growth or change in imaging appearance at follow-up. Similarly, the patient's sex was not a significant predictor of growth or change in imaging characteristics.

Conclusions.—Follow-up imaging and neurosurgical evaluation may be considered optional in older children with pineal cysts.

▶ Cystic-appearing pineal glands are common findings on imaging studies. Up to 10% of routine magnetic resonance (MR) scans show 'pineal cysts' on careful examination, and they are found in 20% to 40% of autopsy cases. Most (80%) are less than 1 cm in diameter and most are asymptomatic, found incidentally at imaging. Most are hyperintense to cerebrospinal fluid (CSF) on T1-weighted images (T1WIs) and do not suppress on fluid-attenuated inversion recovery (FLAIR). Rim (less commonly nodular) enhancement can be seen. Although cystic-appearing pineal glands are common in adults, their prevalence in children is less well defined. The authors of this article previously reported approximately 2% prevalence of pineal cysts identified on imaging studies in children. Whether they are found in children or adults, they are often of concern to radiologists and referring clinicians and may be referred to neurosurgeons or—perhaps more commonly—subjected to repeated imaging over time to distinguish them from pineocytoma. This nice article followed a large group of young patients (all less than 25 years) with pineal cysts. Over 90% (98 of 106 patients) had no change in size or imaging appearance at a mean interval of 3 years following initial imaging. Perhaps most important was their finding that younger age was associated with change in appearance. Cysts with "atypical" imaging were no more likely to increase in size or appearance than classic-appearing pineal cysts. They reported one case of surgically-proven pineocytoma, a WHO grade I pineal parenchymal tumor that typically grows very slowly. In a related article that appeared in the same journal describing internal structure in pineal cysts on high-resolution MR,[1] the authors found that although the presence of a thin wall supports the diagnosis of "benign pineal cyst," fine internal septations or small internal cysts in pineal glands are common and are not an evidence for underlying neoplasm. The newly-described "pineal parenchymal tumor of intermediate differentiation" or PPTID, a WHO grade II or III neoplasm intermediate in malignancy between pineocytoma and pineoblastoma may account for the so-called "atypical pineocytomas" that in the past seemed to follow a modestly more aggressive pattern. In our experience, most PPTIDs occur in middle-aged and older patients. The authors' conclusion, that is, early follow-up imaging and neurosurgical evaluation should be considered optional (not mandatory), especially in older children, seems reasonable. Repeat imaging at yearly intervals generally will not find any interval change in these lesions whether they are in children or adults.

A. G. Osborn, MD

Reference

1. Pastel DA, Mamourian AC, Duhaime AC. Internal structure in pineal cysts on high-resolution magnetic resonance imaging: not a sign of malignancy. *J Neurosurg Pediatrics*. 2009;4:81-84.

Utilization, Diagnosis, Treatment and Cost of Migraine Treatment in the Emergency Department

Friedman D, Feldon S, Holloway R, et al (Univ of Rochester – Neurology, NY; Univ of Rochester – Ophthalmology, NY; et al)
Headache 49:1163-1173, 2009

Objective.—To determine the percentages of patients receiving migraine-specific therapy and to estimate the rate of unnecessary neuroimaging studies in the emergency department (ED).

Methods.—A retrospective study was conducted analyzing medical records and hospital charge data of ED visits for migraine during 2005 in 2 university-affiliated hospitals. Following a preliminary review of 23 randomly selected ED charts selected to determine the reliability of the coding process, 172 other charts were selected to include 1 visit per patient with a primary discharge diagnosis code of 346.0, 346.1, or 346.9. The diagnosis of migraine was confirmed using predefined criteria. Demographic information, treatment strategies, laboratory and neuroimaging tests, response to therapy, discharge planning, and charge data were evaluated.

Results.—Of 156 patients with completed visits, neuroimaging studies were performed in 36 patients (23%), and only 4 patients had no documented justification for obtaining imaging studies. Seventy-eight patients (50%) had a potential contraindication to receiving migraine-specific therapy. Nine patients (11.5% of eligible patients) received migraine-specific therapy. Most patients were treated with a combination of parenteral antiemetics, narcotics, or ketorolac.

Conclusion.—This analysis supports previous studies indicating the underutilization of migraine-specific treatment in the ED, and suggests that the ED is generally used as a "last resort" when the patient's home medication fails. Because of various contraindications, migraine-specific medications may not be a treatment option in up to 50% of patients seen in the ED. Although almost all of the neuroimaging studies were justified, the radiology charges were a major contributing factor to the overall financial burden of emergency migraine care.

▶ This article comes from a journal (*Headache*) that few radiologists read—but perhaps more (including myself) should. Why? Headache is a pretty common presenting complaint in emergency rooms (ERs). In ours, patients with headache are triaged to radiology for non-contrast head CT scan—even before they have seen a physician. This article states that "acute headache" accounts for up to 4% of all ER visits and that up to 60% may suffer from migraine.

The article also concludes that although almost all of the neuroimaging studies in these patients were "justified," radiology charges are a major contributing factor to the overall financial burden (which is not insignificant) of emergency migraine care. In this article, imaging was the single largest cost associated with ER migraine treatment.

So what does "justified" mean? By their a priori definition, it means first or worst headache, change in previous pattern of migraine, headeaches increasing in frequency or severity, known diagnosis of cancer or HIV/AIDS, and so forth. It was "inappropriate" if the ER physician's differential diagnosis suggested migraine with normal examination, tension-type headache, sinus headache, and so forth. But what if the ER doctor hasn't even seen the patient yet? NECT scans are used as a screening study to improve ER patient throughput. Is that appropriate? It doesn't certainly seem to be good medicine to me.

In this study, 23% of the migraine patients had neuroimaging studies performed; 89% were considered "justified." So how many were actually abnormal? Turns out: None. Neuroimaging results did not change the management of any patients in the study cohort.

A. G. Osborn, MD

Brain: Neoplasms

White matter tract involvement in brain tumors: a diffusion tensor imaging analysis

Yen PS, Teo BT, Chiu CH, et al (Buddhist Tzu Chi General Hosp and Tzu Chi Univ, Hualien, Taiwan)

Surg Neurol 72:464-469, 2009

Background.—Characterization of WM alteration using MR imaging is important in the pre- and intraoperative assessment of brain tumors.

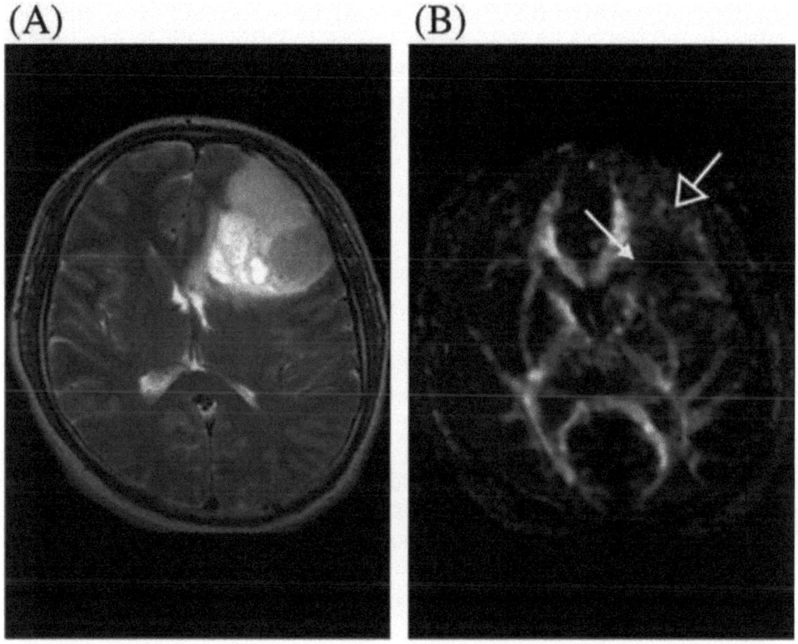

FIGURE 1.—Diffusion tensor imaging directionally encoded color map in a 34-year-old patient who presented with generalized seizure. The T2-weighted image (A) revealed a left frontal mass with marked perifocal hyperintensity. The DTI directionally encoded color map (B) showed markedly decreased hue of the anterior limb of the internal capsule (AL) (solid arrow), although an infiltrative or destructive process cannot be differentiated. The hollow arrow may represent displaced fiber and the center of the lesion may represent disruption. (Reprinted from Yen PS, Teo BT, Chiu CH, et al. White matter tract involvement in brain tumors: a diffusion tensor imaging analysis. *Surg Neurol.* 2009;72:464-469, with permission from Elsevier.)

This study characterizes the extent and severity of WM tract alterations near brain tumors using DTI in an effort to determine preoperative viability or resectability of the adjacent WM tracts. Fractional anisotropy is an important DTI-derived metric of MR imaging.

Methods.—Twenty-one patients underwent MR DTI. Eighty-six WM tracts composed of 43 WM lesions paired with 43 contralateral WM hemispheric controls were categorized using FA. Neuroradiologists categorized the WM tracts as edematous, displaced, disrupted, or infiltrated with tumor using directionally encoded color maps. A mixed model analysis was used to compare FA.

Results.—Of the lesioned tracts, 5 were scored as edema, 9 as infiltration, 18 as displacement, and 11 as disruption. A significant $\Delta FA_{\%}$ was found between the lesioned and contralateral hemispheres only in WM disruption ($P = .0056$). Both edema FA and disruption FA are significantly less than displacement FA ($P < .05$). The FA change ($\Delta FA_{\%} = [FA_{lesion} - FA_{normal}]/FA_{normal} \times 100\%$) on the lesioned side was calculated. A ΔFA% less than −30% is likely to be associated with WM disruption. A positive ΔFA% is likely to be associated with edema or displacement, and a ΔFA% between 0% and −30% is likely to be associated with WM displacement or infiltration.

Conclusions.—Quantitative analysis of DTI data may provide insight as to whether WM tracts are salvageable preoperatively.

► We all know (or should) that diffusely infiltrating astrocytomas (the most common group of primary brain neoplasms) typically involve brain well beyond detectable margins on conventional magnetic resonance (MR) imaging: beyond any enhancement for sure, with viable neoplastic cells found in the "edema" extending peripheral to the enhancing margins. I teach our residents that malignant astrocytomas, especially glioblastoma multiforme, could almost be considered "holobrain" tumors. It does not matter whether or not the margins look well delineated, they are not.

Now diffusion tensor imaging (DTI) comes along. Initially used to evaluate white matter tract displacement around tumors preoperatively, the authors demonstrate that actual infiltration (not just displacement) can be detected using DTI (Fig 1). Whether or not this will alter treatment is questionable as tumor debulking, even if resection is subtotal, is standard neurosurgical practice. But this use of DTI looks promising, especially if it helps achieve what the authors describe as "the delicate balance between maximal resection of tumor tissue and minimal injury to surrounding normal white matter."

A. G. Osborn, MD

Brain: Stroke and Intracranial Hemorrhage

Distal hyperintense vessels on FLAIR: An MRI marker for collateral circulation in acute stroke?

Lee KY, Latour LL, Luby M, et al (Natl Insts of Health, Bethesda, MD; et al)
Neurology 72:1134-1139, 2009

Background.—Hyperintense vessels (HV) on fluid-attenuated inversion recovery imaging are frequently observed in acute ischemic stroke patients. However, the exact mechanism and clinical implications of this sign have not yet been clearly defined. The features of HV and its relevance to other imaging factors are presented here.

Methods.—Prominence and location of HV were documented in 52 consecutive patients with middle cerebral artery (MCA) territory infarction, before treatment with IV recombinant tissue plasminogen activator. Pretreatment ischemic lesion volume, perfusion lesion volume, and vessel occlusion were determined in addition to recanalization status and ischemic lesion volume on follow-up imaging. NIH Stroke Scale (NIHSS) was used as a measure of clinical severity.

Results.—HV distal to arterial occlusion was observed in 73% of patients; more frequent in proximal than distal MCA occlusion patients. Among the 38 patients with proximal MCA occlusion, initial perfusion lesion volume was comparable among patients with different grade distal

HV. However, patients with more prominent distal HV had smaller initial, 24-hour, and subacute ischemic lesion volumes and lower initial NIHSS scores.

Conclusions.—The presence of distal hyperintense vessels before thrombolytic treatment is associated with large diffusion-perfusion mismatch and smaller subacute ischemic lesion volumes in patients with proximal middle cerebral artery occlusion.

▶ Fluid-attenuated inversion recovery (FLAIR) hyperintense vessels in acute cerebral ischemia-infarction can be dramatic and have been termed the "dot sign" (ie, seeing FLAIR hyperintense middle cerebral artery [MCA] branches "end on" in the sylvian fissure). The assumption was that these represented thrombus (less likely) slowed antegrade or retrograde collateral flow across the pial watershed.

The authors found FLAIR hyperintense vessels distal to MCA occlusion in almost 75% of cases, which is higher than what many other investigators have found. We have used FLAIR hyperintense vessels as an ancillary sign of acute ischemia/infarction, but the authors demonstrate that this finding has broader implications, that is, it may indicate a large diffusion-perfusion "mismatch." Sounds good, right? Especially as these patients may be good candidates for thrombolysis. My big question is, does this really make sense? Are nearly three-quarters of patients with proximal MCA occlusions candidates for thrombolysis? Probably not. As the authors admit, the major limitation of this study is the lack of an established association between this finding (distal vessel hyperintensity on FLAIR) and any independent measures of collateral blood flow. The obvious one, digital subtraction angiography, was not studied.

Nevertheless, this study does point out the importance of looking at the vessels themselves—not just the brain parenchyma—on MR scans in patients with suspected acute cerebral ischemia (Fig 2B in the original article).

A. G. Osborn, MD

MRI Detection of New Microbleeds in Patients With Ischemic Stroke: Five-Year Cohort Follow-Up Study

Gregoire SM, Brown MM, Kallis C, et al (UCL Inst of Neurology, London, UK; London School of Hygiene and Tropical Medicine, UK)
Stroke 41:184-186, 2010

Background and Purpose.—Little is known about the development of cerebral microbleeds in patients with ischemic stroke. We studied the incidence of new microbleeds in a cohort of patients with ischemic stroke or transient ischemic attack screened for microbleeds at baseline.

Methods.—Twenty-one surviving patients with ischemic stroke or transient ischemic attack were followed up after a mean interval of 5.5 years with repeat MRI and clinical assessment. Predictors of new microbleeds were tested in logistic regression.

Results.—Of patients with microbleeds at baseline, 50% had new microbleeds at follow-up compared with 8% of those without baseline microbleeds ($P = 0.047$). The presence of microbleeds at baseline predicted new microbleeds (OR, 12; 95% CI, 1.02 to 141.34; $P = 0.048$), as did mean systolic blood pressure (OR, 1.28 per unit increase; 95% CI, 1.23 to 1.33; $P < 0.001$). One patient had a stroke (intracerebral hemorrhage) during follow-up.

Conclusions.—Patients with ischemic stroke or transient ischemic attack are at risk of developing new microbleeds over 5.5 years, despite most surviving patients remaining clinically stable. Systolic blood pressure is the strongest predictor of microbleed development; better control of hypertension may help prevent new microbleed formation.

▶ As the authors note, brain "microbleeds" are increasingly being detected in patients with ischemic stroke when T2* sequences are added to magnetic resonance (MR) scans in this clinical setting. Some authors have considered microbleeds as a contraindication to thrombolysis, although this is controversial.

The 2 most common causes of cerebral "microbleeds" in middle-aged and older adults are chronic hypertensive encephalopathy and amyloid angiopathy. What the authors looked at is longitudinal data on whether systemic hypertension increases the risk of developing new microbleeds. The long and short of it was a firm "yes," especially as baseline systolic blood pressure was above 150 mm Hg in all patients in whom new microbleeds developed.

The bottom line is simply more evident for something we have known for a long time, that is, controlling blood pressure is very important. What I found especially interesting is that although their at-risk group of patients did indeed develop additional new microbleeds at follow-up imaging(Fig 1 in the original article), they were clinically stable. What does that mean? They noted that the only "vascular events" were one intracerebral hemorrhage and one nonfatal myocardial infarction. It is possible—in fact, I think highly likely—that the damage microbleeds do is not gross vascular events, but progressive damage that is more neurocognitive. That was not addressed.

A. G. Osborn, MD

Reperfusion Is a More Accurate Predictor of Follow-Up Infarct Volume Than Recanalization: A Proof of Concept Using CT in Acute Ischemic Stroke Patients

Soares BP, Tong E, Hom J, et al (Univ of California, San Francisco; et al)
Stroke 41:e34-e40, 2010

Background and Purpose.—The purpose of this study was to compare recanalization and reperfusion in terms of their predictive value for imaging outcomes (follow-up infarct volume, infarct growth, salvaged penumbra) and clinical outcome in acute ischemic stroke patients.

Material and Methods.—Twenty-two patients admitted within 6 hours of stroke onset were retrospectively included in this study. These patients underwent a first stroke CT protocol including CT-angiography (CTA) and perfusion-CT (PCT) on admission, and similar imaging after treatment, typically around 24 hours, to assess recanalization and reperfusion. Recanalization was assessed by comparing arterial patency on admission and posttreatment CTAs; reperfusion, by comparing the volumes of CBV, CBF, and MTT abnormality on admission and posttreatment PCTs. Collateral flow was graded on the admission CTA. Follow-up infarct volume was measured on the discharge noncontrast CT. The groups of patients with reperfusion, no reperfusion, recanalization, and no recanalization were compared in terms of imaging and clinical outcomes.

Results.—Reperfusion (using an MTT reperfusion index >75%) was a more accurate predictor of follow-up infarct volume than recanalization. Collateral flow and recanalization were not accurate predictors of follow-up infarct volume. An interaction term was found between reperfusion and the volume of the admission penumbra >50 mL.

Conclusion.—Our study provides evidence that reperfusion is a more accurate predictor of follow-up infarct volume in acute ischemic stroke patients than recanalization. We recommend an MTT reperfusion index >75% to assess therapy efficacy in future acute ischemic stroke trials that use perfusion-CT.

▶ Accurately predicting infarct volume is key to acute stroke management. The dictum, "time is brain," is probably better stated "time is penumbra" (according to Gil Gonzalez of Massachusetts General Hospital, who gives some of the best stroke talks I have ever heard). Of course, the idea is to "rescue" at-risk brain as contrasted with "doomed" brain in the densely ischemic core of an infarct.

The authors show that reperfusion, not collateral flow and recanalization, is the most accurate predictor of final infarct volume (Fig 2 in the original article). They established a decrease of 75% or greater in the mean transit time (MTT) volume defined reperfusion. They reported this in a relatively small group of patients, but this finding looks promising. Stay tuned.

A. G. Osborn, MD

Brain: Trauma

Diffusion Tensor Imaging Abnormalities in Patients With Mild Traumatic Brain Injury and Neurocognitive Impairment

Lo C, Shifteh K, Gold T, et al (Montefiore Med Ctr and Albert Einstein College of Medicine, Bronx, NY; et al)

J Comput Assist Tomogr 33:293-297, 2009

Objective.—To determine if diffusion tensor imaging can differentiate patients with chronic cognitive impairment after mild traumatic brain injury (TBI) from normal controls.

Methods.—Ten patients with persistent cognitive impairment after mild TBI were evaluated at least 2 years after injury. Fractional anisotropy (FA) and apparent diffusion coefficient (ADC) were measured at white matter regions susceptible to axonal injury after TBI. Comparison was made to 10 normal controls.

Results.—Fractional anisotropy was significantly lower (4.5%; $P = 0.01$) and ADC higher (7.1%; $P = 0.04$) in patients at the left side of the genu of the corpus callosum. The mild TBI group also demonstrated a significant increase in FA within the posterior limb of the internal capsule bilaterally (left, 5.1%; $P = 0.03$; right, 1.9%; $P = 0.04$).

Conclusions.—These results demonstrate low FA and high ADC in the genu of the corpus callosum of mild TBI patients with persistent cognitive impairment, suggesting that permanent white matter ultrastructural damage occurs in mild TBI, and that such damage may be associated with persistent cognitive disability. Further longitudinal studies are warranted to elucidate the full importance of the findings.

▶ I've often wondered why patients with mild traumatic brain injury (TBI) sometimes don't do well, reporting persistent difficulties and cognitive impairment that may last years after the initial injury. Part of the problem may be in the definition of "mild" TBI. In many studies, patients with Glasgow Coma Scores (GCS) of 14-16 are often grouped together as "mild" TBI. Well, there's a big difference between a 16 (completely normal, no loss of consciousness, etc) and a 14. Maybe a GCS of 14 or even 15 is a whole lot different than a 16.

Before the advent of diffusion tensor imaging (DTI), the presence of brain microhemorrhages on T2* scans was used as a surrogate marker of axonal injury. Now, with the increasingly widespread use of DTI, it is possible to demonstrate subtle white matter abnormalities directly. That's what the authors did and found that in a small group of 10 patients with persistent cognitive impairment (admittedly on nonstandardized neuropsychological examination), fractional anisotropy (FA) was lower and apparent diffusion coefficient (ADC) higher in the corpus callosum splenium suggesting what they called "white matter ultrastructural damage."

The splenium of the corpus callosum, together with the internal capsule and subcortical white matter fiber tracts, is a common site of diffuse axonal injury. But areas other than the splenium didn't show significant changes in FA. The authors attributed this to the "extremely mild" TBI in their group. They may be justified in calling their group extremely mild as only 1 of the 10 patients had any abnormalities on conventional magnetic resonance (MR) (a small focal area of lobar gliosis). Specifically, they identified no evidence of hemorrhage on any of their subjects. Had they used susceptibility-weighted imaging (SWI), it is conceivable they might have detected microhemorrhage that wasn't apparent on the GRE scans.

Bottom line: This is a very interesting, very small, and very preliminary study that shows intriguing findings, that is, objective evidence of diffuse axonal injury that isn't detectable on standard MR imaging (or, for that matter, on the DTI images by visual inspection). Stay tuned (you can bet the personal injury lawyers will be!).

A. G. Osborn, MD

Brain: Vascular Malformations

Intraventricular cerebral cavernomas: a series of 12 patients and review of the literature

Kivelev J, Niemelä M, Kivisaari R, et al (Helsinki Univ Central Hosp, Finland)
J Neurosurg 112:140-149, 2010

Object.—Intraventricular cavernomas (IVCs) occur in only 2–10% of patients with cerebral cavernomas. Reports concerning IVC are scarce and are limited mostly to sporadic case reports. In this paper, the authors present a series of 12 patients with IVCs that were treated at a single neurosurgical department. In addition, the authors reviewed the literature.

Methods.—All clinical data were analyzed retrospectively. Follow-up questionnaires were sent to all patients. Outcome was assessed using the Glasgow Outcome Scale. The authors also conducted a PubMed search and found 77 cases of IVC.

Results.—The patients' median age was 47 years, and the male/female ratio was 2:1. A cavernoma occurred in the lateral ventricle in 6 patients, in another 5 it was in the fourth ventricle, and 1 had a lesion in the third ventricle. Almost all patients presented with acute headache on admission and in more than half, the symptoms were related to cavernoma bleeding. In total, 8 rebleedings occurred in 5 patients during a median of 0.4 years. Three patients with a cavernoma of the fourth ventricle presented with a cranial nerve deficit. In 8 cases, a cavernoma was surgically treated an average of 1.3 years after the diagnosis. Only 1 patient underwent surgery in the acute phase after a major intraventricular/intracerebral hemorrhage. The median follow-up time was 2 years. No patient was lost to follow-up, and no patient died. In total, on follow up 9 patients improved and 3 had a persistent neurological deficit, of which 2 existed before surgery.

Conclusions.—In the present series, the IVCs had a high tendency for rehemorrhage. Surgery is advocated when hemorrhages are frequent, and the mass effect causes progressive neurological deficits. Microsurgical removal of the IVC is safe, but in the fourth ventricle it can carry increased risk for cranial nerve deficits.

▶ Mass in the ventricle? Most radiologists do not think of cavernous malformations (CMs or, as the authors termed them, "cerebral cavernomas"). The most common lateral ventricle masses in adults are choroid plexus cysts (xanthogranulomas) followed by meningioma and metastasis. In children under the age of 5 years, it is probably choroid plexus papilloma. Masses are relatively rare in the third ventricle; choroid plexus papillomas can be seen in the fourth ventricle, especially in adults.

Cavernous malformation? Way down the list. But it is an important entity. As the authors point out, only 5% to 10% of CMs occur in the lateral ventricles. But when they do occur here, they can be mistaken for more ominous pathology (eg, hemorrhagic metastases). In this series, 7 of the 12 reported cases were in the lateral ventricle (not always in the atrium or choroid plexus), whereas 5 were in the fourth ventricle. The youngest patient was 15 years old, so most were middle-aged adults.

We have seen several CMs and when they have the typical "popcorn ball" appearance with multiple locules of blood with fluid-fluid levels, the diagnosis is not difficult. But when they present with acute hemorrhage, and especially if they are in a location like the frontal horn, the diagnosis is more problematic. Surgery may provide the only definitive answer.

My recommendation: If you see a hemorrhagic intraventricular mass on magnetic resonance (MR) (Fig 4 in the original article), do a T2* sequence (GRE or SWI). That may disclose the multilocular nature of the mass surrounded by a hemosiderin rim that suggests the diagnosis. It will also identify other CMs in the parenchyma, which is extremely important. If even 1 additional lesion is found, the patient has multiple CM syndromes and a mutation in one of the CCM genes.

A. G. Osborn, MD

Head and Neck

A Prospective Study of Paranasal Sinus Osteomas in 1,889 Cases: Changing Patterns of Localization

Erdogan N, Demir U, Songu M, et al (Izmir Ataturk Res and Training Hosp, Turkey)

Laryngoscope 119:2355-2359, 2009

Objectives/Hypothesis.—We developed the largest paranasal sinus computed tomography (CT) scan study so far by including 1,889 cases to investigate the prevalence, localization, age distribution, and the secondary complications of paranasal sinus osteomas.

Study Design.—Prospective study.

Methods.—A prospective study was performed on 1,889 consecutive adult individuals who underwent paranasal sinus CT examinations with suspected sinus disease.

Results.—The prevalence of paranasal sinus osteomas was determined to be 3%. Osteomas were located most frequently in the ethmoid sinuses. The size of the osteomas varied from approximately from 2 to 30 mm. Of the osteomas found, 37% were accompanied by pathological sinonasal findings.

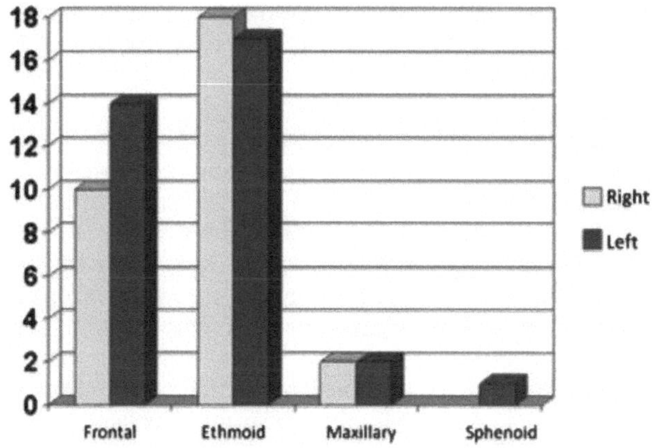

FIGURE 1.—Distribution of osteomas according to sinuses. (Courtesy of Erdogan N, Demir U, Songu M, et al. A prospective study of paranasal sinus osteomas in 1,889 cases: changing patterns of localization. *Laryngoscope*. 2009;119:2355-2359.)

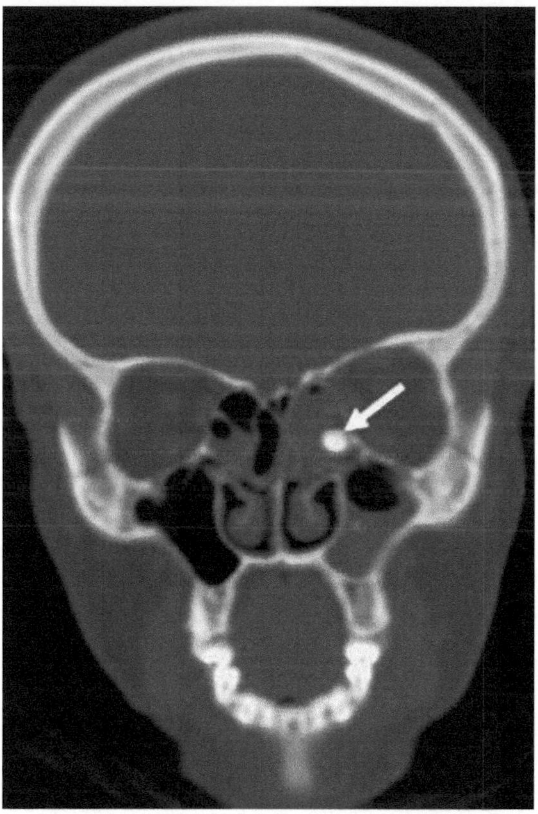

FIGURE 5.—Focal inflammation due to an osteoma in the ethmoid region. (Courtesy of Erdogan N, Demir U, Songu M, et al. A prospective study of paranasal sinus osteomas in 1,889 cases: changing patterns of localization. *Laryngoscope*. 2009;119:2355-2359.)

Conclusions.—The prevalence of sinonasal osteoma in the paranasal sinus region detected by CT scan is supporting the literature, whereas the localization pattern challenges it.

▶ This interesting study both confirms and belies the "conventional wisdom" (CW) concerning osteomas. These benign slowly growing bone-forming tumors are found almost exclusively in the craniofacial skeleton with the paranasal sinuses being the most common site. In the sinuses, the CW holds that approximately 80% are found in the frontal sinuses, followed by 20% in the ethmoid sinuses. The authors found the overall prevalence of paranasal sinus (PNS) osteomas to be about 3% of patients undergoing sinus CT scans. In this largest known study of its kind, they found that the most common site was actually the ethmoid sinuses, representing 55% of PNS osteomas. Frontal sinus osteomas represented 37.5% with 6% in the maxillary and only 1.5% in the sphenoid sinuses (Fig 1). Most were confined to the sinuses; the largest osteomas tended to be found in the frontal sinus with all ethmoid lesions less than 20 mm. Most PNS osteomas were asymptomatic, although 37% had associated pathologic findings, such as obstruction and inflammation, especially when located in the ethmoids (Fig 5). Peak prevalence was in the third decade with a modest male predominance.

A. G. Osborn, MD

Incidental head and neck ^{18}F-FDG uptake on PET/CT without corresponding morphological lesion: early predictor of cancer development?

Heusner TA, Hahn S, Hamami ME, et al (Univ of Duisburg-Essen, Germany)
Eur J Nucl Med Mol Imaging 36:1397-1406, 2009

Purpose.—To retrospectively determine whether increased/asymmetric FDG uptake on PET without a correlating morphological lesion on fully diagnostic CT indicates the development of a head and neck malignancy.

Methods.—In 590 patients (mean age 55.4 ± 13.3 years) without a head and neck malignancy/inflammation FDG uptake was measured at (a) Waldeyer's ring, (b) the oral floor, (c) the larynx, and (d) the thyroid gland, and rated as absent (group A), present (group B), symmetric (group B1) or asymmetric (group B2). Differences between groups A and B and between B1 and B2 were tested for significance with the *U*-test ($p < 0.05$). An average follow-up of about 2.5 years (mean 29.5 ± 13.9 months) served as the reference period to determine whether patients developed a head and neck malignancy.

Results.—Of the 590 patients, 235 (40%) showed no evidence of enhanced FDG uptake in any investigated site, and 355 (60%) showed qualitatively elevated FDG uptake in at least one site. FDG uptake values (SUV_{max}, mean ± SD) for Waldeyer's ring were 3.0 ± 0.89 in group A ($n = 326$), 4.5 ± 2.18 in group B ($n = 264$; $p < 0.01$), 5.4 ± 3.35 in group B1 ($n = 177$), and 4.1 ± 1.7 in group B2 ($n = 87$; $p < 0.01$). Values for the oral floor were 2.8 ± 0.74 in group A ($n = 362$), 4.7 ± 2.55 in group B ($n = 228$; $p < 0.01$), 4.4 ± 3.39 in group B1 ($n = 130$), and 5.1 ± 2.69 in group B2 ($n = 98$, $p = 0.01$). Values for the larynx were 2.8 ± 0.76 in group A ($n = 353$), 4.2 ± 2.05 in group B ($n = 237$; $p < 0.01$), 4.0 ± 2.02 in group B1 ($n = 165$), and 4.6 ± 2.8 in group B2 ($n = 72$; $p = 0.027$). Values for the thyroid were 2.4 ± 0.63 in group A ($n = 404$), 3.0 ± 1.01 in group B ($n = 186$; $p < 0.01$), 2.6 ± 0.39 in group B1 ($n = 130$), and 4.0 ± 1.24 in group B2 ($n = 56$; $p < 0.01$). One patient developed a palatine tonsil carcinoma (group B1, SUV_{max} 3.2), and one patient developed an oral floor carcinoma (group B1, SUV_{max} 3.7).

Conclusion.—Elevated/asymmetric head and neck FDG accumulation without a correlating morphological lesion can frequently be found and does not predict cancer development. In populations in which goitre is endemic, FDG uptake by the thyroid is common and not associated with thyroid cancer.

▶ As positron emission tomography/computed tomography (PET/CT) becomes more widespread, "incidentalomas" seen as uptake in the head and neck becomes an issue. Is uptake truly incidental and of no clinical significance? Does it presage cancer development? These important questions were addressed by Heusner et al. Although this was a retrospective study, it did point out some significant and highly interesting findings. First of all, a significant majority (60%) of patients had elevated [^{18}F]-fluoro-D-glucose (FDG) uptake in at least 1 site—sites such as oral cavity and the larynx—where head and neck cancers are likely to occur (Fig 1 in the original article). Out of the 355 patients who had elevated uptake, only 2 developed carcinomas over the study period (follow-up was 1 year, a potential weakness of the article).

FDG accumulation is frequent, not specific for tumor, and in the absence of a correlating morphologic lesion, does not predict cancer development. Of special note is their conclusion that FDG uptake by the thyroid is common and usually associated with goiter, not cancer.

A. G. Osborn, MD

Maxillary Mucosal Cyst Is Not a Manifestation of Rhinosinusitis: Results of a Prospective Three-Dimensional CT Study of Ophthalmic Patients

Kanagalingam J, Bhatia K, Georgalas C, et al (Royal Natl Throat Nose and Ear Hosp, London, UK; Moorfields Eye Hosp, London, UK; Academic Med Ctr, Amsterdam, The Netherlands)

Laryngoscope 119:8-12, 2009

Background.—Mucosal cysts in the maxillary antrum (MMC) are a common finding in imaging of the paranasal sinuses. Their significance remains in doubt and their prevalence in the general nonrhinitic population is unknown.

Objectives.—To establish the prevalence of MMCs in patients without nasal complaints and identify any association with putative causative factors. To identify any correlation between objective radiologic evidence of chronic rhinosinusitis or dental disease and the presence of cysts.

Study Design.—Prospective case series.

Methods.—Computed tomography (CT) images from 257 patients undergoing scans of the orbits for ophthalmic reasons were entered into this study. Patients responded to a questionnaire prior to scanning inquiring about nasal complaints, treatment for nasal disorders, previous nasal injury, allergy, asthma, recent upper respiratory tract infection, and dental root disease of the upper jaw.

Results.—Overall 35.6% of patients studied had radiologic evidence of at least one maxillary mucosal cyst. There was no association between the presence of cysts and subjective or objective evidence of sinus or dental disease. Specifically, there was no correlation between the presence of cysts and the total and ostiomeatal complex Lund-Mackay radiologic scores.

Conclusions.—Maxillary mucosal cysts are prevalent in nonrhinitic patients and do not reflect sinus or dental disease. Their presence should not be an indication for sinus surgery.

▶ Maxillary mucosal cysts (MMCs) are a common incidental finding on routine head CTs and MRs, let alone on dedicated imaging of the nose and paranasal sinuses. The key question: So what? We always comment on their presence but does this really mean anything?

The authors prospectively studied orbital CT scans in "nonrhinitic" patients and found that over one-third of all scans demonstrated imaging evidence of at least one maxillary mucosal cyst. Half were located on the floor of the maxillary sinus with one-quarter found on the lateral wall. They found these MMCs had no relationship to nasal complaints, infection, dental root disease, or sinus symptoms. They concluded that MMCs are not pathologic, do not reflect either sinus or dental disease, and should not be an indication for sinus surgery. There was also no relationship between MMCs and ostiomeatal complex disease.

So here's the question: If it doesn't mean anything, should we report such findings when present? Practically speaking, probably not...but we gotta call 'em like we see 'em. So I will continue to report them.

A. G. Osborn, MD

Partially Cystic Thyroid Nodules on Ultrasound: Probability of Malignancy and Sonographic Differentiation

Lee M-J, Kim E-K, Kwak JY, et al (Yonsei Univ, Seoul, Korea)
Thyroid 19:341-346, 2009

Background.—Thyroid nodules are very common and a large portion are mixed echoic, with both solid and cystic areas. There are few studies regarding the reliability of ultrasonographic criteria for evaluating solid nodules as they apply to mixed echoic nodules. The object of this study was to evaluate the frequency of malignancy in mixed echoic thyroid nodules and ascertain the ultrasound findings that help distinguish benign from malignant nodules.

Methods.—Among 1056 thyroid nodules undergoing ultrasound with fine-needle aspiration (FNA) biopsy, 392 nodules (37.1%) were mixed echoic. From this group of 392, the nodules that were read as benign or malignant on histopathology examination after surgery and the nodules that were not resected but were considered to be benign or malignant on cytology were analyzed for their ultrasonographic features. The nodules were divided into the following three groups. Group 1 ($n = 93$) included nodules in which the solid portion was <50%; group 2 ($n - 216$) comprised nodules in which the solid portion was >50%; and group 3 ($n = 26$) included mixed echoic (spongy) nodules. We also analyzed features of the solid portion of the nodule, namely, its position (eccentric or not), shape, margin, and echogenicity, and whether there were micro/macrocalcifications.

Results.—In the FNA sample, 52 nodules were inadequate for cytological diagnosis, but the remaining 340 (86.7%) were adequate. Eighteen of the nodules were malignant and 317 were benign, yielding a malignancy rate of 5.4%. By group, the malignancy rate was 2.2% (2/93) in group 1, 7.4% (16/216) in group 2, and 0% (0/26) in group 3. There were more malignancies in group 2 than the other groups ($p = 0.040$). Among sonographic findings, eccentric placement ($p = 0.007$) and the presence of microcalcifications ($p < 0.001$) were significantly correlated with malignancy.

Conclusions.—About 5% of partially cystic nodules in our series were malignant. When more than 50% of the nodule is solid and the solid portion of the nodule is eccentric, the risk of malignancy is greater. As has been noted for completely solid nodules, microcalcifications are associated with an increased risk of malignancy.

► Such a pesky organ, the thyroid... and it's even more bothersome now that multidetector-row CT scans of the cervical spine have become routine in trauma patients. It's become increasingly rare to get a "brain only" CT scan even in patients with only minor injury. And why not? CT is more accurate in detecting spine injury than plain films. And, as these patients often get a "screening brain" anyway, why not add CT of the cervical spine too (well, there are lots of economic and medical practice issues, but that's another story)?

As we see many more cervical spine CTs rolling through our reading rooms, we are, of course, seeing lots more imaging of patients' necks. In our institution, both bone and soft tissue algorithm reconstructions are used for both the source and reformatted images. The result: That pesky thyroid is seen and "stuff" is "found." To phrase it like a comment in the American Journal of Neuroradiology by Tony Mancuso did, "Oh #*$%#! Another pesky incidental thyroid nodule!"[1] Or, as one author noted, "Thyroid incidentaloma: the ignorant in pursuit of the impalpable."[2]

TABLE 3.—Diagnostic Index of Malignant Ultrasound Features

US Features	Sensitivity	Specificity	PPV	NPV	Accuracy
a. Solid portion ≥50%	16/18 (88.9)	117/317 (36.9)	16/216 (7.4)	117/119 (98.3)	133/335 (39.7)
b. Eccentric solid position	8/18 (44.4)	264/317 (83.3)	8/61 (13.1)	264/274 (96.4)	272/335 (81.2)
c. Containing microcalcifications	7/18 (38.9)	312/317 (98.4)	7/12 (58.3)	312/323 (96.6)	319/335 (95.2)
a or b	18/18 (100)	84/317 (26.5)	18/251 (7.2)	84/84 (100)	102/335 (30.4)
a or c	18/18 (100)	113/317 (35.6)	18/222 (8.1)	113/113 (100)	131/335 (39.1)
b or c	12/18 (66.7)	261/317 (82.3)	12/68 (17.6)	261/267 (97.8)	273/335 (81.5)
a or b or c	18/18 (100)	82/317 (25.9)	18/253 (7.1)	82/82 (100)	100/335 (29.9)

Percentages in parentheses.
US, ultrasound; PPV, positive predictive value; NPV, negative predictive value.

With an autopsy prevalence of up to 50%, we are, for sure, going to find thyroid nodules. What to do?

The clinical significance of thyroid "incidentalomas" is indeed controversial. Some authors such as Jin et al[3] report a relatively high prevalence of malignancy in thyroid nodules (11%) found on CT. Some recommend ultrasound, which, in the authors' article, showed that only 5% of partially cystic nodules were malignant. Investigating thyroid nodules is expensive.[4] Maybe if thyroid lesions appear cystic they may not need further evaluation? If, as the authors note, more than 50% of the nodule is solid and/or eccentrically placed, the risk of malignancy is higher (Table 3). Or maybe if microcalcifications are present, they deserve further imaging? Which needs fine-needle aspiration biopsy (an expensive procedure) and which doesn't? I think there is abundant evidence that there is no CT feature that reliably distinguishes benign from malignant lesions in thyroid gland.[5] So what to do?

Please e-mail me your ideas and comments. Let's continue the dialogue of this controversial subject next year.

A. G. Osborn, MD

References

1. Mancuso AA. Oh #*$%#! Another pesky incidental thyroid nodule! *AJNR Am J Neuroradiol.* 2005;26:2444-2445.
2. Topliss D. Thyroid incidentaloma: the ignorant in pursuit of the impalpable. *Clin Endocrinol (Oxf).* 2004;60:18-20.
3. Jin J, Wilhelm SM, McHenry CR. Incidental thyroid nodule: patterns of diagnosis and rate of malignancy. *Am J Surg.* 2009;197:320-324.
4. Mehanna HM, Jain A, Morton RP, Watkinson J, Shaha A. Investigating the thyroid nodule. *BMJ.* 2009;338:b733.
5. Shetty SK, Maher MM, Hahn PF, Halpern EF, Aquino SL. Significance of incidental thyroid lesions detected on CT: correlation among CT, sonography, and pathology. *AJR Am J Roentgenol.* 2006;187:1349-1356.

Sphenoid Septations and Their Relationship With Internal Carotid Arteries: Anatomical and Radiological Study

Fernandez-Miranda JC, Prevedello DM, Madhok R, et al (Univ of Pittsburgh School of Medicine, PA)

Laryngoscope 119:1893-1896, 2009

Objectives/Hypothesis.—We sought to investigate the anatomical relation of the intrasphenoid septations to the internal carotid artery (ICA).

Methods.—Twenty-seven preoperative high-resolution computed tomography angiographic (CTA) scans with 1 mm of separation acquisition were examined. In addition, an endoscopic endonasal approach and high-resolution computed tomography were done on 27 fresh-frozen cadaveric heads. The number of intrasphenoid septa and their relation to the ICAs were analyzed endoscopically and radiologically. Complete and incomplete septations were included in the analysis. A total of 54 sphenoid sinuses were studied.

Results.—Out of 27 sphenoid sinuses radiologically studied from real patients, 23 (85%) and 11 (41%) had at least one or two septa, respectively, touching one of the ICA. Out of 27 sphenoid sinuses endoscopically examined from cadavers (excluding one conchal type), 24 (89%) had at least one septation inserted in the ICAs. Two or more septations were inserted in the carotid prominence in 48% of sphenoid sinuses. The radiological examination of the anatomical specimens revealed similar results, with discrepancy in just one case (1/27) where it did not identify an incomplete septation inserting at ICA. No significant differences were found between the groups. From a total of 54 sphenoid sinuses studied, 47 (87%) had at least one septum related to the ICA, and only 13% presented a typical isolated midline septation.

Conclusions.—As demonstrated both radiographically and endoscopically, most intrasphenoidal septa insert at the parasellar or paraclival carotid prominence. As such, extreme care should be taken when identifying and removing these septations intraoperatively (Fig 1).

▶ The relationship between osseous septations in the sphenoid sinus and the position of the cavernous internal carotid arteries is critical to transsphenoidal/endonasal approaches to the skull base, especially for the pituitary fossa. As the authors note, intrasphenoidal septae are common, with 85% of patients having at least 1 internal septation. In cadaver studies, nearly 90% had at least 1 septation that inserted into the internal carotid artery (ICA) prominence and nearly 50% had 2 (Fig 1C,D). As these septations often must be reduced in the surgical approach to the pituitary fossa, such septations should not be regarded lightly as most insert into the parasellar or paraclival carotid prominence.

So what does this mean for radiologists? When reading MRs or CTs in patients who may undergo endonasal approach to the skull base, the presence

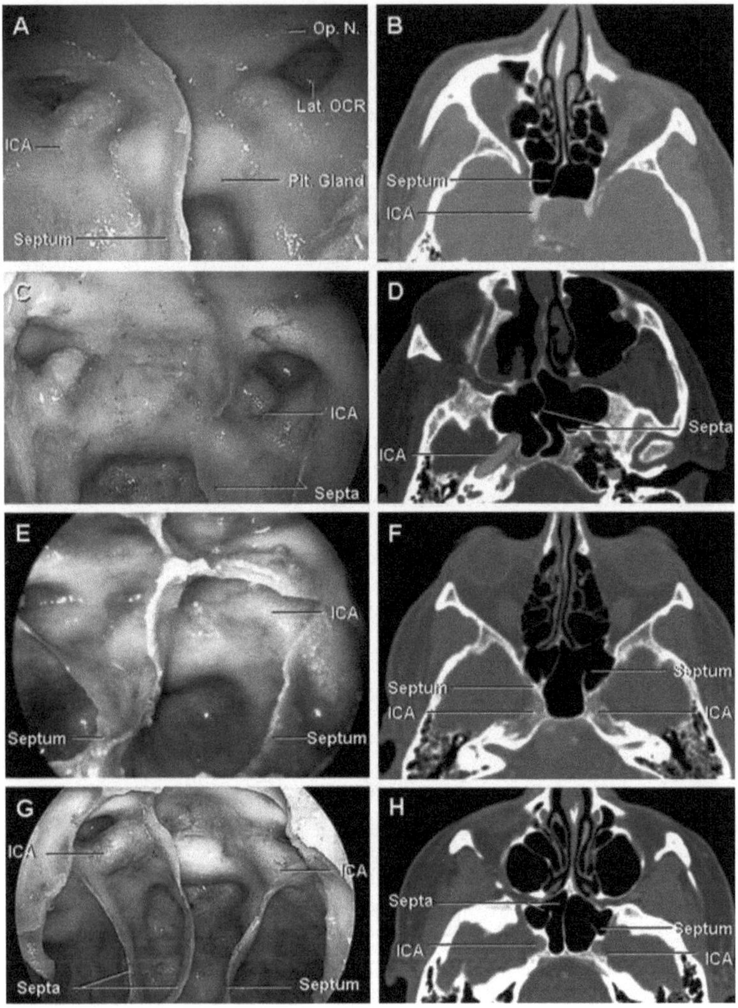

FIGURE 1.—Examples of sphenoid septations and their anatomical relationship to the internal carotid artery (ICA) as observed in endoscopic endonasal dissections from anatomical specimens (A, C, E, and G), and preoperative CTA studies from surgical patients (B, D, F, and H). (A) A single septum is related to the ICA at its parasellar and paraclival segments. The optic nerves (Op. N.) and a pair of very well-pneumatized lateral optico-carotid recesses (Lat. OCR) are identified. The pituitary gland (Pit. Gland) is located behind the sellar bone. (B) A single septum is identified inserting into the carotid prominence at the parasellar segment of the ICA. (C) Two different septations are observed related to the ICA at the parasellar and paraclival segments. (D) Two different septations are observed, and only the left one seems to be related to the ICA on that specific axial level. (E) In this case two septations are identified. The left one clearly relates to the ICA, whereas the right one bifurcates into two separate septations, one of them being related to the ICA at the parasellar segment. (F) Again, two different septations are observed related to the ICA. Interestingly, the right septation bifurcates into two separate septations. (G) Three independent septations are observed in this specimen. All of them are in contact to the ICA at some point of its trajectory. (H) Three independent septations are observed in this patient. The one in the midline does not contact the ICA, but both left and right septations insert into the carotid prominence. (Courtesy of Fernandez-Miranda JC, Prevedello DM, Madhok R, et al. Sphenoid septations and their relationship with internal carotid arteries: Anatomical and radiological study. *Laryngoscope* 2009;119:1893-1896.)

of internal sphenoid septations should be noted in the report and wherever possible potential contact with the ICA identified.

A. G. Osborn, MD

Spine and Spinal Cord

Cervical myelopathy in patients with ossification of the posterior longitudinal ligament

Mochizuki M, Aiba A, Hashimoto M, et al (Numazu City Hosp, Shizuoka, Japan; et al)

J Neurosurg Spine 10:122-128, 2009

Object.—The authors assessed the clinical course in patients with a narrowed cervical spinal canal caused by ossification of the posterior longitudinal ligament (OPLL), but who have no or only mild myelopathy. Additionally, the authors analyzed the factors contributing to the development and aggravation of myelopathy in patients with OPLL-induced spinal canal stenosis.

Methods.—Between 1997 and 2004, the authors selected treatments for patients with cervical OPLL in whom the residual space available for the spinal cord was ≤ 12 mm. Treatment decisions were based on the severity of myelopathy at presentation. Twenty-one patients with no or mild myelopathy (defined as a Japanese Orthopaedic Association [JOA] scale score ≥ 14 points) received conservative treatment, with a mean follow-up period of 4.5 years. In 20 patients with moderate or severe myelopathy (JOA scale score < 14 points), the authors performed surgery via an anterior approach. The clinical course in these patients was assessed with the JOA scale and the OPLL types were classified. The authors evaluated the range of motion between C-1 and C-7, the developmental segmental sagittal diameter, the percentage of spinal canal diameter occupied by the OPLL (% ratio), and the residual space available for the spinal cord on cervical radiographs; T2-weighted MR images were examined for high signal changes (HSCs).

Results.—In the conservative treatment group, 8 patients showed improvement, 12 remained unchanged, and 1 patient's condition became slightly worse during the observation period. Fifteen patients in this group had mixedtype, 3 had continuous-type, 2 had localized-type, and 1 had a segmental-type OPLL. In the surgically treated group, there were 12 patients with segmental-type, 10 patients with mixed-type, and 1 with localized-type OPLL. The mean range of motion at C1-7 was 36.4° in the conservatively treated group and 46.5° in the surgical group ($p < 0.05$). No significant difference was seen between the groups in terms of developmental segmental sagittal diameter, % ratio, or residual space available for the cord. No HSCs were noted in the conservative group, while 17 patients in the surgical group had HSCs ($p < 0.05$).

Conclusions.—In the present study, the authors demonstrate that the mobility of the cervical spine and the type of OPLL are important factors contributing to the development and aggravation of myelopathy in patients with OPLL-induced spinal canal stenosis. The authors advocate conservative treatment in most patients with OPLLs who have no or only mild myelopathy, even in the presence of spinal canal narrowing.

▶ Ossification of the posterior longitudinal ligament (OPLL) is seen as "flowing," multilevel ossification posterior to the vertebral bodies. Most common in the midcervical region, it can—when severe—cause canal narrowing, cord compression, and myelopathy. The "conventional wisdom" regarding OPLL is that if the canal diameter is under 6 mm or stenosis exceeds 60%, the risk of developing progressive myelopathy increases. The authors add additional important information, that is, mobility of the cervical spine and the type of OPLL (classified as continuous, mixed, segmental, or localized) are important factors contributing to the development and aggravation of myelopathy. Perhaps most importantly, they advocate conservative treatment in patients with OPLL who have no or mild myelopathy even if the canal appears significantly narrowed on imaging studies. One caveat: If patients have spinal canal narrowing AND high signal intensity in the adjacent spinal cord at magnetic resonance (MR) imaging (Fig 4 in the original article), the risk of myelopathy developing or increasing is high, and more aggressive treatment (ie, surgical decompression) is justified.

A. G. Osborn, MD

Heavily T2-weighted MR myelography vs CT myelography in spontaneous intracranial hypotension

Wang Y-F, Lirng J-F, Fuh J-L, et al (Taipei Veterans General Hosp, Taiwan; et al)

Neurology 73:1892-1898, 2009

Objective.—To assess the diagnostic accuracy of heavily T2-weighted magnetic resonance myelography (MRM) in patients with spontaneous intracranial hypotension (SIH).

Methods.—Patients with SIH were recruited prospectively, and first underwent MRM and then computed tomographic myelography (CTM). The results of MRM were validated with the gold standard, CTM, focusing on 1) CSF leaks along the nerve roots, 2) epidural CSF collections, and 3) high-cervical (C1-3) retrospinal CSF collections. Comparisons of these 3 findings between the 2 studies were made by κ statistics and agreement rates. Targeted epidural blood patches (EBPs) were placed at the levels of CSF leaks if supportive treatment failed.

Results.—Nineteen patients (6 men and 13 women, mean age 37.9 ± 8.6 years) with SIH completed the study. MRM did not differ from CTM in the detection rates of CSF leaks along the nerve roots (84% vs 74%, $p = 0.23$), high-cervical retrospinal CSF collections (32% vs 16%, $p = 0.13$), and epidural CSF collections (89% vs 79%,

$p = 0.20$). MRM demonstrated more spinal levels of CSF leaks (2.2 ± 1.7 vs 1.5 ± 1.5, $p = 0.011$) and epidural collections (12.2 ± 5.9 vs 7.1 ± 5.8, $p < 0.001$) than CTM. The overall level-by-level concordance was substantial for CSF leaks along the nerve roots (C1-L3) ($\kappa = 0.71$, $p < 0.001$, agreement = 95%) and high-cervical retrospinal CSF collections (C1-3) ($\kappa = 0.73$, $p < 0.001$, agreement = 92%), and moderate for epidural CSF collections (C1-L3) ($\kappa = 0.47$, $p < 0.001$, agreement = 72%). Ten of the 14 patients (71%) receiving targeted EBPs experienced sustained symptomatic relief after a single attempt.

Conclusions.—Heavily T2-weighted magnetic resonance myelography was accurate in localizing CSF leaks for patients with spontaneous intracranial hypotension. This noninvasive technique may be an alternative to computed tomographic myelography before targeted epidural blood patches.

▶ Since its initial description of MR brain scans in the mid-1990s, spontaneous intracranial hypotension (SIH) has been recognized as an uncommon but very important cause of headaches. SIH is typically caused by spontaneous spinal cerebrospinal fluid (CSF) leaks and has also been reported after lumbar puncture, shunting, decompressive surgery for Chiari I malformation, and other procedures.

In many cases, epidural blood patch (EBP) is performed empirically without localization of the precise site of the CSF leaks although, as the authors point out, there is some evidence that "targeted" EBPs are more effective than untargeted EBPs.

If patients respond to an untargeted EBP, swell. But many—up to two-thirds of patients in some series—do not. While a "high-volume" EBP can be performed in the case of initial failure, why not target the EBP if possible? Contrast myelography with CT (CTM) may provide information as to the site of CSF leak. But it may not. And, as an invasive procedure, it may in some cases cause symptoms to worsen.

The ideal scenario would be a noninvasive study aimed at detecting the CSF leakage site. While the study group in this case was relatively small, they did succeed in using MR myelography (MRM) to target EBP in about 70% of cases. Typical MRM and CTM findings of CSF leakages were nicely demonstrated (Fig 2 in the original article).

One rather startling finding: Multiple leaks (a bit over 40%) were common in their group. Bottom line: MRM looks promising and is probably worth trying in difficult cases. Many institutions, including our own, will probably continue to use empiric EBPs because of the expense (and time required) to obtain total spine MRM.

A. G. Osborn, MD

Outcome of percutaneous rupture of lumbar synovial cysts: a case series of 101 patients

Martha JF, Swaim B, Wang DA, et al (New England Baptist Hosp, Boston, MA; et al)
Spine J 9:899-904, 2009

Background Context.—Lumbar facet joint synovial cysts are benign degenerative abnormalities of the lumbar spine. Previous reports have supported operative and nonoperative management. Facet joint steroid injection with cyst rupture is occasionally performed, but there has been no systematic evaluation of this treatment option.

Purpose.—To profile the role of facet joint steroid injections with cyst rupture in the treatment of lumbar facet joint synovial cysts.

Study Design/Setting.—Retrospective chart review and long-term follow-up of patients treated for lumbar facet joint synovial cysts.

Patient Sample.—One hundred one patients treated for lumbar facet joint synovial cysts with fluoroscopically guided corticosteroid facet joint injection and attempted cyst rupture.

Outcome Measures.—Oswestry Disability Index and numeric rating scale score for back and leg pain.

Methods.—A retrospective review and a subsequent interview were conducted to collect pre-treatment and posttreatment pain and disability scores along with details of subsequent treatment interventions. Group differences in pain and disability scores were assessed using paired *t* test. Multiple clinical factors were analyzed in terms of risk for surgical intervention using logistic regression modeling and Cox proportional hazards modeling.

Results.—Successful cyst rupture was confirmed fluoroscopically in 81% of cases. Fifty-five patients (54%) required subsequent surgery over a period averaging 8.4 months because of inadequate symptom relief. All patients reported significant improvement in back pain, leg pain, and disability at 3.2 years postinjection, regardless of their subsequent treatment course ($p<.0001$ in all groups). There was no significant difference in current pain between patients who received injections only and those who underwent subsequent surgery.

Conclusions.—This study presents the largest clinical series of nonsurgical treatment for lumbar facet joint synovial cysts. Lumbar facet joint steroid injection with attempted cyst rupture is correlated with avoiding subsequent surgery in half of treated patients. Successful cyst rupture does not appear to have added benefit, and it was associated with worse disability 3 years postinjection. Long-term outcomes are similar, regardless of subsequent surgery.

▶ Cysts arising from degenerating facet joints are common in patients over the age of 60 years. When small, they are probably incidental findings but can get large enough to cause spinal stenosis, nerve root compression, and significant mass effect on the thecal sac.

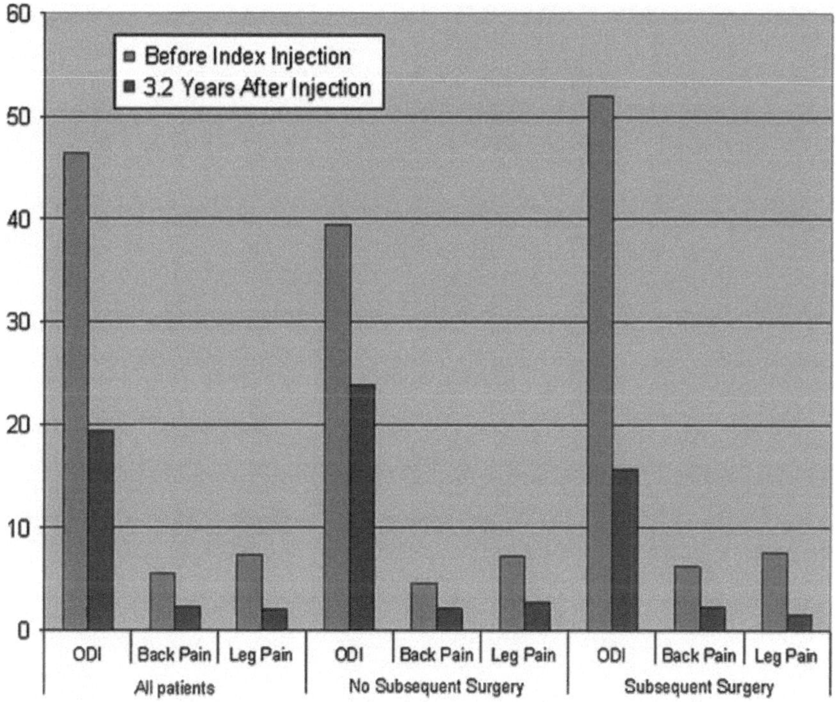

FIGURE 1.—A comparison of patient-reported preinjection and postinjection outcomes for all patients and by surgery grouping. There were significant differences between retrospective preinjection and postinjection scores on the Oswestry Disability Index (ODI), back pain, and leg pain in all patients ($p<.0001$) and by surgery grouping. Patients who underwent surgery reported higher levels of disability ($p=.01$) and back pain ($p=.03$), but similar levels of leg pain ($p=.43$) on the retrospective pretest. (Reprinted from Martha JF, Swaim B, Wang DA, et al. Outcome of percutaneous rupture of lumbar synovial cysts: a case series of 101 patients. *Spine J*. 2009;9:899-904, with permission from Elsevier.)

How to treat these benign but bothersome abnormalities has been controversial. Until the current article, there has been no systematic evaluation and comparison of treatment options. The authors report the largest group of patients with nonoperative treatment as the initial approach. Injecting the facet joints themselves with steroids was successful in about half the cases, and outcome was the same as with patients who had subsequent surgery (which was eventually done in 55% of the patients) (Fig 1). Of interest: Successfully "rupturing" the cyst did not appear to be necessary and even contraindicated as it was associated with worse long-term disability.

A. G. Osborn, MD

8 Cardiac Imaging

Introduction

The past year has been one where new technology, new applications of cardiac CT, and a substantial amount of "hard data" regarding cardiac CT and MRI have emerged. Several larger extramurally funded trials, multicenter trials, and registry-based studies have been completed and published. This follows the initial years of many smaller single-center studies. To reduce the danger of bias, I have asked leading cardiac MRI and CT experts to rank the articles that they feel had the greatest impact. I gratefully acknowledge the input from Drs Stephan Achenbach, Michael Shapiro, Allen Taylor, Udo Hoffmann, Ron Blankstein, Jack Ziffer, and Ricardo Cury. Although the absolute number of articles mentioned by this group was quite high, there was a smaller number of articles that we all felt had (or will have) great impact on the field. You will find these on the following pages.

Recently, a new scanner that enables high pitch (3.0 compared to standard of ~0.2) cardiac imaging has entered the market and an initial analysis was published that showed that in select patients (nonobese, low steady heart rate) coronary CT is feasible at incredibly low radiation doses (less than 1 mSv) with high image quality and diagnostic accuracy. Several emergency room–based trials have been published, including the ROMICAT I trial that showed that cardiac CT could be an effective triage tool for patients with acute chest pain who have not been ruled in for myocardial infarct (MI) in initial evaluation but otherwise would be admitted to "rule out MI." A registry-based study by Raff et al demonstrated that educating physicians with regard to dose savings tools results in significantly lower patient radiation doses in a real-life setting. A new application of cardiac CT has been evaluated by Blankstein and others, who showed feasibility of pharmacologic stress and rest perfusion CT can be performed at reasonable radiation doses and with accuracies comparable to radionuclide imaging.

An MRI delayed-enhancement study has demonstrated the prognostic significance of MRI of patients with and without left ventricular dysfunction. Another MRI study has the potential to change how we manage patients with suspected cardiac hemochromatosis; it showed that MRI can identify individuals who will develop heart failure within 1 year from the scan.

Lastly, due to the heterogeneity of technology and acquisition options, guidelines have emerged that delineate the minimum standards for performance and interpretation and reporting of cardiac CT.

I hope YEAR BOOK readers will find the selected articles informative and applicable to their practice and will enjoy reading them.

Suhny Abbara, MD

Adenosine-Induced Stress Myocardial Perfusion Imaging Using Dual-Source Cardiac Computed Tomography

Blankstein R, Shturman LD, Rogers IS, et al (Massachusetts General Hosp and Harvard Med School, Boston; et al)

J Am Coll Cardiol 54:1072-1084, 2009

Objectives.—This study sought to determine the feasibility of performing a comprehensive cardiac computed tomographic (CT) examination incorporating stress and rest myocardial perfusion imaging together with coronary computed tomography angiography (CTA).

Background.—Although cardiac CT can identify coronary stenosis, very little data exist on the ability to detect stress-induced myocardial perfusion defects in humans.

Methods.—Thirty-four patients who had a nuclear stress test and invasive angiography were included in the study. Dualsource computed tomography (DSCT) was performed as follows: 1) stress CT: contrast-enhanced scan during adenosine infusion; 2) rest CT: contrast-enhanced scan using prospective triggering; and 3) delayed scan: acquired 7 min after rest CT. Images for CTA, computed tomography perfusion (CTP), and single-photon emission computed tomography (SPECT) were each read by 2 independent blinded readers.

Results.—The DSCT protocol was successfully completed for 33 of 34 subjects (average age 61.4 ± 10.7 years; 82% male; body mass index 30.4 ± 5 kg/m^2) with an average radiation dose of 12.7 mSv. On a per-vessel basis, CTP alone had a sensitivity of 79% and a specificity of 80% for the detection of stenosis ≥50%, whereas SPECT myocardial perfusion imaging had a sensitivity of 67% and a specificity of 83%. For the detection of vessels with ≥50% stenosis with a corresponding SPECT perfusion abnormality, CTP had a sensitivity of 93% and a specificity of 74%. The CTA during adenosine infusion had a per-vessel sensitivity of 96%, specificity of 73%, and negative predictive value of 98% for the detection of stenosis ≥70%.

Conclusions.—Adenosine stress CT can identify stress-induced myocardial perfusion defects with diagnostic accuracy comparable to SPECT, with similar radiation dose and with the advantage of providing information on coronary stenosis.

▶ This study addresses an important issue related to the myocardial perfusion defects that can be seen on coronary CT. Clinically and hemodynamically significant stenoses that may warrant treatment range from approximately 70% to 100%. However, most of these, perhaps 70% to 98%, stenoses will not produce perfusion defects at rest. The hemodynamic significance will only result in perfusion defects if myocardial blood flow demand is raised, namely during exercise or pharmacologically induced stress. This study has shown that within a reasonable radiation dose limit it is possible to show coronary morphology, myocardial cardiac function, and myocardial stress and rest perfusion – all with one easy to tolerate quick test, cardiac CT. Although interpretation of perfusion defects remains difficult, the diagnostic accuracy for stress-rest perfusion mismatches in this study was similar to SPECT at comparable radiation doses; however, CT delivered also detailed anatomic information, valve, and myocardial function. The presence of myocardial calcification and subendocardial fatty metaplasia are powerful signs indicating remote, possibly nontransmural, occasionally otherwise missed myocardial infarcts.

S. Abbara, MD

Cardiac T2* Magnetic Resonance for Prediction of Cardiac Complications in Thalassemia Major

Kirk P, Roughton M, Porter JB, et al (Univ College London, UK)

Circulation 120:1961-1968, 2009

Background.—The goal of this study was to determine the predictive value of cardiac T2* magnetic resonance for heart failure and arrhythmia in thalassemia major.

Methods and Results.—We analyzed cardiac and liver T2* magnetic resonance and serum ferritin in 652 thalassemia major patients from 21 UK centers with 1442 magnetic resonance scans. The relative risk for heart failure with cardiac T2* values <10 ms (compared with >10 ms) was 160 (95% confidence interval, 39 to 653). Heart failure occurred in 47% of patients within 1 year of a cardiac T2* <6 ms with a relative risk of 270 (95% confidence interval, 64 to 1129). The area under the receiver-operating characteristic curve for predicting heart failure was significantly greater for cardiac T2* (0.948) than for liver T2* (0.589; $P < 0.001$) or serum ferritin (0.629; $P < 0.001$). Cardiac T2* was <10 ms in 98% of scans in patients who developed heart failure. The relative risk for arrhythmia with cardiac T2* values <20 ms (compared with >20 ms) was 4.6 (95% confidence interval, 2.66 to 7.95). Arrhythmia occurred in 14% of patients within 1 year of a cardiac T2* of <6 ms. The area under the receiver-operating characteristic curve for predicting arrhythmia was significantly greater for cardiac T2* (0.747) than for liver T2* (0.514; $P < 0.001$) or serum ferritin (0.518; $P < 0.001$). The cardiac T2* was <20 ms in 83% of scans in patients who developed arrhythmia.

Conclusions.—Cardiac T2* magnetic resonance identifies patients at high risk of heart failure and arrhythmia from myocardial siderosis in thalassemia major and is superior to serum ferritin and liver iron. Using cardiac T2* for the early identification and treatment of patients at risk is a logical means of reducing the high burden of cardiac mortality in myocardial siderosis.

Clinical Trial Registration.—URL: http://www.clinicaltrials.gov. Unique identifier: NCT00520559.

▶ This study is included because of its potential to change management of patients with cardiac siderosis. Primary hemochromatosis (also known as hereditary or "classic" hemochromatosis) is a genetic disorder that results in increased iron absorption, whereas secondary hemochromatosis may be caused by frequent blood transfusions used to treat patients with thalassemia and other anemias, although several other underlying causes exist. Myocardial hemosiderosis is a sequelae of iron deposition in the myocardium and is responsible for high mortality due to heart failure and arrhythmias in affected patients.

In this multicenter study, the authors examined 652 patients with thalassemia major with MRI at 1442 time points, looking for potential presence and severity of myocardial siderosis by using myocardial T2* value measurements.

They found that T2* values of less than 10 ms were highly sensitive and specific for the prediction of development of heart failure within 1 year from the examination. The risk increased with shorter T2* values. Nearly half of the patients with T2* of less than 6 ms developed heart failure within a year, whereas only 2 out of more than 1100 scans with T2* of greater than 10 were followed by development of heart failure within 1 year.

T2* measurements had a tremendous incremental value over the traditional metrics of total body iron, liver iron, and serum ferritin, which in fact performed poorly in predicting heart failure. Risk determination is very important because of the high mortality and the presence of a therapeutic option, namely cardiac chelation therapy.

Please also note the informative and interesting editorial.[1]

S. Abbara, MD

Reference

1. Wood JC. History and current impact of cardiac magnetic resonance imaging on the management of iron overload. *Circulation.* 2009;120:1937-1939.

Coronary computed tomography angiography with a consistent dose below 1 mSv using prospectively electrocardiogram-triggered high-pitch spiral acquisition

Achenbach S, Marwan M, Ropers D, et al (Univ of Erlangen, Germany)
Eur Heart J 31:340-346, 2010

Aims.—We evaluated the feasibility and image quality of a new scan mode for coronary computed tomography angiography (CTA) with an effective dose of less than 1 mSv.

Methods and Results.—In 50 consecutive patients (body weight ≤ 100 kg, sinus rhythm ≤60 b.p.m. after pre-medication), coronary CTA was performed using a dual-source CT system with 2 × 128 × 0.6 mm collimation, 0.28 s rotation time, a pitch of 3.2 or 3.4, 100 kV tube voltage and current of 320 mA s. Data acquisition was prospectively triggered at 60% of the R–R interval and completed within one cardiac cycle. Image quality was evaluated using a four-point scale (1 = absence of any artefacts to 4 = uninterpretable). In all 50 patients, imaging was successful. Mean duration of data acquisition was 258 ± 20 ms. Mean dose-length product was 62 ± 5 mGy cm, the effective dose was 0.87 ± 0.07 mSv (0.78–0.99 mSv). Of the 742 coronary artery segments, 94% had an image quality score of 1, 5.0% a score of 2, 0.9% a score of 3, and 4 segments (0.5%) were 'uninterpretable'.

Conclusion.—In non-obese patients with a low and stable heart rate, prospectively ECG-triggered high-pitch spiral coronary CTA provides excellent image quality at a consistent dose below 1.0 mSv.

▶ This article was selected because it is the first study that was conducted on a new generation of dual source CT scanner using a high pitch spiral CT approach. In the past year, retrospectively gated, tube-current modulated, spiral acquisitions as well as prospectively triggered, axial, stop-and-shoot acquisition techniques have been refined, and remarkable dose reductions could be achieved while maintaining diagnostic accuracy. This generation scanner allowed the authors to get diagnostic images in all of their 50 patients with unprecedented low radiation doses of less than 1 mSv. However, it is important to notice that this patient population was selected based on body weight less than 100 kg and a heart rate of 60 or less. This resulted in exclusion of nearly 30% of the evaluated consecutive patients. However, it remains a tremendous advance even if this technology allows us to acquire coronary CT at 1 mSv in only 70% of our patients. It remains to be demonstrated how well this technology will perform at higher heart rates and in larger patients.

S. Abbara, MD

Coronary Computed Tomography Angiography for Early Triage of Patients with Acute Chest Pain: The ROMICAT (Rule Out Myocardial Infarction using Computer Assisted Tomography) Trial

Hoffmann U, Bamberg F, Chae CU, et al (Massachusetts General Hosp and Harvard Med School, Boston; et al)

J Am Coll Cardiol 53:1642-1650, 2009

Objectives.—This study was designed to determine the usefulness of coronary computed tomography angiography (CTA) in patients with acute chest pain.

Background.—Triage of chest pain patients in the emergency department remains challenging.

Methods.—We used an observational cohort study in chest pain patients with normal initial troponin and nonischemic electrocardiogram. A 64-slice coronary CTA was performed before admission to detect coronary plaque and stenosis (>50% luminal narrowing). Results were not disclosed. End points were acute coronary syndrome (ACS) during index hospitalization and major adverse cardiac events during 6-month follow-up.

Results.—Among 368 patients (mean age 53 ± 12 years, 61% men), 31 had ACS (8%). By coronary CTA, 50% of these patients were free of coronary artery disease (CAD), 31% had nonobstructive disease, and 19% had inconclusive or positive computed tomography for significant stenosis. Sensitivity and negative predictive value for ACS were 100% (n = 183 of 368; 95% confidence interval [CI]: 98% to 100%) and 100% (95% CI: 89% to 100%), respectively, with the absence of CAD and 77% (95% CI: 59% to 90%) and 98% (n = 300 of 368, 95% CI: 95% to 99%), respectively, with significant stenosis by coronary CTA. Specificity of presence of plaque and stenosis for ACS were 54% (95% CI: 49% to 60%) and 87% (95% CI: 83% to 90%), respectively. Only 1 ACS occurred in the absence of calcified plaque. Both the extent of coronary plaque and presence of stenosis predicted ACS independently and incrementally to Thrombolysis In Myocardial Infarction risk score (area under curve: 0.88, 0.82, vs. 0.63, respectively; all $p < 0.0001$).

Conclusions.—Fifty percent of patients with acute chest pain and low to intermediate likelihood of ACS were free of CAD by computed tomography and had no ACS. Given the large number of such patients, early coronary CTA may significantly improve patient management in the emergency department.

▶ Several million Americans are admitted each year from emergency departments to cardiac units, without being ruled in or ruled out for acute coronary syndromes (acute myocardial infarction or unstable angina pectoris), which result in cost exceeding 8 billion US dollars. This well designed blinded study showed that CT may at minimum reliably exclude acute coronary syndrome (ACS) in a substantial fraction of patients without acute coronary syndrome and allow save discharge in more than 50%. This is important evidence that together with that from other studies will help redesign paradigms of management of patients with acute chest pain, which will allow rapid discharge of patients with non-ischemic etiologies of chest pain and may allow faster treatment initiation in those who do have ACS. The Rule Out Myocardial Infarction using Computer Assisted Tomography (ROMICAT) study is to be followed by a similarly designed multicenter trial. Another study worth reviewing examined 283 patients with potential acute coronary syndrome and with a low Thrombolysis In Myocardial Infarction (TIMI) risk score and absence of ischemia on initial EKG.[1] The authors found that 84% of patients could be safely discharged based on the negative coronary CT result without any death or myocardial infarction during the 30-day follow-up period.

S. Abbara, MD

Reference

1. Hollander JE, Chang AM, Shofer FS, McCusker CM, Baxt WG, Litt HI. Coronary computed tomographic angiography for rapid discharge of low-risk patients with potential acute coronary syndromes. *Ann Emerg Med.* 2009;53:295-304.

Prognostic Significance of Delayed-Enhancement Magnetic Resonance Imaging: Survival of 857 Patients With and Without Left Ventricular Dysfunction

Cheong BYC, Muthupillai R, Wilson JM, et al (Texas Heart Inst at St Luke's Episcopal Hosp, Houston; et al)
Circulation 120:2069-2076, 2009

Background.—Left ventricular ejection fraction is a powerful independent predictor of survival in cardiac patients, especially those with coronary artery disease. Delayed-enhancement magnetic resonance imaging (DE-MRI) can accurately identify irreversible myocardial injury with high spatial and contrast resolution. To date, relatively limited data are available on the prognostic value of DE-MRI, so we sought to determine whether DE-MRI findings independently predict survival.

Methods and Results.—The medical records of 857 consecutive patients who had complete cine and DE-MRI evaluation at a tertiary care center were reviewed regardless of whether the patients had coronary artery disease. The presence and extent of myocardial scar were evaluated qualitatively by a single experienced observer. The primary, composite end point was all-cause mortality or cardiac transplantation. Survival data were obtained from the Social Security Death Index. The median follow-up was 4.4 years; 252 patients (29%) reached one of the end points. Independent predictors of mortality or transplantation included congestive heart failure, ejection fraction, and age ($P < 0.0001$ for each), as well as scar index (hazard ratio, 1.26; 95% confidence interval, 1.02 to 1.55; $P = 0.033$). Similarly, in subsets of patients with or without coronary artery disease, scar index also independently predicted mortality or transplantation (hazard ratio, 1.33; 95% confidence interval, 1.05 to 1.68; $P = 0.018$; and hazard ratio, 5.65; 95% confidence interval, 1.74 to 18.3; $P = 0.004$, respectively). Cox regression analysis showed worse outcome in patients with any DE in addition to depressed left ventricular ejection fraction (<50%).

Conclusion.—The degree of DE detected by DE-MRI appears to strongly predict all-cause mortality or cardiac transplantation after adjustment for traditional, well-known prognosticators.

▶ This study was included because it is one of the first to deliver good prognostic data on presence or absence of delayed hyperenhancement (DE) at cardiac MRI. It is a large single center study that included all patients within a 3-year window who had DE and left ventricular (LV) function sequences performed, except patients with hypertrophic or infiltrative cardiomyopathy, and myocarditis.

A Kaplan Meier analysis showed clear separation of the curves for those patients with DE compared with those without DE, and transplant-free survival at 7 years after the MRI was 83% of those patients who had no delayed hyperenhancement, compared with only 63% in those who were positive for abnormal DE.

In the original article, Fig 4 shows transplant-free survival curve after adjustment for all confounding variables. Patients are stratified by left ventricular ejection fractions (LVEF) greater or less than 50% and by the presence or absence of DE. The curves show that in addition to the prognostic value of LVEF, there is significant incremental prognostic value contributed by DE analysis.

S. Abbara, MD

Radiation Dose From Cardiac Computed Tomography Before and After Implementation of Radiation Dose–Reduction Techniques

Raff GL, Chinnaiyan KM, Share DA, et al (William Beaumont Hosp, Royal Oak, MI; Univ of Michigan Health System; et al)

JAMA 301:2340-2348, 2009

Context.—Cardiac computed tomography angiography (CCTA) can accurately diagnose coronary artery disease, but radiation dose from this procedure is of concern.

Objectives.—To determine whether a collaborative radiation dose–reduction program would be associated with reduced radiation dose in patients undergoing CCTA in a statewide registry over a 1-year period and to define its effect on image quality.

Design, Setting, and Patients.—A prospective, controlled, nonrandomized study conducted during a control period (July-August 2007), an intervention period (September 2007-April 2008), and a follow-up period (May-June 2008) at 15 hospital imaging centers participating in the

Advanced Cardiovascular Imaging Consortium in Michigan, which included small community hospitals and large academic medical centers. A total of 4995 sequential patients undergoing CCTA for suspected coronary artery disease were enrolled; 4862 patients (97.3%) had complete radiation data for analysis.

Intervention.—A best-practice CCTA scan model was used, which included minimized scan range, heart rate reduction, electrocardiographic-gated tube current modulation, and reduced tube voltage in suitable patients.

Main Outcome Measures.—Primary outcomes included dose-length product and effective radiation dose from all phases of the CCTA scan. Secondary outcomes were image quality assessed by a 4-point scale (1 indicated excellent; 2, good; 3, adequate; and 4, nondiagnostic) and frequency of diagnostic-quality scans.

Results.—Compared with the control period, patients' estimated median radiation dose in the follow-up period was reduced by 53.3% (dose-length product decreased from 1493 mGy × cm [interquartile range {IQR}, 855-1823 mGy × cm] to 697 mGy × cm [IQR, 407-1163 mGy × cm]; $P < .001$) and effective dose from 21 mSv (IQR, 12-26 mSv) to 10 mSv (IQR, 6-16 mSv) ($P < .001$). The greatest reduction in dose occurred at low-volume sites. There were no significant changes in median image quality assessment during the control period compared with the follow-up period (median image quality of 2 [images rated as good] vs median image quality of 2; $P = .13$) or frequency of diagnostic-quality scans (554/620 patients [89%] vs 769/835 patients [92%]; $P = .07$).

Conclusion.—Consistent application of currently available dose-reduction techniques was associated with a marked reduction in estimated radiation doses in a statewide CCTA registry, without impairment of image quality.

Trial Registration.—clinicaltrials.gov Identifier: NCT00640068.

► This study was included because of its very powerful message. The Prospective Multicenter Study on Radiation Dose Estimates of cardiac CT Angiography in Daily Practice I (PROTECTION I) trial, which was reviewed in 2009,[1] had shown us that there is a wide range of radiation dose across imaging centers and CT vendors. In this study by Raff et al, several sites that reported radiation doses to a state-wide registry before and after they were educated as to appropriate radiation reduction strategies. Sites were urged to adhere to the ALARA principles and use appropriate radiation reduction strategies tailored to individual patients. The figure (Fig 2 in the original article) shows 2 graphs that illustrate the fraction of patients (y -axis) who had received certain radiation doses (x -axis). The line with the solid dots shows a clear shift of the distribution of radiation doses toward lower doses. The authors also report higher percentages of patients who had the target doses of 15 mSv or less. This study illustrates that awareness of radiation savings techniques and tailoring cardiac CT parameters in individual patients will result in decreased radiation doses with maintained diagnostic values.

S. Abbara, MD

Reference

1. Hausleiter J, Meyer T, Hermann F, et al. Estimated radiation dose associated with cardiac CT angiography. *JAMA*. 2009;301:500-507.

SCCT guidelines for performance of coronary computed tomographic angiography: A report of the Society of Cardiovascular Computed Tomography Guidelines Committee

Abbara S, Arbab-Zadeh A, Callister TQ, et al (Society of Cardiovascular Computed Tomography, Washington, DC)

J Cardiovasc Comput Tomogr 3:190-204, 2009

The increasing use of coronary computed tomographic angiographic (CTA) requires the establishment of standards meant to ensure reliable practice methods and quality outcomes. The Society of Cardiovascular Computed Tomography Guidelines Committee was formed to develop recommendations for acquiring, interpreting, and reporting of these studies in a standardized fashion. Indications and contraindications for specific services or procedures are not included in the scope of these documents. These recommendations were produced as an educational tool for practitioners to improve the diagnostic care of patients, in the interest of developing systematic standards of practice for coronary CTA based on the best available data. Because of the highly variable nature of individual medical cases, an approach to scan performance that differs from these guidelines may represent an appropriate variation based on a legitimate assessment of an individual patient's needs.

The Society of Cardiovascular Computed Tomography Guidelines Committee makes every effort to avoid any actual or potential conflicts of interest that might arise as a result of an outside relationship or a personal interest of a member of the Guidelines Committee or either of its Writing Groups. Specifically, all members of the Guidelines Committee and of both Writing Groups are asked to provide disclosure statements of all such relationships that might be perceived as real or potential conflicts of interest relevant to the document topic. The relationships with industry information for Committee and Writing Group members are published in the appendices of the document. These are reviewed by the Guidelines Committee and will be updated as changes occur.

▶ This document aims to set the minimum acquisition standards for cardiac CT. It is based on expert consensus and on currently available evidence. The document covers several aspects of cardiac CT acquisition, including physician and technician competencies, patient screening and preparation, contrast injection protocols, cardiac CTA acquisition, techniques to reduce radiation, and others. After each section there are bulleted single sentence recommendations. One recommendation asks for biannual internal review of each site's radiation levels and comparison with published references followed by internal protocol review and potential protocol optimization, if appropriate. A number of radiation savings techniques that mandate that CT parameters be adjusted to individual patients are emphasized. For example, tailoring the tube voltage and current for each individual patient to a given patient's characteristics and clinical indication was recommended, and use of low voltage (100 kVp) is recommended to reduce radiation dose for nonobese patients. The use of ECG triggered tube current modulation or better prospective ECG triggering was recommended in every patient with low and regular heart rate in whom diagnostic image quality can be expected. A second guideline for the interpretation and reporting of cardiac CT was published by the same society.[1] This document aims to standardize reporting and to establish basic elements and optional elements that each cardiac gated CT report should contain. This document also contains a 'CT friendly' 18-segment model of the coronary artery anatomy to aid reporting.

S. Abbara, MD

Reference

1. Raff GL, Abidov A, Achenbach S, et al. SCCT guidelines for the interpretation and reporting of coronary computed tomographic angiography [published online ahead of print January 29 2009]. *J Cardiovasc Comput Tomogr.* 2009;3:122-136.

Article Index

Chapter 1: Thoracic Radiology

Chapter 2: Breast Imaging

Chapter 3: Musculoskeletal System

Chapter 4: Pediatric Radiology

Chapter 5: Economics, Research, Education, and Quality

Chapter 6: Interventional Radiology

Chapter 7: Neuroradiology

Chapter 8: Cardiac Imaging

Author Index

G

H

I

J

K

L

M

N

O

P

R

FSC
www.fsc.org
MIX
Papier | Fördert
gute Waldnutzung
FSC® C083411

Kolibri 360 GmbH
Ferdinand-Jühlke-Straße 7
99095 Erfurt, Deutschland
produktsicherheit@kolibri360.de